T0212997

Technological Innovation as an Evolutionary Process

Technological artefacts and biological organisms 'evolve' by very similar processes of blind variation and selective retention. This analogy is explored systematically, for the first time, by a team of international experts from evolutionary biology, history and sociology of science and technology, cognitive and computer science, economics, psychology, education, cultural anthropology and research management. Do technological 'memes' play the role of genes? In what sense are novel inventions 'blind'? Does the element of design make inventions 'Lamarckian' rather than 'Darwinian'? Is the recombination of ideas the essence of technological creativity? Can invention be simulated computationally? What are the entities that actually evolve – artefacts, ideas or organizations? These are only some of the many questions stimulated and partially answered by this powerful metaphor. By its practical demonstration of the explanatory potential of 'evolutionary reasoning' in a well-defined context, this book is a ground-breaking contribution to every discipline concerned with cultural change.

JOHN ZIMAN is well-known internationally for his many scholarly and popular books on condensed-matter physics and on science, technology and society. He was born in 1925, and was brought up in New Zealand. He took his DPhil at the University of Oxford and lectured at the University of Cambridge before becoming Professor of Theoretical Physics at Bristol University in 1964. His research on the electrical properties of metals earned his election to the Royal Society in 1967. After voluntary early retirement from Bristol University in 1982 he devoted himself to the systematic analysis and public exposition of various aspects of the social relations of science and technology, on which he is a recognized world authority. In 1997, as Convenor of the Epistemology Group, which studies the evolution of knowledge and invention, he invited leading scholars from a number of disciplines to work together on this book.

Technological Innovation as an Evolutionary Process

Edited by JOHN ZIMAN

on behalf of *The Epistemology Group*

CAMBRIDGE
UNIVERSITY PRESS

PUBLISHED BY THE PRESS SYNDICATE OF THE UNIVERSITY OF CAMBRIDGE
The Pitt Building, Trumpington Street, Cambridge, United Kingdom

CAMBRIDGE UNIVERSITY PRESS
The Edinburgh Building, Cambridge CB2 2RU, UK
40 West 20th Street, New York NY 10011–4211, USA
477 Williamstown Road, Port Melbourne, VIC 3207, Australia
Ruiz de Alarcón 13, 28014 Madrid, Spain
Dock House, The Waterfront, Cape Town 8001, South Africa

http://www.cambridge.org

© Cambridge University Press 2000

This book is in copyright. Subject to statutory exception
and to the provisions of relevant collective licensing agreements,
no reproduction of any part may take place without
the written permission of Cambridge University Press.

First published 2000
First paperback edition 2003

Typeface Swift 9.5/14pt   System 3B2   [CE]

A catalogue record for this book is available from the British Library

Library of Congress Cataloguing in Publication data

Technological innovation as an evolutionary process / edited by John Ziman
on behalf of the Epistemology Group
    p.   cm.
Includes bibliographical references and index
ISBN 0 521 62361 8 hardback
1. Technological innovations - Social aspects. 2. Evolution (Biology)
1. Ziman, J. M. (John Michael), 1925-   .
T173.8.T363   2000
600-dc21   99-15474   CIP

ISBN 0 521 62361 8 hardback
ISBN 0 521 54217 0 paperback

In memory of

DONALD THOMAS CAMPBELL

1916 – 1996

Contents

vii

Contributors

Dr JANET DAVIES BURNS
Department of Educational Studies and
    Community Support,
Massey University,
Private Bag 11 035,
Palmerston North,
New Zealand.

Prof. W. BERNARD CARLSON
Division of Technology, Culture and
    Communications,
University of Virginia,
Charlottesville VA 22901,
USA.

Prof. EDWARD W. CONSTANT II
Department of History,
Carnegie Mellon University,
Pittsburgh PA 15213–3890,
USA.

Prof. PAUL A. DAVID
All Souls College,
Oxford OX1 4AL,
England.

Mr GERARD FAIRTLOUGH
22 Holst Court,
65 Westminster Bridge Road,
London SE1 7JQ,
England.

Prof. JAMES FLECK
The Management School,
University of Edinburgh,
7 Bristo Square,
Edinburgh EH8 9AL,
Scotland.

Mrs SARAH HARRISON JP
Department of Social Anthropology,
University of Cambridge,
Free School Lane,
Cambridge CB2 3RF,
England.

Dr EVA JABLONKA
The Cohn Institute for the History and
    Philosophy of Science and Ideas,
Tel Aviv University,
Tel Aviv 69978,
Israel.

Prof. ALAN MACFARLANE
Department of Social Anthropology,
University of Cambridge,
Free School Lane,
Cambridge CB2 3RF,
England.

Mr T.L. MARTIN
Springlands Cottage,
Sandy Lane,
Henfield,
Sussex BN5 9UX,
England.

Dr GEOFFREY F. MILLER
ESRC Research Centre for Economic Learning
    and Social Evolution,
University College London,
Gower Street,
London WC1E 6BT,
England.

Prof. JOEL MOKYR
Department of Economics,
Northwestern University,
2003 Sheridan Road,
Evanston IL 60208–2400,
USA.

Prof. RICHARD NELSON
School of International and Public Affairs,
Columbia University,
420 W 118th Street,
New York NY 10027,
USA.

Prof. DAVID PERKINS
Harvard Graduate School of Education,
323 Longfellow Hall,
Appian Way,
Cambridge MA 02138,
USA.

Prof. JOAN SOLOMON
Centre for Science Education,
The Open University,
Walton Hall,
Milton Keynes MK7 6AA,
England.

Prof. RIKARD STANKIEWICZ
Research Policy Institute,
Box 2017, S-220 02 Lund,
Sweden.

Dr DAVID TURNBULL
School of Social Inquiry,
Deakin University,
Geelong,
Victoria 3217,
Australia.

Prof. WALTER G. VINCENTI
Department of Aeronautics and Astronautics,
Stanford University,
William F. Durand Building,
Stanford CA 94305,
USA.

Prof. JOHN ZIMAN
27 Little London Green,
Oakley,
Aylesbury,
Bucks HP18 9QL,
England.

Preface

'Isn't technological innovation rather like biological evolution?' This was the question that Gerry Martin asked me, one day in the autumn of 1993. Like all his questions, it turned out to be more penetrating than it mildly suggested. It certainly seemed very plausible. People had asked it before – but why had so few stayed longer for an answer? After five years, many opinions, much talk, and several megabytes of text, here is a provisional reply.

Quite clearly, this enquiry could not be confined to a single academic discipline. Backed generously by the Renaissance Trust, we conjured up *The Epistemology Group*, a market-place for ideas about 'the Evolution of Knowledge and Invention'. Scholars from different intellectual traditions debated these in a series of informal seminars (1994–95, at the Royal Society of Arts in London), presented some of their findings in a public Forum (June 1996, at the London School of Economics), and hammered them out together at an International Workshop (January 1997, at Goring, Oxfordshire).

This volume, then, is not just a compilation of diverse opinions on a puzzling theoretical question. Whether they mutually agree or disagree, the authors of the various chapters speak for themselves. But they do so out of a shared experience, as conscientious participants in a collective intellectual enterprise, addressing the same set of problems, locating their responses in the same conceptual frame and yet remaining mindful of the issues on which they knew they were divided. This is why the production of this book has been such an agreeable task, to which all the authors have contributed harmoniously.

My editorial work has mostly been to cut out near-duplicate accounts of some of the general themes and then to tack the various chapters together with cross-references. These take the form '(Vincenti, ch.13)', or, more cryptically, '(§13.3)' to indicate that this theme is also dealt with at length by Walter Vincenti in chapter 13, or more particularly in section 3 of that chapter. Superscript numbers (e.g. '7') indicate notes listed at the end of the book, in which literature citations are given in 'Harvard' form (e.g. 'Campbell 1964'). Full details for these references are set out in a comprehensive bibliography at the end of the book. This bibliography also operates as an author index, in that at the end of each reference there is a code in square brackets which indicates where that reference was cited in the text. For example [13, 7] means that this work is cited as note 7 in chapter 13.

The shared framework of concepts, problems and controversies within which we worked is outlined in chapter 1. Let me emphasize, however, that this introductory chapter is not the usual artful synthesis of the multiplicity of perspectives actually presented in this volume: it represents the situation as we had already come to see it *before* the Goring meeting. This book thus owes a great deal to the many other participants in the Epistemology Group activities where these perceptions had evolved. In particular, let me thank the following:

Prof. Barry Barnes, Department of Sociology, University of Exeter; Dr Kerstin Berminge, Department of Theory of Science, University of Göteborg, Sweden; Sir James Black FRS, Prof. of Analytical Pharmacology, King's College Hospital Medical School, London; Prof. Margaret Boden FBA, School of Cognitive and Computer Science, University of Sussex; Prof. Werner Callebaut, Department of Philosophy, University of Limburg, Maastricht, Netherlands; Prof. Mihaly Csikszentmihalyi, Department of Psychology, University of Chicago, Chicago IL, USA; Dr Manuela Delpos, Konrad Lorenz Institut, Altenberg-Donau, Austria; Prof. Daniel Dennett, Centre for Cognitive Science and Philosophy, Tufts University, Medford MA, USA; Prof. Richard Gregory CBE FRS, Department of Psychology, University of Bristol; Dr Rob Iliffe, Department of Humanities, Imperial College, London; Prof. Peter Munz, Victoria University of Wellington, New Zealand: Prof. Keith Pavitt, Science Policy Research Unit, University of Sussex; Prof. Henry Plotkin, Department of Psychology, University College, London; Dr André Pomiankowski, Galton Laboratory, University College, London; Dr Emma Rothschild, King's College, Cambridge; Dr Simon Schaffer, Department of History and Philosophy of Science, University of Cambridge; Dr Elizabeth Van Meer, Department of Applied Philosophy, Wageningen Agricultural University, Wageningen, Netherlands; Dr Richard Webb, Centre for the Philosophy of the Natural and Social Sciences, London School of Economics.

As can be seen from our numerous references to his writings, we all owe a great deal to the work of Professor Donald Campbell of Lehigh University, who was a very active founding member of the Epistemology Group, and who was to have attended the Goring Workshop and contributed a chapter to this volume. His death on 6 May 1996 not only brought to an end a most distinguished and creative scholarly career but took from us an inspiring and much loved colleague and friend.

In the end, however, our main debt of gratitude is to Gerry Martin. We have not really answered his question yet, but his stimulus and support has made us think about interesting things, which is what we all most enjoy. And we will go on searching.

John Ziman
(Convenor)

The Epistemology Group (reg. charity No.1057502)
27 Little London Green, Oakley, Aylesbury, Bucks, HP18 9QL

I
EVOLUTIONARY THINKING

1

Evolutionary models for technological change

JOHN ZIMAN

1.1   The biological analogies

Go to a technology museum and look at the bicycles. Then go to a museum of archaeology and look at the prehistoric stone axes. Finally, go to a natural history museum and look at the fossil horses. In each case, you will see a sequence, ordered in time, of changing but somewhat similar objects. The fossils, we know, are sampled from the history of a family of biological organisms. They are similar because they are related by reproductive descent. But here – and quite generally throughout this book – when we say that they have *evolved* we mean more than that they have 'developed gradually'.[1] We are indicating that this development has occurred through genetic variation and natural selection, sometimes, in outwardly static circumstances, apparently spontaneously, sometimes as an adaptive response to a changing external environment. Can technological innovation be explained in similar terms? Do *all* cultural entities 'evolve' in this sense – that is, change over time by essentially the same mechanism?

These museum displays, and the similarities between them, are, of course, highly contrived.[2] But the basic analogy between biological and cultural evolution has often been remarked.[3] From the middle of the nineteenth century onwards, it was noted more or less independently by such eminent scholars as William Whewell, Karl Marx, Thomas Henry Huxley, Ernst Mach, William James and Georg Simmel. The basic idea has been extended by later authors such as Jean Piaget, Konrad Lorenz, Donald Campbell, Karl Popper and Jacques Monod. Sometimes it is presented as just one aspect of a general principle of 'Evolutionary Epistemology', which interprets the whole story of human social, intellectual and material development as the continuation of organic evolution by other means.[4] But it is a simple idea that makes obvious sense in its own right – for example, in accounting for the immense variety of artefacts that are invented and put on the market, and for the superior utility of the few that eventually survive.[5]

What is more, it is an idea that can easily be developed in considerable detail. Try it out in conversation around the coffee table. Most people nowadays know enough about how Darwin explained 'the origin of species' to apply the same reasoning to 'the origin of inventions'. They regularly refer to technological concepts in quasi-biological terms – 'fitness', 'survival', 'niche', 'hybrid', 'genealogy', etc. – as if the analogy needed no further explanation. This usage is so convenient that we shall employ it throughout wherever it is not misleading. Indeed, the theme of this book is perfectly exemplified by the usefulness of the technical language of evolutionary biology for summing up succinctly a variety of cultural phenomena.

First of all: it is very easy to point to *structural* analogies between certain biological processes and the processes involved in technological innovation. One can immediately think of mechanisms whereby material artefacts – indeed, also, less tangible cultural entities, such as scientific theories, social customs, laws, commercial firms, etc. – undergo *variation* by *mutation* or *recombination* of characteristic *traits*. Many different variants are put on the market (or published, or practised, or adjudicated, or invested in, etc. as the case may be). There they are subjected to severe *selection*, by customers and other users (or competing groups, courts of appeal, banks, and so on). The entities that *survive* are *replicated*, *diffuse* through the population and become the predominant type.

Further thought reminds us that *mutualistic* relationships are very common, as between pens and inks, or between bombers and radar systems. Indeed, technical innovations in an industry such as car manufacture are so interrelated that one might describe it as a whole *ecological system* of *coevolving* artefacts. As the selective environment changes, so do such systems evolve and *adapt* to it. On the other hand, isolated subpopulations – *demes* – may separate and evolve independently in different directions for long periods before recombining. And so on.

What is more, the history of technology provides numerous episodes that are remarkably similar to well-known biological phenomena. Some of these *phenomenological analogies* will be discussed in later chapters of this book. Overall they make an impressive list. In my own limited reading I have come across suggested technological analogues of what an evolutionary theorist would term *diversification, speciation, convergence, stasis, evolutionary drift, satisficing fitness, developmental lock, vestiges, niche competition, punctuated equilibrium, emergence, extinctions, coevolutionary stable strategies, arms races, ecological interdependence, increasing complexity, self-organization*, *unpredictability*, *path dependence*, *irreversibility* and *'progress'*. Admittedly, some of these suggested similarities are very questionable, so that it would take us too far afield to cite, decode and try to justify them in detail. Indeed, the whole biological analogy is often dismissed as naive. But the mere fact that such a list can be compiled at all shows just how many quite specialized 'evolutionary' characteristics are apparently common to both technological and biological systems.

1.2   The technological 'disanalogies'

The directness and diversity of these analogies strongly suggest the possibility of transforming the notion of 'technological evolution' from an evocative *metaphor* into a well-formed *model*. But before trying to set up such a model, we must look at the flip side of the comparison. Unfortunately, as many students of the subject have pointed out, there are many 'disanalogies' to take into account. Technological systems are not like biological systems in a number of important ways.

The most obvious difference is that novel artefacts are not generated randomly: they are almost always the products of conscious *design*. In the language of neo-Darwinism,[6] they are not 'Weismannian': the variations that are presented for selection are not produced by a mechanism that is entirely blind to their ultimate fate. Inventors learn by experience and experiment, and visualize their creations before they make them. Their inventions thus acquire characteristics which are deliberately handed on to the next generation. Technological innovation thus has 'Lamarckian' features, which are normally considered to be forbidden in biology.

Another major difference is that there is no strict technological equivalent of a biomolecular *gene*. To sustain the overall analogy, it is convenient at times to talk about technological systems in terms of 'memes'[7] – elementary concepts that endure over long periods, replicate themselves and shape the actual artefacts. But this terminology is abstract and metaphorical. 'Memes' are not operationally equivalent to the indivisible entities hypothesized by Mendel and made flesh by Crick and Watson. The characteristic features of an artefact cannot be analysed uniquely into precisely defined design elements that endure unchanged for long periods. Thus, all bicycles have wheels, but these are so varied in design and construction that it is not very useful to regard them as manifestations of a 'wheel meme' that persists from type to type.

'Meme' language is instructive, of course, in emphasizing the heritability of technological traits, as distinct from the physical survival of the artefacts that exhibit these traits. Biological theorists make much of the relationship between the *phenotypes* that are actually subjected to selection, and the *genotypes* that encode them. This is much simpler and more precise than the relationship between artefacts and their 'memotypes'. The design of a novel artefact is often analysed and revised many times before any engineering work begins. Technological memes can be transmitted, stored, revived, varied and selected independently of the actual artefacts to which they might apply. It would be possible, for example, to construct a workable modern 'penny-farthing' bicycle solely on the basis of an old photograph or patent specification.

But here again, there are serious 'disanalogies'. The genome of a biological organism is a 'recipe' for its development from conception, rather than a 'blueprint' of its adult form. In technological evolution, 'memes' from distant lineages often recombine, and 'multiple parentage' is the norm. No biological organism is like, say, a computer chip, which combines basic ideas, techniques and materials from a variety of distinct fields of chemistry, physics, mathematics and engineering. Does the differentiation of organisms into separate *species*, which Charles Darwin made central to evolutionary theory, truly apply to inventions? The 'cladogram' of a technological artefact usually looks more like a neural net than a family tree![8]

1.3　Is 'evolution' compatible with 'design'?

At first sight, then, the evolutionary metaphor for technological innovation is very appealing. In many respects, both the underlying mechanisms and the broad patterns of technological change are quite reminiscent of those found in biological evolution. But the idea of turning these structural and phenomenological analogies into a realistic model soon meets obstacles. Indeed, these obstacles have seemed so daunting that only the most daring scholars of the subject – notably the late Donald Campbell – have tried to surmount them. The viability of the project embodied in this book is thus seriously in question!

To proceed further, it is essential to understand what we are up against. The first major obstacle has deeper roots than the Lamarckian heresy. 'Design' is central to modern technology.[9] How can that be reconciled with 'evolution', which both Darwin and Lamarck explained as a process through which complex adaptive systems emerge *in the absence* of design? We may well agree that technological change is driven by variation and selection – but these are clearly not 'blind' or 'natural'. This work is being done largely by conscious human effort, without apparently needing guidance from any 'hidden hand', whether of Nature, the market, or God.

An evolutionary model incorporating intentional factors, such as memory and mental imagery, thus seems self-contradictory. Should we abandon the project altogether, in favour of some other theoretical paradigm, such as 'self-organization', or 'social constructivism'? To avoid this, advocates of 'universal Darwinism'[10] rightly point out that human cognition is the product of natural selection, and operates on selectionist principles.[11] They thus maintain the 'blindness' of the whole process by locating all the action in lower-level neural events whose causes might as well be considered random for all that we can find out about them. Alternatively, the effects of design are assimilated notionally into the selection stage of the cycle, where already-achieved wisdom is used, as in computer problem-solving, to reduce the search space.[12] But these are reductionist strategies that complicate the model far beyond any hope of practical application. Throughout this book, therefore, we shall accept these intentional factors in the way that we ordinarily understand them – for example, as reported to us in everyday psychological terms by 'creative' persons engaged in inventive activities.[13]

But does the 'random variation' that is such a fundamental element of an evolutionary mechanism really have to be as 'blind' as, say, the mutation or recombination of molecules of DNA in sexual reproduction? Remember Darwin's insight that it is *populations* that evolve, not individual organisms. All that may be required is that there should be a stochastic element in what is actually produced, chosen and put to the test of use.

In practice, design processes are always imperfect and indeterminate. As we all know, the 'best-laid plans' of inventors, engineers, research managers, market analysts, company directors, etc. 'gang aft agley'. What is more, there is usually so much uncertainty and disagreement on so many significant points that a wide range of artefacts is made available for selection. Again, the criteria by which technological innovations are selected are not universally agreed, so that artefacts with similar purposes may be designed to very different specifications and chosen for very different reasons. In other words, this apparently deep-rooted obstacle has little substance. There is usually enough diversity and *relatively* blind variation in a population of technological entities to sustain an evolutionary process.

1.4   Artefacts as cultural constructs

'Design' denotes more than rational construction. It indicates *purpose*. A technological artefact is defined in terms of its practical use. Unless novel specimens are made completely mindlessly – and that would not be true even of the stereotypical stone axe – their variant features are bound to be correlated with their intended role in the lives of their makers. After all, this is what distinguishes an 'artefact' from a 'useful object', such as a pebble picked up for throwing.

A technology is not really separable from the *culture* in which it is embedded. Material artefacts encode, embody, convey or transmit whole systems of immaterial ideas and behavioural patterns. It is only for brevity that we talk about them as if they were specimens in a museum, identified only by a name and acquisition date, as if unaware of the invisible cultural aura that gives each object its meaning.

Indeed, the concept of 'technology' – 'the application of practical sciences to industry or commerce'[14] – is not restricted to material objects. It also includes a whole variety of systematic technical procedures, such as farming routines or medical therapies, where the material instruments are not the centre of attention. The spectrum of *social constructs* stretches without a break from concrete objects such as stone axes, bicycles and aspirin pills, to the most abstract entities such as commercial contracts, legal precedents and economic theories.

In other words, a comprehensive model of technological innovation would have to cover almost *every* aspect of *cultural* change. In this book, we focus, for simplicity, on the evolution of tangible cultural objects such as swords, cathedrals, turbojets and pharmaceutical products. But the social context in which and for which they are produced is not merely a passive environment to which they must adapt. Many intangible features of the surrounding culture – for example, military techniques and commercial practices – change over time, hand in hand with the changes in the artefacts to which they are connected.

Thus, if technological entities (in the narrow sense) are deemed to 'evolve', then this interpretation must surely extend to the social entities with which they interact. In default of an alternative theory of a socio-cultural change, we have to include them in our evolutionary model along with their technological counterparts. This book is necessarily much concerned with the complications arising from this widening of the basic metaphor.

1.5   Institutions, roles and behaviour

The above argument can be turned on its head. One could well say that we are interested primarily in the evolution of *cultural* entities in general,[15] and choose to study material artefacts because they have the useful property of being concrete, relatively stable physical objects. They are amongst the few socially meaningful entities that can be preserved unchanged for centuries and are not dissipated by close scrutiny. Their evolutionary trajectories ought to be much easier to investigate and understand than those of less tangible cultural entities, such as languages, rituals, organizations or ideologies. Thus, the study of technological innovation, seemingly so marginal to the humanistic endeavour, could eventually lead right into its centre.

In other words, our starting point might have been evolutionary interpretations of cultural change as such, rather than the analogies between technological and biological evolution. Indeed, there is an extensive, if rather incoherent literature on this subject,[16] which will be referred to in detail at various points in this book. But much of this literature is not really relevant to our theme, since it mainly derives its conceptual framework from *sociobiology*. That is to say, it is concerned primarily with the evolutionary interaction between the biological traits of human beings – in particular, hereditable traits such as sexual preferences, linguistic capabilities, affective responses, etc. – and their social behaviour.[17] This *coevolutionary* process was obviously fundamental to the emergence of modern humans as social beings.

But the biological engine of sociobiology seems to have run out of steam many tens of thousands of years ago.[18] On the other hand, cultural evolution has continued at an ever-increasing pace. Indeed, quite enough technological innovation has taken place since then to provide ample material for our study. We may confidently assume, therefore, that all the 'inventions' that we are concerned with in this book originated amongst people who were physically, intellectually and emotionally 'just like us'.

What is more, we can discard the 'methodological individualism' intrinsic to sociobiology and evolutionary psychology. That is, instead of trying to reduce social action to patterns of individual *behaviour*, we can analyse it in terms of the *institutions* that shape it and give it meaning. Here, of course, we are adopting a much disputed sociological stance. But it does allow us to talk intelligibly about the evolution of organizations, social roles, cultural practices, languages, symbols, concepts, etc. without being committed to any particular opinion about whether such entities are 'really real'.[19]

Indeed, as we noted incidentally above, 'Darwinian' processes of variation and selection can be observed at work amongst commercial firms, social customs, laws, scientific theories, etc. For example, *evolutionary economics*[20] focusses on industrial firms, treating them as social institutions driven by market forces to adapt to changing technological regimes. One of the obvious features of technological innovation in an advanced industrial society is that it involves the coevolution of marketable artefacts, scientific concepts, research practices and commercial organizations. The transistor, for example, was a novel engineering device conceived theoretically by solid-state physicists working in the research and development laboratory of a telephone company. Each of these elements not only contributed to the process of innovation, but was itself changed by its participation.

But here again, by starting with entities at the material end of the spectrum of cultural entities, we do not run head-on into the fog of indefinability surrounding more general theories of 'cultural Darwinism'. Even the most systematic of these theories[21] have encountered very serious conceptual difficulties in trying to identify the units of variation and selection in, say, the abstract world of scientific theories. The application of evolutionary theory to narrowly 'technological' change promises to avoid some of these difficulties, or at least to meet them in a different order, with different weights, and sometimes in simpler forms.

1.6   Selectionism versus instructionism

We can now see that a realistic evolutionary model of technological change must be more complicated than its biological counterpart. It has to incorporate 'design' as well as 'selection', and must extend into the domains of social institutions and abstract ideas. Biologists often complain about the misconceptions of non-biologists about the nature of organic evolution, and about their ignorance of the diversity of its mechanisms.[22] But the processes at work in technological innovation are extremely heterogeneous, and change radically from era to era. As a result, the analogies and disanalogies between biology and technology look different according to what we choose to place at each end of the comparison.

There is a temptation to leap over these messy complications by proposing ever more abstract versions of Darwinism. But a watered-down model designed to meet all such objections would be weaker even than the basic metaphor. The standard neo-Darwinian account of biological evolution has a logical coherence and proven explanatory power which is hard to match. The challenge is to retain and exploit these virtues in the cultural domain.

In particular, what are we to make of the striking phenomenological similarities of biological and technological change? Perhaps these phenomena are not really sensitive to the structural details of the system. Perhaps they are common to all systems that evolve by mechanisms that include stages of partially random variation, selection and replication.[23]

This line of argument is confirmed by the results of computer simulations on very simple models. The burgeoning literature on artificial life, genetic algorithms, cellular automata, etc.[24] contains instances of almost all the phenomena common to technological and biological evolution. For example, some forms of artificial life clearly exhibit 'punctuated equilibrium': they evolve almost imperceptibly for long periods, with sudden episodes of radical change.

Thus, instead of lumping technology and biology together into the same species, we should perhaps treat them as distinct members of a larger genus of *complex systems*.[25] Rather than insisting that our ideas about evolutionary processes should conform to strictly 'Darwinian', or 'neo-Darwinian' principles, we should be exploring the properties of a more general *selectionist* paradigm.[26] We could then give up such Procrustean exercises as trying to make industrial firms look just like organisms, and design concepts just like genes, and concentrate on the actual structural relations between the entities that make up each type of system. In other words, our evolutionary model of technological innovation need not be quasi-biological, just as our evolutionary model of the biological world need not be quasi-cultural, even though they have many general features in common

1.7   Understanding innovation

That, in outline, is the realm of thought that opens up behind the evolutionary metaphor for technological innovation. What do we hope to gain by exploring it further?

In the first place, improved understanding of cultural change is not irrelevant to evolutionary biology. Evolutionists tend to take biology as the standard model, as if all evolutionary processes had to conform to its peculiarities. The exploration of an alternative system throws into relief those features that are specific to biology, such as nearly permanent genes and sexual reproduction, and suggests limits to their evolutionary functions. Would a 'Lamarckian' factor necessarily alter the nature of a 'Darwinian' mechanism? Might it not just improve the efficiency of the search for a higher peak in fitness space? Could it perhaps facilitate self-organization and damp out some of the random fluctuations as the system – life itself – approaches the edge of chaos?

Less hypothetically, technological change is one of the most striking features of our present-day civilization. And yet, in spite of much research effort, it still escapes elucidation. It would overwhelm this book to go through all the different theories, models and metaphors that have been proposed to explain how it occurs. A 'selectionist' approach to this puzzling problem area promises to show more clearly the relative roles of apparently contrary factors, such as 'creativity' and 'design'. Should one think of material artefacts or conceptual 'memes' as the entities that evolve? Historians, ethnographers and prehistorians of technology might then consider whether there has been a progressive move away from selection towards design in the invention or improvement of artefacts. We shall probably never know how much systematic thinking went into the production and selection of stone axes; it is still unclear how much random trial and error will really be needed in the creation of the next generation of 'designer pharmaceuticals'.

Indeed, technological innovation is of such enormous social importance that it is worth pursuing such questions further into the interstices of scientific, industrial and commercial life. What is the actual relationship between 'selection' and 'design' in the research and development divisions of industrial firms? Does this relationship differ from firm to firm, or from industry to industry – and if so, why? Is the overall process speeded up by more feedback (and feed forward) between the various stages, or does each system evolve at its own characteristic rate? What are the real selection criteria in various types of market? An evolutionary perspective should yield new insights into such practical matters.

This exploration of a plausible proposition thus starts with many more questions than it can ever expect to answer. There is evidently plenty of conceptual and empirical space into which it can expand and evolve. That, surely, is a blessed state for any worth-while intellectual enquiry!

2

Biological evolution: processes and phenomena

EVA JABLONKA AND JOHN ZIMAN

2.1   Darwin's theory today

As we shall argue throughout this book, an 'evolutionary' model of cultural change need not be expected to imitate biological evolution in every detail. Nevertheless, any discussion of this subject is bound to draw on the theory of bio-organic evolution for most of its concepts and terminology. The rest of the book would make no sense without a sound basic understanding of the processes and phenomena covered by this theory.

This is even more necessary today than it was 25 years ago. Since the early 1970s, ideas in evolutionary biology have been changing more than outsiders usually realize. An ongoing, slow expansion in the concept of biological heredity has meant that evolutionary biology has begun to encompass phenomena that were traditionally divorced from it. Evolutionary thinking is also being applied to different levels of biological organization. Consequently there is now a greater focus on the evolution of complexity and organization, and an emphasis on processes as well as structures as the objects of natural selection.

Despite all these changes, the basic Darwinian theoretical framework of evolution by natural selection has been retained. In fact, Darwin's theory is more influential and fruitful today than it has been for a long time.[1] We shall use this framework as the basis of our discussion and, although we cannot cover the huge ground that evolutionary biology encompasses, we shall try to illustrate its incredible richness, and its potential relevance for studies of cultural evolution.

As we have seen in the previous chapter, Darwin's selection theory goes beyond the strict biological realm. According to Maynard Smith,[2] for Darwinian evolution to occur we need to have a population of entities with the following three properties: (1) the entities must be able to *multiply*, that is, one entity gives

13

rise to two; (2) there must be *variation* within the population with respect to various characteristics of the entities, including those that influence their chances of multiplying; (3) some of the variations must be *hereditary*, so that when an entity of type A multiplies it usually gives rise to entities of type A, and when an entity of type B multiplies it usually gives rise to type B entities. If an entity of type B changes to a new type C, the new type C should also breed true and give rise to more entities of type C. Natural selection is a necessary and inevitable emergent outcome for a population of entities with such properties. Given enough time, evolution – a change in the frequency and/or the nature of the constituent entities in the population – will necessarily occur.

This formulation stresses the population nature of the theory, and therefore the difference between an evolutionary process and a developmental process. A developing organism changes over time. Suppose that a developing organism does not naturally age, and goes on changing indefinitely. Should we consider this an evolutionary process? In a sense we could, but this evolving entity is doomed, because deleterious variations cannot be selectively eliminated without ending the life of the ever-developing organism. It is only in a population of entities that selective elimination, though involving the death of individuals, can lead to changes that are perpetuated. The population structure of Darwinian theory is one of its most important features, and one of those distinguishing it from previous evolutionary theories.[3]

Maynard Smith's criteria lead to another insight which is fundamental to Darwin's theory: the insight that members of populations and species are likely to be related by descent. As Darwin himself emphasized, his theory assumes a common origin from which members of a species have descended. Entities can be grouped together on the basis of their being descendants from the same parent/s. Closely related members of the population constitute a *lineage* that can evolve or go extinct. Usually lineages that have evolved separately from each other diverge and are more different from each other than lineages that are closely related. Since hereditary similarity and relatedness are typically associated, the classification of entities into groups according to the similarities among them is also a historical reconstruction of the branching, diverging process of evolution. However, lineages can sometimes come together after diverging from each other and form a new lineage, through a process of *hybridization*.

For the theory of evolution by natural selection to be more than a general and somewhat abstract algorithm, we must fill it with precise content. What are the units of hereditary variations? What is the nature of the processes that bring about variation? How does the hereditary process work? What are the objects of selection? The population of what is evolving? How does the structure of the population affect evolution? Biologists can give very precise answers to parts of these questions, and the great success of genetics in the twentieth century derives from the elucidation of the nature of a fundamental unit of hereditary variation, the gene, and the processes that lead to its inheritance. Some aspects of the organization of genes on chromosomes, the regulation of gene expression, the structure of DNA and the transfer of genetic information to proteins are understood in great detail, and this understanding has led to *biotechnology*, based on genetic engineering of DNA structures. But hereditable variation involves more than genes and DNA.

2.2   Heritable variation

As most people know, genes are made of DNA, a molecule made up of a string of nucleotides, whose sequence can be transformed, through complex biochemical processes, into RNA and proteins. Parts of DNA sequence also participate in the regulation of this transformation. A gene can be thought of as a chunk of DNA responsible for an organismic function. A large part of the cellular machinery is devoted to the controlled transfer of information from the DNA nucleotide sequence to various types of proteins and RNA molecules. The sum total of the DNA is regarded as the sum total of the hereditary potential of the organism, its *genotype*. However, the genotype only has meaning through its specific realization, its expression as it interacts with the environment, through its *phenotypic* manifestation.

The interaction of genotype and environment is usually quite active. Many organisms from plants and earthworms to beavers and human beings construct their own environment, and hence influence the way that their genotype interacts with the environment. The multiple ways in which the same genotype can be expressed are most clearly seen in plants. Plant cuttings that share the same genotype can be grown in different environments and develop different phenotypes.

According to the present, genic neo-Darwinian version of Darwinism, however, this interaction goes only one way. The genotype, the genetic potential, the DNA base sequence, is transmitted across generations, while the phenotype, its context-dependent realization, reproduces but is not inherited. Thus, we would expect that phenotypic variations of genetically identical individuals that develop in different environments will not be inherited. The only variations that are heritable should be variations in DNA base sequence. But as we shall see (Jablonka, ch.3), changes in DNA are not the only source of heritable variation, so this feature of neo-Darwinism needs to be modified. For the moment, however, let us look at the way that variations in DNA base sequence are generated.

Molecular studies have revealed the existence of an elaborate enzymatic machinery that unwinds the DNA, replicates it, and repairs most of the changes in base sequence that occur before replication, or as a result of replication mistakes or other failures of the machinery. Some changes in DNA base sequence, however, are not repaired or restored, and are the raw material on which selection ultimately acts. The term *mutation* now covers many different types of variations in DNA base sequence, including changes in a single base, additions or deletions of bases, changes in the position and number of sequences, and changes in the number of chromosomes and chromosome sets.

In organisms that reproduce sexually, the amount of variation between individuals within a population is enormously increased through the regular processes of chromosomal *segregation* and *recombination* as gametes form and fuse to form new individuals. These processes allow the variations in genes, originally present in different individuals, to appear within a single individual, and lead to the genetic uniqueness of sexually reproducing individuals.

The variations in DNA base sequence brought about by mutation and by the sexual processes are often random, in that the variations are not 'informed' about the environment in which they are later selected. In other words, the processes that generate variations are not adaptive responses to the life experiences of the organism that produced them, and do not anticipate the life experiences of the organism that inherits them. According to the neo-Darwinian version of Darwinism, the environment selects the most appropriate variations but it does not preferentially induce them. However, recent research on microorganisms suggests that, in some conditions, some variations in DNA are induced by the environment in a way that results in a more adaptive response than would be predicted if chance alone was responsible for adaptive value. Although the mechanisms leading to such processes are not well under-stood, and much controversy surrounds this active area of research, it is quite clear that to regard all variations in DNA base sequence as 'blind mistakes' is wrong. It seems that, during evolution, mechanisms have evolved that target mutations to particular subsets of genes, under particular types of environ-mental stress.

The DNA system is a system with *digital* information, that is, information is coded in a decomposable sequence of units drawn from a standard set. In such a system, changing one unit into another (e.g. a change of the nucleotide adenine into the nucleotide cytosine at some point in a chunk of DNA, or a change of one letter for another in a written word) leads to the inheritance of the change. The machinery that replicates or re-generates the information is indifferent to the precise linear organization of the units and to the functional meaning of the information. A change in a unit can sometimes lead to a change in meaning: in a word a change of a letter can make the word into nonsense (BED becomes BQD) or transform it into another word (BED becomes BAD); similarly a change of nucleotide in DNA can alter the translated protein into one that is non-functional or one which has altered activity in the cell.

Nevertheless, although the digital system of organizing information is the most familiar to us, not all heritable information is digital. For example, the information in a self-maintaining autocatalytic network or a cycle (for example a neural network, or a chemical cycle) resides in the dynamic activity and the regulatory organization of this network or cycle. Such *analogue* information is not readily decomposable, and the information can be thought of as residing in the dynamic organization of the system as a whole. Many heritable patterns of behaviour in animals and man have an analogue, network, organization. Only a few human activities, especially some rationally controlled activities, can be manipulated in a combinatorial manner characteristic of digital information systems (see Stankiewicz, ch.17).

An analogue system has much less potential for producing heritable varia-tions than a digital one (Perkins, ch.12), because few changes in a network structure will lead to the formation of an alternative, stable, self-maintaining autocatalytic network. In contrast, a DNA sequence of only 100 nucleotides in length has 4^{100} heritable variants! If this DNA sequence codes for a protein, a substantial subset of the 4^{100} variants will alter the functional properties of the protein within the cell. Most function-changing variants will be deleterious, and some will make no difference to the function of the protein in its present context, so they are selectively neutral. But a minority will be beneficial. How many of the vast number of variants will be deleterious, neutral or beneficial depends on the precise structure of the protein and its interactions with the cellular and organismal environment. The total amount of DNA in a typical mammal (a length of 3×10^9 nucleotides) gives the DNA system a practically unlimited amount of heritable variation.

Of course, for any particular species only a very tiny subset of all possible genetic variations has been realized. Quite apart from the relatively short time since the species first appeared, there are other limitations. Some heritable changes may occur more easily than others because of the physical and chemical properties of the molecule and its cellular milieu. Moreover, natural selection sometimes has resulted in systems that lead to the preferential generation of some heritable variations, to the evolution of locus-specific rates of mutation and to increased recombination under some situations of environmental stress. Thus, there are limitations and biases in the generation of variation.

2.3   Multiplication and heredity

The process of multiplication can take several forms. An entity can multiply by division, by fragmentation, by budding, by a process of division followed by fusion (sex), and by various processes of reproduction where elements of the environment are converted into elements that actively participate in an autocatalytic cycle. The outcome of multiplication is growth in the number of identical or similar units.

Most of our assumptions about heredity and multiplication are derived from what we know about DNA replication on the one hand, and about technological processes of copying on the other hand. DNA replication is a very simple process in principle, although an enormously complex one in practice. Essentially, from one original, double-stranded molecule of DNA we get two double-stranded molecules of DNA each made of one old, original, strand and a new complementary strand. DNA replication is therefore termed *semi-conservative*, for the process results in two daughter molecules in each of which only half of the original molecule has been physically conserved. A process of photocopying, like most imitation processes, is *conservative*, since the original entity from which copies are made remains intact.

But there are less straightforward processes that elude traditional definitions in terms of replication or copying. For example, in an autocatalytic cycle, the activity of the cycle results in the formation of elements that are part of the cycle. The number of each part of the cycle, and in a sense the number of identical cycles, grows with time. However, the distinction between growth and heredity in such analogue systems only becomes clear when multiplication occurs, when the original entity fragments, divides or buds, and each resulting 'daughter' unit contains a complete cycle.

Heredity also becomes less straightforward in a sexual system. The sexual process leads to the formation of mixtures of offspring genotypes, and the genotypic hereditary similarity between parent and offspring is, on the average, 50% rather than nearer to 100%, as in a non-sexual system. However, offspring are more similar to their parents than to the average of the population, and the constituent elements of the genotypes (the genes), although they may change their sequence during the processes of recombination, often retain their overall structure for long periods of time.

2.4   The objects of selection

Selection is an inevitable consequence of a system in which a population of entities has the heritable variation that affects reproductive success (*fitness*).

However, selection can have limited and not very interesting effects if the number of heritable variants is limited.[4] For example, in a system with only four possible variants, selection can only lead to cycling between four states according to conditions, but can never lead to anything transcending them. But in a system such as the DNA system, where each of many genes has many possible variants, several variations can accumulate in an individual entity. The number of selectable variants is effectively unlimited.

By definition, the objects of selection are the entities whose multiplication (their survival, reproduction, or both) is affected. The units of variation are not necessarily synonymous with the objects of selection. For example, although a gene is a unit of heritable variation, it is the organism (e.g. a bacterium) which lives or dies, or divides more or less rapidly. The way that genes affect the fate of organisms usually cannot be accounted for in terms of the effects of individual genes. A particular gene may sometimes have a positive effect on an organism's ability to live and reproduce, sometimes a negative effect, depending on the genetic combination in which it is found and the environment in which it is expressed. When considered in isolation, the gene's effect may be on the average selectively neutral, so the gene itself is not a direct object of selection.

It is possible to talk about genes as direct objects of selection when an *allele* – that is one of the variant forms of a particular gene – can be shown to directly 'compete' with another allele. For example, when genes are self-replicating DNA or RNA molecules in a test-tube (or on a primitive earth), these molecules are both units of hereditary variation and objects of selection. Similarly, genes are also direct objects of selection when alleles compete with each other during the formation of sex cells, a relatively uncommon process called *meiotic drive*.

Genes and organisms are not the only objects of selection. The cells that make up the body of a multicellular organism can also be objects of selection during development, as indeed we now know happens in the immune system of mammals.[5] It also happens in cancer. The objects of selection can also be higher-order levels of organization – for example, if different species have different rates of extinction or of speciation, those that survive and speciate more readily can be said to have been positively selected.

The way the products of multiplication behave after they have been produced is relevant to whether or not we define the entity under study as an object of selection. If daughter entities stay together after being produced, the local population is composed of close relatives, and this may influence the way selection would affect this group of relatives. As Hamilton showed in the early 1960s, in some conditions, close relatives are expected to behave altruistically towards each other. The degree of relatedness reflects the probability that a particular individual and its kin share the same gene: parents and offspring are related to each other by 50%, and have a 50% chance of sharing a given gene; grandchildren and grandparents are related by 25%, and have 25% chance of sharing the same gene, and so on. If a shared gene influences the probability of altruistic behaviour, and if this behaviour leads to an increase in the *overall* reproductive success within the kin group, the altruism-affecting gene will be favoured by selection and will increase in frequency. This may happen even if the cost to *some* individual carriers in the kin-group is very high; indeed the altruism-inducing gene may spread even if most of the carriers relinquish reproduction altogether, as do the workers in some ultrasocial honey bee and ant species! Hamilton's insight makes us consider simultaneously three levels of organization: the gene, whose spread in the population is monitored; the individual, whose reproductive success in the kin-group is followed (some reproduce a lot, like queens in honey bee colonies; some, like workers, do not reproduce); and the kin-group whose relative success with respect to other kin-groups is compared.[6]

Notice, moreover, that many higher organisms can learn certain types of behaviour simply by learning from their parents or other associates. Distinctive patterns of social behaviour – even 'altruism' – can thus be passed on from generation to generation in populations of birds and mammals to form social units with distinct traditions without genetic correlates.[7] Again discriminatory group behaviour that leads to its own spread is not limited to humans: crows will treat aggressively an individual that does not know their group dialect, while they treat more tolerantly one that has learnt it and shares it.[8] Relegating 'cultural' evolution to man alone is mistaken, and calling it non-biological begs the question.

The selection processes at different levels of organization are simultaneous, and need not always be in the same direction. An example in the mouse is illuminating. There is a meiotic drive system in mice that leads to a mutant allele of a particular gene being present in 90% of the gametes of male mice (instead of the normal 50%). So there is positive selection for the mutant allele at the gene level. However, this same mutant allele causes male sterility in individuals that have two copies of it, so there is selection against mice with the two alleles at the level of individual mice. But there is more to it than that: mice are socially organized, so usually only a single dominant male mates with the females in the group. If this male (as well as the majority of other males in the group) has two copies of the mutant allele and is therefore sterile, the group will probably become extinct, so there is selection against groups in which the allele is common. The frequency of the mutant allele in the population is the outcome of positive selection at the gene level, and negative selection at the individual and group level.

Even selection at a single level, for example at the level of individual organisms, can often be very complex. Selection for survival and selection for mating success sometimes pull in opposite directions: those individuals most attractive to the other sex need not be those who have the best survival strategies. In some circumstances, there is a trade-off between survival ability and mating success which precludes successful selection for both maximal survival and maximal attractiveness.

Whatever the levels at which selection is acting, the structure of the population, such as its size and its isolation from other populations, may have overwhelming evolutionary effects. In a small population the effects of natural selection can sometimes be outweighed by sampling effects – for example, suboptimal genotypes can reach appreciable frequencies (and occasionally even become fixed). Chance effects of this kind can be of great importance for adaptive evolution. When the effects of genes are non-additive, it may only be possible to reach a very superior genotype through a series of 'construction' steps involving individuals of low fitness. In a large population such intermediate genotypes will be winnowed, but *random genetic drift* in a small population may allow them (as Sewell Wright put it), to cross the 'valleys' in the fitness landscape (see Perkins, ch.12) and reach a higher 'adaptive peak'. Thus, in small populations, the combination of chance and selection can lead to evolutionary outcomes that neither alone can attain.

2.5   Adaptation

The Natural Theologians of the early nineteenth century explored the engineering design of living beings and explained natural design as a proof of God's creative intelligence. Evolution by natural selection was Darwin's powerful alternative to this prevalent view. He showed how its cumulative effects over long periods of time, in different ecological niches, could explain the most striking phenomena in nature: the diversity of life forms, their exquisite functional designs, and the near perfection of their adaptation to their environment. For example, organisms moving habitually through water evolve hydrodynamic body structure, organisms moving habitually through air evolve aerodynamic structures, organisms exposed to heat for very long periods evolve adaptation to heat, and so on. The directional selection pressure exerted by the stable physical properties of such abiotic environments leads to the accumulation of variations that make the organism experiencing it better adjusted to the environment.

Despite the fact that there are several different ways of responding to a given selection pressure, there are many cases that illustrate that good engineering solutions tend to recur. The effects of stable and general aspects of the environment on adaptations can be best seen through the examples of *convergence*: the evolution of functionally similar structures and processes in different lineages as an adaptation to the same kind of environment. For example, the hydrodynamic structure of the fish, the dolphin, and the penguin are *analogous*, owing to convergence, to the same selection pressure exerted by the need to move in water. The similarity is not a consequence of common descent and has arisen independently in these distinct lineages.

On the other hand, organisms may have formally similar structures but very different functions, such as the wing of the bat, the hand of man, and the flipper of the whale. These are *homologous structures*, whose similarity is explained by their common descent rather than by their common function. Another example of homologies are the non-functional *vestiges* of once functional structures, such as the human duodenal appendix. These are degenerate structures whose traces have not been eliminated by natural selection, which are related by descent to functional structures in other species. Such non-functional homologies are striking demonstrations of the historical 'footprints' left by the commonality of descent. On the other hand, important innovations often result from *exaptations*: modifications of a homologous structure that performed a different function in ancestors, such as the flight feathers of birds that are thought to have been originally used only for thermoregulation.[9]

The 'environment' which exerts selection pressure, and to which organisms adapt, is clearly a very complex and dynamic concept and includes changing biotic as well as abiotic aspects. The environment of a parasite is the body of its host while, for the host, the parasite is part of its own environment. As the parasite evolves to exploit its host, the host evolves to defend itself and reject the parasite, and the two are entwined in a close evolutionary war of attrition. *Symbiosis*, which is thought to be an important source of evolutionary innovations, is another important example of closely linked *coevolution*, in this case a partnership with mutually advantageous effects for both partners. Coevolution involves all kinds of interactions among organisms: the prey is an important aspect of the environment of the predator and vice versa; some of the environment of the female is the male, and of the male the female; the social set-up in which organisms live is a major aspect of their environment. All the organisms with which an individual interacts are part of its environment: those belonging to its family and population as well as those belonging to other species. Closely interacting organisms coevolve and adapt to each other.

Some of this coevolution involves antagonistic interactions, as when prey and predator, or parasite and host, coevolve. Other interactions are cooperative as those between host and symbiont, or among individuals in a social group. When the interactions are antagonistic we may often find that coevolution leads to a dynamic *arms race* where every side has to go on evolving, just to keep in the same relation relative to its evolving natural enemy, like Alice's Red Queen who had to keep running to remain in the same place.

As we have seen, the concept of the environment cannot always be usefully separated from that of the organism. The environment, as Lewontin emphasized, is not external to the organism, but is to a smaller or to a larger extent constructed by it.[10] This construction can have long-term evolutionary effects. The most famous example is the formation of an oxygen-rich atmosphere by the activity of oxygen fixing organisms, that has become the environment to which their descendants need to adapt. By their activities organisms create the selection pressures that affect subsequent generations. This niche-construction by the organism has important effects on the way that evolution proceeds. As a population evolves, both the biotic and abiotic aspects of the environment coevolve with it.[11]

However, not every conceivable adaptation can actually evolve by natural selection. First, only a subset of variations is functionally compatible with what already exists and functions. Second, the organization of living organisms renders many new variations selectively neutral: the dynamic organization of organismal functions has been selected to compensate for and neutralize the effects of most new variations. Many biological systems are therefore highly channelled (or *canalized*), *locked in* by developmental patterns, so that it is very difficult for them to change. The functional and organizational constraints are most dramatically evident in the persistence of the same *bauplan* – a basic structural pattern (such as the vertebrate limb) – in many lineages over long stretches of time. There are therefore strong constraints on the visibility of heritable variations to natural selection, and hence on the evolution of new adaptations.

2.6   Speciation and macroevolution

In addition to adaptation, the other great problem facing Darwin was the diversity of species. It is estimated that the number of species currently living on Earth is something between 10 million to 100 millon. We no longer try to explain this diversity by assuming that they were all created by God. So how do species originate?

There is still much debate among biologists about how exactly a species should be defined. It is clear, however, that the living world is differentiated into more or less coherent groups of organisms (though with rather fuzzy edges!), with morphological, physiological, genetic and behavioural similarities amongst individuals in such groups. Many evolutionary biologists have defined species of sexually reproducing organisms in terms of *reproductive isolation*: members of different species usually do not mate and, if they do, their offspring are commonly sterile or impaired in some other way.

The apparent discontinuities between species make *speciation* a special problem, for most evolutionary processes are assumed to occur continuously and we therefore expect a gradual change, and hence continuity rather than discontinuity. It is usually assumed that geographic isolation between populations, combined with gradual adaptation to their physically different and separate niches, can lead, over thousands of generations (a brief instance on the geological time-scale), to sufficient accumulation of hereditary adaptations to result in speciation. The common occurrence of *endemic species* (restricted to a unique locality) on oceanic islands inspired and supported this position. The island species are often similar to species on the nearby continent, and descended from individuals belonging to continental species. The island population began diverging and adapting to its new island environment, until it became sufficiently different to be considered a separate species. Such populations are often reproductively isolated from the original continent population.

This type of speciation, initiated by geographic isolation (*allopatric speciation*), is probably very common. In sexual populations, an allopatric distribution enhances the chance of local genetic divergence, and speciation is often the by-product of such divergence. Divergence can happen, as Darwin suggested, through adaptation to new conditions, but mere isolation followed by inevitable random genetic drift over a long period can lead to a new species.

In addition to speciation that is initiated through geographic isolation, members of a population living in the same place can become reproductively isolated from other members (this type of speciation is called *sympatric speciation*). In a contiguous population, an accidental chromosomal change, or a change in ecological circumstances which affect sexual interactions can lead to reproductive isolation between parts of the population. For example, when mating occurs at the feeding site, developing a taste for a new type of food may lead to isolation between two parts of a population and eventually to speciation. Many mechanisms potentially able to cause to sympatric speciation are now recognized.[12] Speciation through hybridization is also common in some groups, and in some plant taxa it is probably the major route for the formation of new species.

Where distinct species can be delineated, it is possible to look at properties such as the rate of speciation and *extinction*, properties that are not reducible to those of the individuals of which species are composed. This perspective demands a consideration of the interactions among populations belonging to different species within an ecological community. The dynamic rules of organization within communities, which have repercussions on evolution within populations, must be investigated.

One of the striking features of 'the theory of evolution' is that it provided a rationale for biological *systematics* – the overall classification of organisms, fossil and extant. The higher *taxa*, such as 'genera', 'families', 'orders', etc., are not operational entities like species, but the observed hierarchy of shared morphological characteristics is clear evidence of their history. Indeed, the use of new molecular genetic techniques permits the arrangement of biological lineages into unique clusters showing how they have divided and diversified in the course of evolution.

When evolution is studied over a long time-scale – over tens and hundreds of millions of years – large-scale *macro-evolutionary* patterns can be discovered. Patterns such as *stasis*, the visible lack of evolutionary change over very long periods of time, followed by relatively abrupt morphological changes which may reflect speciation events (*punctuations*), can be discerned. Evolutionary trends can be followed, and transitions between major groups can be investigated. However, to understand processes and phenomena at this level, simple extrapolation from the genetic level is not sufficient. The genetic level usually constrains rather than specifies phenomena at higher levels, so both lower-level constraints and higher-level determinants have to be studied.

2.7   Progress and its ambivalences

Throughout its history, the concept of evolution as a scientific theory has been entwined with its social and normative aspects.[13] While the idea of 'progress' has long been potent in the social sciences and technology, the notion of 'evolutionary progress' is fraught with problems. On the one hand, simple living forms still constitute the majority of species and the great bulk of living matter on earth.[14] On the other hand, it is impossible to ignore the historical emergence of increasingly more complex organisms.[15] The concept of 'progress' is clearly meaningless without reference to a specific criterion, but most criteria, such as the amount of coding DNA or the number of cell types in a multicellular organism, are very crude, whereas others, such as adaptability, are very, very ambiguous.

However, over the whole history of organic evolution, new levels of biological organization and individuality emerged in succession. The type of information involved, its quantity, its storage, and its transmission has evolved as organismic complexity increased. For example, the evolution of multicellular organisms is associated with the evolution of cellular communication and the processing and maintenance of transmitted epigenetic information. The increase in the type and quantity of information a system carries is a symptom of a major transition to a new level of organization and individuality.[16] If this increase in complexity is our measure of 'progress', then the appearance of a hominid species capable of vastly enlarging the type and quantity of information by 'cultural' means, can surely be considered one of the greatest of these upwards transitions.

Acknowledgement. We are very grateful to Marion Lamb for her helpful comments on this chapter.

3

Lamarckian inheritance systems in biology: a source of metaphors and models in technological evolution

EVA JABLONKA

3.1 The genic model

No one can doubt that genetics has played a key role in the development of biology and medicine in the twentieth century. However, the way genetics itself developed has, I believe, restricted thinking about heredity and has led to a very narrow view of the nature of information transmission in the evolution of organisms and cultures. If the biological analogy for technological evolution is to be explored further, then this mental bottle-neck must be unblocked. This is the theme of this chapter.

The conceptual foundations of genetics – the definitions of the gene, the genotype, and the phenotype – were laid at the end of the first decade of the twentieth century by Wilhelm Johannsen.[1] Johannsen emphasized the difference between what is transmitted to the next generation and what is not. He distinguished between the genotype – which represents the heritable potentialities – and the phenotype, which is the realization of these potentialities. He saw the genotype as the essential part of heredity, and defined the gene as the unit of heritable information, the basic unit of the genotype.

The concepts Johannsen developed are closely related to the ideas elaborated by Weismann in the second half of the nineteenth century. Weismann suggested that although the hereditary substance in the nucleus of somatic cells changes during development, that in germ-line cells does not. This idea, in a mutated form, became known as 'Weismann's doctrine'. According to this doctrine, the hereditary information that is passed from generation to generation in the germ-cells is not altered by developmental or environmental stimuli, which affect only somatic cells. Johannsen's genotype became identified with the unchanging and immortal information passed through the germ line, while the phenotype was identified with the expression of the genotype as it occurred during the development of the soma.

A gulf was thus established between heredity and development. The gulf was deepened in the 1920s, when, in order to get to grips with the transmission of traits, Morgan and other students of heredity, like Mendel himself, chose those organisms and characters that showed clear segregation of phenotypes, and made a positive decision not to worry too much about what went on between the gene and the character.

By the time the modern synthesis of evolutionary biology was made in the 1940s, the dominance of Mendelian genetics and the way in which embryology had developed meant that development was almost entirely left out of the synthesis.[2] The view of heredity that was incorporated into the synthesis was that of chromosomes carrying linear arrays of discrete genes that pass from one generation to the next, unaffected by environmental factors. Mutation of these genes was thought to be rare and random.

This view provided the framework for the successful expansion of theoretical and experimental population genetics, for quantitative genetics and for ecological genetics. It even enabled physiological adaptation to be given a role in evolutionary theory: the work of Schmalhausen and Waddington showed how, through selection for combinations of genes that allowed adaptability, new adaptations could evolve by the process Waddington called genetic assimilation.[3]

The arrival of molecular biology, and the proclamation of the central dogma that information flows from DNA to proteins and not in the reverse direction, reinforced the centrality of the gene in heredity and evolution. Weismann's doctrine that information in the germ line could not be changed by external influences seemed to be confirmed. As Mayr repeatedly and emphatically proclaimed, heredity is 'hard'; there was no place in evolutionary theory for 'soft' inheritance, the inheritance of induced modifications.[4] Such 'Lamarckian' inheritance is impossible. It seemed that the nature of biological heredity, as well as the origin of heritable variation, had been completely and fully deciphered.

Modern (or genic) neo-Darwinism applied this conception of heredity and the origin of variation to the general Darwinian idea of evolution by natural selection. Darwin's basic idea is deceptively simple: natural selection is a logical necessity in a world with interacting entities that display multiplication, heredity and heritable variation.[5] Adaptive, cumulative changes can occur through the selection of individual entities with heritable differences. In this general form, the theory says nothing about the nature and the origin of the hereditary variation, nor about the mechanisms of heredity and of multiplication. In fact, Darwin believed that some of the hereditary variations are acquired during the individual's lifetime.

Genic neo-Darwinism – the molecular-genetics version of the new synthesis – makes definite and specific assumptions about the origin of variation (random or blind), the type of heritable information (DNA sequence variations, digital information) and the modes of heredity and multiplication. Despite its general acceptance as the exclusive model of biological evolution, genic neo-Darwinism is really only one specific application of the Darwinian principle.

3.2   Dysfunctions of the genic model in cultural evolution

With such a historical and ideological background, it is perhaps not surprising that the prototype model for cultural evolution is the genic model. The type of information implied is digital information (§2.2) – cultural muliplication and heredity are modelled on the replication of DNA, and cultural variation is modelled on variation in DNA sequence resulting from recombination and segregation and non-directed mutations. Following Johannsen, the cultural potential (ideas, concepts, etc.) is assumed to be inherited, while the cultural phenotype, its actualization, is not.

This fundamental distinction between genotype and phenotype became central to the concept of biological heredity and has been carried over by philosophers of science to cultural evolution. Cultural variations are considered to be 'blind' or 'random', as in classical genetics. This seems to correspond to the intuition that, to be new, there must be an element of randomness in the production of variations. It is also assumed that information must be digital, as in a DNA sequence or in a written text. When information is *non-digital*, however, it cannot be localized in a sequence of signs but is integral to the activity of an autocatalytic cycle or an overall structure.[6] With *analogue* information the distinction between genotype and phenotype does not apply, except as a rational abstraction.

The genic perspective is insufficient not only as a metaphor for thinking about and modelling cultural evolution, but also for understanding biological evolution. The simple rules of the Darwinian 'evolutionary selection game' mask the complexity of the processes underlying it, because evolution is a game that evolves its own rules. During evolution, information storage and transfer have evolved. The nature of heritable information has changed and diversified. New systems of transmitting information – the epigenetic systems and the behavioural systems – have been added to the inheritance system based on nucleic acids.

Epigenetic inheritance systems underlie the transmission of the functional states of genes and of cellular structures in cell lineages, in cell groups, and sometimes between organisms.[7] Behavioural inheritance systems enable the transfer of learnt, behavioural information through social learning in mammals and birds[8] and symbolic language in humans. Through epigenetic inheritance and behavioural inheritance, information can be transmitted from one generation to the next in ways other than through the base sequence of DNA, and these transmission systems allow the inheritance of environmentally induced, acquired and learnt variations. During Darwinian evolution, mechanisms allowing the inheritance of acquired variations – 'Lamarckian' mechanisms – have evolved.

Even variation in DNA sequences, generally assumed to be the consequence of random changes, is now receiving more scrutiny. Studies of bacteria and eukaryotes have shown that some changes in DNA sequence may not be random: in some conditions, the mutations that occur are gene-specific and even function-specific. The molecular mechanisms that produce this effect are not clear and are the subject of lively debates.[9] But it has been acknowledged that, through natural selection of non-directed variations, systems could have evolved that preferentially produce or stabilize DNA changes that are adaptive in the conditions that produced them.[10]

Natural selection is a general principle of evolutionary change at many levels, including the cultural level. However, adherence to the genic model of heredity is problematic for students of cultural evolution. Unlike genetic variations, cultural variations can hardly be described as non-directed: individual learning and social learning play a great role as the source of new heritable variation. Furthermore, unlike the genetic system of DNA replication, the 'inheritance' of behaviour patterns and the products of behaviour follows diverse and changing routes, occurs at different levels of social organization, and involves different types of potentially transmissible information.

To apply the basic models of population genetics to cultural evolution, they have to be modified substantially in order to accommodate some of the special features of cultural inheritance.[11] Models of the coevolution of genes and culture, where cultural evolution provides the selection pressures for genetic evolution and vice versa, are also becoming more common.[12] However, since the distinction between genotype and phenotype is not clear in the cultural context, and is certainly not neatly tied up with cultural transmission versus cultural expression (the two are, in fact, often intimately connected), it is frequently not clear what the unit of cultural transmission is. Although these problems are not insoluble,[13] the many discussions about the usefulness or otherwise of the term 'meme' reflect a feeling that a modified version of genic evolution fails to capture some essential features of cultural evolution.

Do we have an alternative to the inadequate model of genic heredity? Although it is clear that the connection between the behavioural inheritance system of birds and mammals and the cultural inheritance of humans is particularly close, the non-human behavioural inheritance system is even less well understood than that of human culture. It cannot at present provide a simple model of cultural evolution. I believe that the more remote but the better understood *epigenetic inheritance systems* (EISs), operating at the cellular level and shared by all extant living organisms, can serve as a more helpful source of analogies for some features of cultural and particularly technological evolution.

EISs are relatively simple inheritance systems, and we understand some of them in biochemical detail. They are more useful as models for cultural evolution than the usual genic models because they have more in common with the cultural inheritance system. Many heritable epigenetic variations are induced or 'acquired', and vertical as well as non-vertical transfer of epigenetic information is common. The epigenetic inheritance systems can also readily demonstrate how evolution by natural selection operates in a system of acquired variations, and elucidate the interactions between instructive and selective processes. Furthermore, different EISs illustrate different types of heritable information, allowing us to go beyond the digital type of information. Since, unfortunately, EISs are not yet as widely known as they deserve to be, I shall briefly review them.

3.3   Epigenetic inheritance systems (EISs)

Epigenesis is the biological process by which an individual organism comes into being through embryonic growth and development. EISs are the systems that enable cells that have different phenotypes but identical genotypes to transmit their phenotypes to their descendants, even when the stimuli that originally induced these phenotypes are no longer present. They perpetuate the determined and differentiated states of somatic cells during the development of multicellular organisms. *Determined* cells breed true: fibroblasts divide to give fibroblasts, and kidney cells divide to give kidney cells. And yet both types of cell may have exactly the same genes on their chromosomes.

Three broad types of EISs that play a role in cell memory have been recognized.[14]

(i) *Steady-state systems* are the systems that perpetuate patterns of metabolic activity, by regulatory feedback-loops. The simplest example is a gene that regulates its own transcription by positive feedback. Once such a gene is turned on by an external inducer, the product of the gene can bind to a control region of the same gene and positively maintain its activity. In this way the activity of a gene that initially depended on the presence of an external inducer is now maintained by self-regulation (fig. 3.1b). If the concentration of the gene product is not too low, and cell division is more or less equal, the regulatory gene-

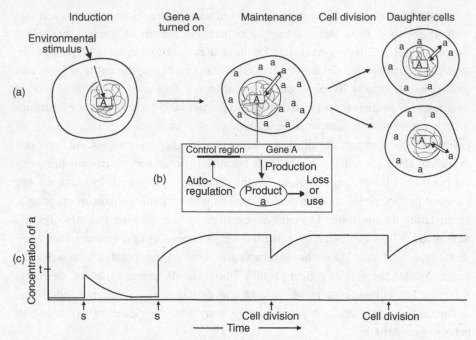

Figure 3.1 A steady-state system showing the perpetuation of an induced active state through cell divisions. (a) After induction, gene A is turned on and its own product, **a**, positively regulates its own activity. (b) The self-regulation of the genetic circuit. (c) The expected behaviour of gene A over time. At the times depicted by the arrows, a stimulus, **s**, inducing the activity of gene A is introduced. Only when the concentration of product **a** exceeds the threshold **t** is gene A turned on by its own product. Once turned on, its activity is maintained by positive feedback. Following cell division, the concentration of **a** drops, but as long as it remains above the threshold the active state will be perpetuated through cell divisions.

product will be transmitted to daughter cells as an automatic consequence of cell division, and the pattern of activity will be inherited with quite high fidelity (fig. 3.1a,c). Moreover, if the regulatory product (or a regulatory molecule that is a product of its activity) diffuses to other cells, these too could be switched to the same heritable activity pattern. In this case there is horizontal transfer of epigenetic information as well as vertical transfer.

Usually steady-state systems are more complicated than the one described, with several gene products and binding regions involved in the self-perpetuating feedback-loop, but the principle remains the same. There are examples from organisms in many different phyla of such self-perpetuating regulatory loops acting at the transcriptional and the post-transcriptional levels.[15]

(ii) *Structural inheritance systems* are the systems responsible for the architectural continuity of cell structures. Pre-existing three-dimensional cell structures

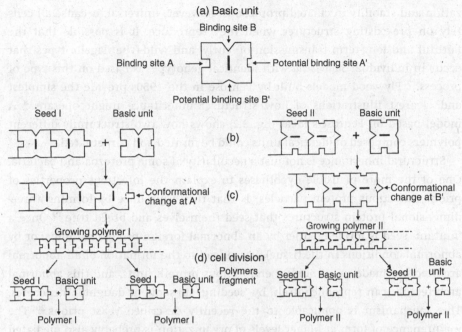

Figure 3.2 The formation and perpetuation of two alternative polymer structures from identical subunits. (a) Basic free units are continuously synthesized and have two active binding sites A and B, and two potential binding sites A' and B'. (b) Two different seeds can be formed from two basic free units – seed I and seed II. The seeds are assumed to pre-exist. The spontaneous formation of seeds of either type is extremely rare, since it involves a spontaneous change of conformation in sites A' and B'. (c) Once a seed is present, it binds a basic unit in a complementary way. Binding site A attaches to seed I, while binding site B attaches to seed II. The binding of a new unit through site A or B induces a change in its conformation in the corresponding potential binding site (A' or B') in the newly attached unit. (d) As a consequence of starting with two different seeds, the structure of the growing polymer is different in the two cells. Following cell division, polymers fragment and the fragments act as seeds for the construction of a new daughter polymer identical to that which was present in the mother cell.

are used as templates for similar new structures in descendant cells. The best examples of this type of inheritance system come from the study of some unicellular organisms, ciliates such as *Paramecium*. Genetically identical cells can show heritable differences in the patterns of ciliary rows on their cell surface, as well as in the organization at a more global level, such as the position and number of their oral parts. Experimentally altered patterns can be transmitted to daughter cells for thousands of cell generations with a fidelity as great as that of the most classical of classical mutations![16]

This system of inheritance, which seems to have reached an extreme specialization and stability in ciliated protozoa is, however, universal, because all cells rely on pre-existing structures when they reproduce. It is possible that the faithful and long-term transmission of 'curly' and wild-type flagella types that occur in individual *Salmonella* with identical genotypes is based on this type of process.[17] Ply-wood models built by Penrose in the 1950s provide the simplest and clearest illustrations of how structural inheritance might operate.[18] A model based on Penrose's ideas (fig. 3.2) shows how two structurally different polymers composed of identical units could be formed and perpetuated.

Structural inheritance is not just a peculiarity of some protozoa and bacteria. One of the more plausible hypotheses to explain the infectious properties of prions (infectious protein particles) is that they reproduce by forming three-dimensional protein structures that seed themselves and breed true.[19] Once a 'mutant' seed is formed (either by an abnormal foreign protein complex, or by abnormal conditions in the tissue), it can lead to the formation of an abnormal architecture made up of normal endogenous protein units, and this abnormal architecture can reproduce itself, by 'seeding' subsequent daughter structures. This mechanism is applicable to the recently described yeast prions.[20] The maintenance of form at higher levels of organization is probably also mediated by processes of structural inheritance, involving the extracellular matrix.[21] Cavalier-Smith has suggested that lipid membranes with their bound protein molecules can perpetuate themselves via structural inheritance, and that the loss and gain of such membranes played a major role in the evolutionary divergence of ancient cells.[22]

(iii) *Chromatin-marking systems* are the systems underlying the inheritance of chromatin structure. Information contained in what we have called *chromatin marks* is carried from one cell generation to the next because it rides with DNA.[23] Marks are binding proteins or chemical groups that are attached to DNA and influence its activity. When DNA is replicated the chromatin marks are reproduced.

The methylation pattern of a gene is a good example of an inherited chromatin mark. In many eukaryotes, some of the cytosines in DNA are methylated. The presence of a methyl group does not affect the coding properties of the codon in which the methylated base participates, but methylation of DNA bases within or around the gene may determine whether or not, or to what extent, the gene is transcribed. The number and pattern of methylated cytosines is related to the functional state of the gene: usually low levels of methylation are associated with potential transcriptional activity, while high levels are associated with inactivity. The same DNA sequence can therefore have several different methylation patterns, each reflecting a different functional state.

Changes in methylation patterns occur during growth and development.

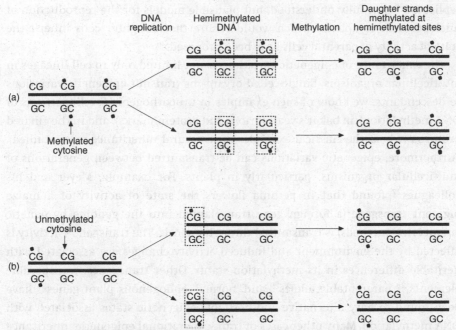

Figure 3.3 The inheritance of alternative patterns, (a) and (b), of DNA methylation.
The dots represent methyl groups. The boxes show hemi-methylated sites following
the replication of DNA. These sites are preferential targets for a methylating enzyme,
which methylates the opposite non-methylated site in the DNA duplex. Different
methylation patterns can therefore be perpetuated through cell division.

Some changes are the results of copying mistakes or accidental removal of the
methyl group, but specific changes in methylation patterns can be induced in
response to particular environmental or developmental stimuli. The alternative
patterns of methylation can be inherited stably through many cell divisions.
Since most of the cytosines that can be methylated occur in CG doublets or CNG
triplets (where N represents any base), CG and CNG sequences on one strand are
partnered, through complementary base pairing, by the same sequences (but in
the opposite direction) in the other strand. Following DNA replication, the
parental DNA strand is methylated, but the daughter strand is not. Maintenance
methylases recognize the asymmetrical, hemimethylated sites and methylate
the cytosines of the new strand.[24] Thus the specific pattern of methylation in a
DNA region is reproduced every cell division, exploiting semi-conservative DNA
replication and the mirror symmetry of CG sites. Figure 3.3 shows the perpetuation of two different methylation patterns imposed on identical DNA sequences.

Chromatin marks involving DNA-associated proteins that affect the maintenance of gene activity can also be transmitted in cell lineages, and such
systems have been described in *Drosophila*.[25] The way these protein marks are

replicated is not fully understood, but plausible models for the reproduction of nucleoprotein complexes which would ensure that daughter cells inherit the state of activity of parental cells have been proposed.[26]

The inheritance of epigenetic variations is not limited only to cell lineages in multicellular organisms. Single-celled organisms transmit epigenetic variations to descendants: we know of such examples in uropathogenic *E. coli* (patterns of DNA methylation), in baker's yeast (prions and protein marks), and in the ciliated protozoa (where the classical examples of structural inheritance were studied). Furthermore, epigenetic variations can be transmitted between generations of multicellular organisms, particularly in plants. For example, Meyer and his colleagues[27] found that in petunia flowers the state of activity of a maize pigment transgene (a foreign gene transplanted into the genome by genetic engineering methods) is transmitted through meiosis. The transgene's activity is affected by the environment and induced activity changes are associated with heritable differences in its methylation status. Other transgenes, transposable elements,[28] paramutable alleles[29] and 'normal' endogenous plant genes[30] have been found to have alternative and heritable epigenetic states associated with DNA methylation. Many other cases of trans-generational epigenetic inheritance in both plants and animals[31] have been reviewed by Jablonka and Lamb (1995) and Jablonka *et al.* (1995). Recently it has been shown that an acquired epigenetic modification in gene expression, which was induced by nuclear transplantation, is inherited through the germ line of mice.[32]

EISs thus have a dual nature: they are both inheritance systems and response systems. While the classical genetic system is a system specialized for the storage and transfer of information, and usually does not directly participate in the response to the environment, the epigenetic inheritance systems are both memory systems and response systems. Specific epigenetic variations are often induced by distinct environmental or developmental changes, and can thus be said to be 'directed' or 'acquired'.

The frequency with which epigenetic variants arise and their rate of reversion vary widely, from 100% to almost zero per cell generation. Furthermore, variations can be produced coordinately at several loci, and the same type of change, leading to a similar functional or structural variation, can be induced in more than one cell within an organism, and in more than one organism. The epigenetic systems may therefore produce rapid, reversible, coordinated, heritable changes.

However, EISs can also underlie random changes, changes that are induced but non-adaptive, and changes that are practically irreversible. The dual nature of epigenetic heritable variations makes them more similar to cultural variations than to the classical genetic variations. EISs also clearly illustrate different types of heritable information: while the chromatin-marking system can be thought of as a digital system, the steady-state and the structural EISs are examples of systems with analogue information.

3.4   Induction and selection of epigenetic variations

It may seem that, since the EISs transmit induced variations, induction is all important, while selection is irrelevant to the ultimate fate of a variant. This is of course a very naive view, because it assumes that induction results in identical epigenetic states in all induced cells. This is often not the case. For example, induction may lead to a changed activity of a gene, altering the pattern and number of methyl groups on the DNA. The extent of gene expression and the stability of this expression as reflected in the pattern of methylation may be variable, and this variation can be cell-heritable and selectable. Furthermore different combinations of genes leading to similar phenotypic effects can be induced and heritably transmitted, and one of the alternatives can be then selected. Thus, induction defines the direction and the general nature of variation, while selection fine-tunes it. An element of randomness is retained, but it is tamed and constrained.

The increasing interest in developmental genetics and evolutionary developmental biology has led to a renewed interest in the idea that somatic selection plays an important role during normal development, and this somatic selection can be based on epigenetic as well as genetic variations in cells.[33] If somatic selection is important, the EISs that underlie cell heredity must play a major role in the process and must have been important determinants of the evolution of ontogeny.

In animals, the best understood case of somatic selection and somatic evolution is seen in the *immune system*. This is based on selection among immunoglobulin-producing cells differing in DNA sequence. A huge pre-existing primary repertoire of antibody-coding genes is established during early development. When the immune system encounters a foreign body – an *antigen* – those cells equipped with antibodies that can best bind this currently prevalent antigen proliferate most efficiently, and are thus selected by the antigen. The proliferation of cells that bind the antigen leads to a high rate of mutation in the region of the gene coding for the binding site in the antibody, then to another round of selection, and to the establishment of a secondary repertoire of antibody-producing cells. These rounds of selection and locus-specific mutation lead to precise adjustment between antibody and antigen, and increase the efficiency of the immune response.[34] The formation of the secondary repertoire – the antigen-induced locus-specific mutations followed by somatic selection – is a good model for what happens when epigenetic variations are induced.

Somatic selection also seems to play an important role in the development of the *nervous system*. Edelman[35] suggested that during early development there is selection for communication among *phenotypically variant* neural cells, leading to the formation of neuronal cell groups. A second round of selection occurs between neuronal groups. In this system, as in the immunoglobulin system, the source of variation is constrained by development and by the direction imposed by induction, but enough redundancy and degeneracy (variation) is left to allow local adjustment, coordination, and global adaptation.

EISs provide the epigenetic variations that can be the basis of many cases of somatic selection. It is known that variations in the pattern of methylation of CG nucleotides, as well as in other potentially heritable features, is extensive in somatic cells. In plants there are numerous examples of somatic selection, and plant developmental biologists believe that somatic selection is a fundamental strategy allowing efficient adaptation to changing environments.[36] It has also been proposed that somatic selection, based on epigenetic variations, is of major importance in plant morphogenesis.[37] Somatic selection for efficient inter-cellular communication (based on epigenetic variations) could also lead to the evolution of what are known as 'costly signals' – in this case seemingly extravagant biochemical molecules used for intra-organismic communication.[38]

The assumption common to all models of developmentally regulated somatic selection is that induction generates variation in specific targets, but it does not specify the precise phenotypic state. That is, induction leads to changes in the cellular phenotype that are in the same general direction, yet are not identical. This induced variation is then sorted by somatic selection. The combination of induction and selection reduces the cost of selection and makes the organism respond to the environment in a cheaper, quicker, and more 'intelligent' way.

Epigenetic inheritance systems are both channels for the transmission of information and part of the regulatory system in cells. To explain their evolution and their evolutionary effects, this dual function has to be taken on board. With EISs, the role of the environment is in inducing, as well as selecting, heritable variation. Selection between cells, and sometimes also between organisms, occurs within the well-defined boundaries set by the inductive processes. This leads to a more focused and efficient selection, both within and between organisms. There is therefore no conflict between selection and induction.

In this the EISs are similar to the higher-level systems of information transfer to behavioural and cultural inheritance systems, where social learning constrains the culturally transmitted variability. However, it is important to study and to model the ways in which the interaction between induction and selection actually occurs, and the constraints imposed on it. To do this it is necessary to define the unit of selection and variation.

3.5   The 'unit' problem

The nature of the 'unit' of selection and variation seems to me the most difficult problem associated with understanding cultural evolution (see Mokyr, ch.5; Fleck, ch.18). Once the 'unit' is agreed upon, it is possible to develop models based on our understanding of other inheritance systems. Technological evolution is somewhat simpler than other domains of culture, because it is easier to define a 'trait' and follow its transformations through time. For example, it is possible to treat the technological history of the bicycle like a very good paleontological record, and use a modified form of models of phenotypic evolution developed for the analysis of paleontological data. This could provide interesting information about rates and patterns of evolution, levels of divergence and convergence, reiterative evolution, and the steadiness of trends. However, if we wish to understand the mechanisms of 'induction' and selection themselves, the psychological and social context of discovery, and the social circumstances of the dissemination of innovations, the 'unit' problem returns.

Dawkins[39] has suggested that the unit of cultural evolution is a 'meme' (§1.2; §18.5). But from the point of view set out here, this presents a problem. The choice of a 'meme' (like that of a gene) as a unit of selection isolates one component of the evolutionary process and leads to the consistent neglect of other inheritance systems and of interactions that lead to the reproduction of phenotypes. It leads to the illusion that the evolution of phenotypes has been explained, whereas it has only been very partially and often misleadingly described.

This problem is more serious than with genes, because the origins of variation in DNA are assumed to be random with respect to the selective effects of environment and there is a single mechanism of DNA replication that is unrelated to the function and structure of the genes it replicates. There are multiple mechanisms underlying the origin and the reproduction of 'memes', and they are not independent of the way the 'meme' is expressed. It is impossible to ignore these mechanisms even for heuristic purposes: they are part of the reproduced 'unit'.

The most effective and useful approach to this problem is to use the developmental framework suggested by several philosophers of biology. Thus, James Griesemer[40] suggests that to truly include development in evolutionary investigations and avoid the danger of explaining evolutionary change only in terms of the DNA inheritance system, we should focus on the analysis of reproduction (or generation) rather than on replication. We should consider all the developmental processes and recurring circumstances that lead to the reproduction of the traits of the organism.

Similarly, Griffiths and Gray suggest that the unit of selection should be the *life-cycle* of the developing individual.[41] They define the developmental process underlying the life-cycle as a series of interactions with developmental resources that display stable recurrence in the lineage. What is reproduced are the interactions, and this occurs through several different inheritance systems as well as through feedback interactions between the organism and the environment which re-produce the initial condition leading to the repetition of the life-cycle. Griffiths and Gray claim that '. . . the individual, from a developmental system perspective, is a process – the life-cycle. It is a series of developmental events which form an atomic unit of repetition in a lineage. Each life-cycle is initiated by a period in which the functional structures characteristic of the lineage must be reconstructed from relatively simple resources. At this point there must be potential for variations in the developmental resources to restructure the life-cycle in a way that is reflected in descendant cycles'.[42]

The resources that have this potential are those that underlie different inheritance systems or that lead to stable feedback interactions between the organism and the environment. According to the life-cycle approach, all inheritance systems and all feedback interactions that lead to the re-occurrence of the life-cycle are part of the 'unit'. This approach is useful for evolutionary theory because it focuses attention on the diverse processes that lead to the generation and reproduction of heritable phenotypes. It is particularly important when discussing cultural evolution (see Fleck, ch.18) because this evolution includes several levels of generation and reproduction of recurring variations.

Explanations of cultural evolution, and even relatively 'simple' examples of technological evolution require an approach that incorporates several levels of organization and focuses on the special processes occurring at each of them. EISs can provide a helpful analogy and are useful for the description of some aspects of technological evolution. They encourage a multi-level and multi-process approach to heredity and information, show that in some cases (when information is analogue as with steady-state systems) the transfer of information is dependent on the specific nature of the information transferred, and illustrate how induced or learnt heritable variations can be selected.

But EISs also have their limitations, because the type of processes that generate and perpetuate cellular epigenetic variations are very different from those occurring during cultural transmission. Adhering too closely to analogies can be hazardous. Nevertheless, I believe that our growing knowledge of EISs may provide historians of technology and evolutionary epistemologists with useful alternative ways of thinking about the history of culture.

Acknowledgement. I am very grateful to Marion Lamb for her invaluable advice and constructive criticism, and her help with the figures.

4

Selectionism and complexity

JOHN ZIMAN

4.1   Adaptation by selection

Evolutionary theories of technological, scientific, economic and other forms of societal change have usually been modelled directly on 'bio-organic' evolution (Jablonka and Ziman, ch.2). But organisms and ecological systems differ significantly from cultural systems in several fundamental features (Ziman, ch.1; Jablonka, ch.3). Quite a lot of thinking and experimentation has been done recently on more generalized 'selectionist' systems which might turn out to be more appropriate models for cultural evolution. But attempts to follow this line of thought[1] tend to take a very schematic view of human society. What we need here is some idea of the defining principles and actual behaviour of these systems before we try to work out their possible cultural or epistemological implications.

The starting point must surely be Donald Campbell's acronym for the essence of Darwinism:[2]

BVSR = Blind Variations Selectively Retained.

In other words, all that is necessary for a population of variable entities to *evolve* is (a) a mechanism for inducing variation, (b) a consistent process of selection, (c) a means for preserving and propagating selected variants. To be precise, this formula requires a recursive factor, such as[3]

G–T–R + T = Generate–Test–Regenerate + Transmit,

to show that evolution is a *cyclic* process, involving more than a single episode of 'trial and error'.

41

The BVSR acronym clearly sums up the basic processes of biological evolution (§2.1). But unlike the widely quoted formula[4]

RIL = Replicator–Interactor–Lineage,

it does not differentiate 'genotypes' from 'phenotypes'. This distinction is characteristic of neo-Darwinism (§3.2), and is often very significant in biological systems. But it was not made by Darwin himself, and is obviously meaningless in the case of many other evolving systems. In natural languages, for example, words and other linguistic entities are varied, selected and replicated through direct interaction with their social environment, without reference to some more basic 'code'.

Historically speaking, the key factor in BVSR is the **S**. What distinguished Darwinism from most preceding theories was that it showed how historical change could be shaped by (natural) *selection* rather than by (divine) *instruction*.[5] But *selectionism* necessarily implies the factors **V** and **R** in equal measure. There must clearly be a sufficiently numerous and diverse range of variants for the selection mechanism to operate on, and an effective means for retaining and multiplying those that are thus preferred.

The **B** factor – 'Blindness' – calls for more discussion. It cannot be equated with being completely 'random', since this would require careful definition and metrication of a *search space* – that is, the hypothetical set of all the variants that might conceivably be put up for selection. In general, as in the neo-Darwinian case, this space is so enormous (§2.2) that it reduces all absolute measures of probability to infinitesimals. What really counts is the *relative* blindness with which variants are generated, especially in relation to their expected adaptive value. This is a logical necessity, since the selection process would be superfluous if it could have been pre-empted by a reliable calculation of its result.

The extent to which variations are 'blind' to their outcomes is thus a significant parameter in any evolutionary process. For example, the efficiency of selectionism as a search strategy (Miller, ch.15) is often greatly improved by reducing the search space to variants of proven adaptive value.[6] It is not helpful, however, to interpret this parameter as a demarcation criterion between 'Darwinism' and 'Lamarckism'. Even in bio-organic systems, which are usually supposed to be perfectly 'Darwinian', we find variations that are subsequent to, and related to, experience of their use (Jablonka, ch.3). Pre-selective 'biases', based on rational extrapolations from individual or collective experience, are prevalent in all forms of cultural evolution. But it is an empirical question in each case whether they are so influential that they effectively *determine* the outcome of the selection process.

In general, a selectionist system, like a 'stencil', requires a large starting population containing a superabundance of 'useless' variants, and leaves behind it a trail of 'failed' entities. It is thus apparently less economical than an instructional mechanism, which can rely on inferences from a limited number of prior observations to design a unique future. But this *superfecundity* (§5.1) or *profligacy* is the price that must be paid (Perkins, ch.12) for close adaptation to changing circumstances and genuine progress into new domains of being.[7] Instructionism is not only unrealistic in its postulation of an all-knowing deity or homunculus.[8] It is also uncreative, in that it seems to guarantee correspondence between knowledge and the world, whereas it actually assumes a pre-categorized world.

The BVSR formula applies to populations of discrete entities, but it does not require that the units of variation and of selection should be the same nor that they should operate at only one level. Indeed, bio-organic evolution itself does not satisfy these requirements (§2.4; §3.5). Nor do the general principles of selectionism make any distinction between 'natural' and 'rational' selection. They do not exclude 'design' in the generation of variants, nor do they exclude 'prudence' in their selection.[9] Once again, these are empirical features of particular systems which do not necessarily have a significant effect on their evolutionary behaviour.

What is more, there is nothing to say that variants need to be selected directly on their own adaptive merits. In practice (§5.3; §12.3; §13.3; §16.4; §20.4), they are more often retained preferentially through the agency of *selectors* using *surrogate* criteria of 'fitness'.[10] For example, an artefact may be preferred for the 'elegance' of its engineering design, with the unstated supposition that it will function with corresponding efficiency. Thus, in cultural evolution, some of the traits of surviving variants may be ill-adapted – or even downright antisocial[11] – for irrelevant reasons. But that means that notions such as 'fitness' and 'progress' are not independent of their context (§15.5): they are only ways of talking about the criteria by which entities are actually selected, or about other qualities that seem to have contributed to their survival.

Selectionism requires a robust mechanism for the Retention, Transmission and Replication (RTR) of selected variants. But this does not necessarily involve a substratum of stable entities – 'memes', say (§1.2) – analogous to bio-organic genes in longevity, fidelity, and fecundity. Under conditions of 'natural' selection, such a mechanism clearly has peculiar evolutionary advantages. In particular, a 'gene pool' can retain 'information' about cyclic phenomena repeated within the intervals between successive replicative generations, and hence, by transmitting genetic instructions across the generations, project into the future with pre-adapted organisms.[12] 'Rational' selection, however, can operate without such a mechanism, since it has access to information stored and transmitted in other forms, such as memories, archives, artefacts and theories.

By its very nature, however, the behaviour of a selectionist system is 'irrational', in that it is *unpredictable* in detail. Its 'blindness', however limited, implies an element of chance in the way that it evolves. This means, moreover, that its historical trajectory is *irreversible*: there is simply no guarantee that a step 'backward' will restore it to its original state. It also means that every outcome is *path-dependent*:[13] there is no in-built mechanism to over-ride the chancy factors so as to arrive eventually at precisely the same point.

Nevertheless, the behaviour of a well-constructed selectionist system is not arbitrary, for it is constrained by its selective environment. And if there is only one rather good way of adapting to these constraints, then this way will most likely be discovered. Thus, evolutionary processes starting along different paths to 'solve' a well-defined practical problem – for example, to design a very fast aircraft – often converge on essentially the same 'solution' (Constant, ch.16). In trivial cases it may be possible to prove analytically that this is actually *the* solution to the problem. But this cannot be assumed in general. One of the characteristic features of all evolutionary processes is that they produce *satisficing* outcomes, not globally optimal ones.

4.2   Other selective systems in biology

Whenever we talk about Darwinism, we refer almost exclusively to its prime example – the evolution of biological organisms through natural selection. But the general selectionist principles discussed above are not mere abstractions with a single domain of exemplification. As this book shows, they operate widely in the realm of human culture. What is more, they can be seen at work in many other aspects of biology.

Ecology, for example, is typically selectionist in principle. The grandest biological application of selectionism must surely be the *Gaia* hypothesis,[14] which characterizes the earth as a self-regulating system where there is extensive feedback between organic evolution and the physical environment. Not only do organisms, species and ecological systems adapt to this environment: the atmosphere and oceans are strongly influenced by changes in the biosphere and the two systems 'coevolve' within a stable envelope which ensures the survival of life itself. Even if we do not buy this argument as a whole, we can appreciate its selectionist rationale.

Selectionism also operates *inside* organisms, as well as in their external relationships. Mutual adaptation may be an internal–internal relationship, as it might be between different bodily organs in the same organism. A list of biological entities involved in *somatic selection systems* would include self-replicating molecules (e.g. DNA), cellular organelles, cells (e.g. lymphocytes), neural groups, and brain functions such as memory.

Complex bodily systems such as the immune system, the nervous system and the brain are not only the products of bio-organic evolution. They each employ 'the memory of selective events' as a basic principle of development and operation.[15] Thus, for example, the *immune system* (§3.3) is almost perfectly 'Darwinian'. This system manufactures an immense variety of 'recognizer molecules' in advance of receiving any information about the shapes to be recognized. But when one of these interacts selectively with, say, a foreign protein, it sets in motion a mechanism for its own replication as a lineage of specific antibodies to this antigen.

Again, in *topobiology*, it is observed that multicellular organisms are given heritable shape by the sequential expression of genes coordinating the growth, rearrangement and specialization of groups of cells. But the genome could not possibly contain enough information to specify the position of each particular cell in the mature organism. It may well be that all organs actually develop by selective retention from a multiplicity of proliferating cells.

The *nervous system*, in particular, is apparently shaped by a BVSR process during the embryonic development of each individual. The neuronal connections are not identical even in identical twins. According to the theory of Neuronal Group Selection, these connections are the outcome of a process of 'neural Darwinism'.[16] Groups of neurons engage in topobiological competition for their positions in the primary neural network, from which functioning circuits are then carved out by the selection of activated synapses. 'Re-entrant signalling' then connects and coordinates the various 'mappings' thus constructed. In other words, selectionism operates at several levels in the self-organization of the brain and its adaptive fit to the world in which the organism is living as it develops.

More metaphorically, the sonar systems of bats and dolphins rely upon the selective reflection and 'retention' of sound pulses emitted 'blindly' by the animal seeking knowledge of its environment.[17] The blind man's stick can be represented in similar terms. And as we move out of biology in the narrower sense into the regions that it shares with the behavioural and human sciences, we find that selectionist *mental* processes are postulated to explain many aspects of memory, consciousness, problem-solving and other forms of intelligent action.[18] These, in turn, shade into the realm of *sociobiology*, where behaviourial traits *coevolve* genetically and culturally.[19] This brings us to the evolution of natural languages,[20] and of the other cultural institutions that are the subject of this book.

4.3   Selectionist methodology

What we need at this point is a general theory of selectionism, backed up by empirical research. The difficulty is, however, that none of these other selectionist systems has been studied as such in great depth. Almost all that we know about them is strongly coloured by direct analogy with bio-organic evolution. It is all too tempting, for example, to see in them taxonomies that mirror 'the Tree of Life', even though they are not constrained by the same rules of inheritance as sexual organisms.

We do know, however, that selectionism *works*, in the sense that a population of entities subjected to repeated BVSR operations evolves towards one that is better adapted to the environment defined by the criteria of selection. As a broad generalization, this is rather obvious, but experiment has demonstrated the speed with which remarkably 'fit' entities can be generated under appropriate circumstances.

This can be seen by the development of practical *selectionist methodologies* in various fields of science and technology (Miller, ch.15). Automatic techniques now make it relatively easy to set up a large, highly diversified population of entities, subject them to artificial selection in favour of some desired property, and then multiply successful entities for practical use, or retain them for a further cycle of variation and selection.

The pharmaceutical industry, for example, has long been using a selectionist methodology. In the search for a new drug, thousands of slightly different chemical entities are synthesized systematically in the laboratory and 'screened' for a particular therapeutic indication. In recent years this process has been simultaneously naturalized and automated. Fast-breeding, rapidly mutating strains of bacteria, into which a particular gene has been spliced, are used to produce millions or billions of variants of the enzyme manufactured by that gene. Mutants that show some activity in performing a desired function, such as catalysing a particular chemical reaction, are then selected and amplified by molecular-biological techniques – and so on. This technology is rapidly growing in scale, scope and industrial application: it has even been used to train RNAs to act as enzymes.[21]

Biological mechanisms are avoided altogether in the use of *genetic algorithms* (GA) to solve technological problems.[22] These work precisely according to the BVSR formula. A great many different computer programs are systematically tested and selected for their efficacy in performing a prescribed task, such as specifying an efficient design for an engineering artefact, then randomly mutated or paired for reselection in an automatic cycle. This is a more subtle methodology than it might seem at first (Miller ch.15), but the heir to thousands of generations of this 'blind watchmaking' is often as good as, or even better than, the most carefully engineered program to perform the desired task.

4.4   Emergent properties of computer models

The most compelling exemplification of selectionism in action is so-called *artificial life* (AL).[23] This slightly pretentious term is applied to a variety of *computer simulations* of natural selection. This activity originated with the invention of self-replicating *cellular automata* (CA), in the form of Q-shaped 'loops' on a computer screen divided into cells that interact with neighbouring cells according to prescribed rules.[24] These computational objects can not only be programmed by a genetic algorithm to change or recombine their 'genetic' elements randomly as they 'reproduce'; they can also be allowed to 'feed' on each other, compete for 'resources', etc. And after each cycle of interaction, they are subjected to the analogue of natural selection – that is, as in real life the criterion of adaptive 'fitness' is 'survival'.

The formula 'AL = CA + GA' does not describe all that is now going on. This is a research field that is itself evolving and proliferating in many directions. It is irrelevant whether AL systems can be said to constitute 'life' as we have hitherto known it. The essential point is that these systems, although very varied in detail, are all designed according to simple, general selectionist principles. The evolutionary potentialities of these principles thus show up in the emergence of characteristic behavioural patterns, even though these have not been consciously preprogrammed, and regardless of whether they are actually observed in real bio-organic or other selectionist systems.

Here again, we cannot avoid describing what happens in 'biomorphic' terms, just as if we were observing living organisms. But we must always remember that this is no more than a convenient and compact language for describing the behaviour of certain patterns on the visual display unit, or of data derived from the inspection of certain memory registers, of an electronic computer. AL models are genuinely *non-biological*. What is more, they can be set up so that they are quite free of specifically bio-organic structural features, such as the distinction between genotype and phenotype or infertility constraints between separated lineages.

In practice, however, AL models tend to be biomorphic in structure, since they have mainly been developed to simulate various stages and levels of the standard Darwinian account of biological evolution. Thus, 'VENUS' uses artificial 'chemistry' to simulate a 'pre-biotic' information soup, where 'organisms' – persistent patterns of computer instructions – emerge and survive.[25] 'Tierra' is modelled on a later stage where digital 'organisms' compete for computer processing time and memory space.[26] Another system follows the evolution of entities labelled laconically '*' that are able to 'feed' from a heterogeneous environment – even up to the level of apparently deferring instant 'gratification'.[27] Other systems are task-oriented, either very abstractly as in GA 'Ramps' competing to solve a formal computational problem,[28] or more pictorially as CA 'ants', which compile 'knowledge' of their environment into their 'genetic structure' as they compete to follow difficult 'trails' across the computer screen.[29]

The most surprising feature of all these models is the degree to which they are reported to have exhibited many familiar characteristics of bio-organic evolution. Some of these characteristics are fairly obvious, such as the way in which 'ants' that have become very efficient at following a certain type of trail prove too 'overspecialized' to diversify and evolve rapidly to perform a new type of task. The emergence of 'parasites' in Tierra might also have been predicted, although it is remarkable that here, as with Ramps, the presence of such parasites actually seemed to enhance the 'fitness' of their 'hosts' in relation to their set tasks, as well as in the 'struggle for survival'. The emergence of parasites evidently leads to much increased, un-programmed diversity in the composition of the population

What is more interesting is that one can identify the mechanisms that are principally responsible for particular evolutionary phenomena. It is found, for example (§15.3), that 'genetic recombination' is sufficient to drive the evolution of GA systems very effectively, with little or no actual 'mutation'. This is because random 'crossovers' tend to preserve important, high fitness, genetic building blocks, once they have emerged, and allow them to carry over into the next generation. In general, this process tends to generate compact but effective building blocks, which would be less likely to be split by recombination.

A feature of all Darwinian systems is that they tend to get caught by 'local peaks' in the adaptive landscape (Perkins, ch.12). 'Ramps' that can exercise the option of 'sexual reproduction' can escape to search for higher peaks, or can even be pushed off such cosy local hills by awkward test cases devised by coevolving 'anti-ramp' parasites, leading in the end to improved, more robust solutions to the computational problem that measured their fitness. Typical coevolutionary 'Red Queen arms races' (§2.5) emerge and can be followed in detail in the interacting genomes.

Another significant feature of some AL systems is that during periods of apparent 'stasis' (§2.6) the underlying genetic make-up of individual entities may in fact be changing and diversifying. The 'gene pool' may accumulate a number of 'recessive alleles' of mutually interdependent genes that cannot be expressed unless all are present. At a certain critical concentration of this 'epistatic' gene complex, a rapid transition to a higher state of fitness may then sweep through the population. In effect, the equilibrium is 'punctuated' by the emergence of novel features associated with the assembly and replication of an interacting 'genetic packet' that seems superficially to behave like a mutant single gene. Long-term stasis in a coevolving population may also be punctuated by the emergence of 'hyperparasites' that divert to themselves some of the metabolism of existing parasites.

An even more subtle observation confirms the *Baldwin effect*, hypothesized over a century ago. A population of organisms that have individually learned a new capability can apparently get evolutionary rewards for traits that contribute to the acquisition of this capability – such as an internal neural net learning system – even if the capability itself was not inherited. On the other hand, this trait may degenerate if the adaptation that it facilitates is too perfect.

I am not citing sources for these reports, nor presenting them here as well-founded scientific facts. In many cases, no doubt, they are strongly coloured by wishful thinking. This is a field where the clear light of scepticism has to shine through swirling mists of imaginative enthusiasm. Nevertheless, such interpretations of the behaviour of simplified selectionist systems are not fanciful in principle, and are obviously of great significance for evolutionary theory. They deserve to be taken seriously, and carefully checked out by more critical research.

As the substance of this book amply demonstrates, cultural systems have non-biological structural features (§1.2) that are very difficult to incorporate plausibly in a comprehensive computer model. But they often satisfy the general principles of selectionism, and may thus be expected to show the sort of evolutionary behaviour observed in most models that incorporate these principles, however schematically. In other words, by relating cultural phenomena directly to the properties of such models we may perhaps escape from the hand-waving rhetoric of the biological 'metaphor' towards a more rigorous theory of cultural evolution.

4.5   Complexity theory

Most systematic studies of selectionist systems have relied on the results of computer simulations. Can this be avoided? Can evolutionary phenomena be analysed by more formal mathematical techniques, such as the solution of coupled differential equations? Certainly, there is a very considerable body of mathematical theory in evolutionary biology and economics which would seem generally applicable to all selectionist systems.

Unfortunately, the results of such applications are seldom very informative. Typically,[30] one may calculate the average effect of the introduction of one or two fitness-changing mutations into an otherwise static population, usually in terms of a few arbitrarily chosen parameters which cannot be directly related to other measurable properties of the system. The results may look perfectly reasonable, but are not normally sufficiently surprising or precise to test against experience.

In reality, the behaviour of BVSR systems is literally incalculable. By definition, they are intrinsically stochastic in their 'blindness' and strongly non-linear in their 'selectivity'. They are not like thermodynamic systems, whose properties are almost completely determined by statistical averages over enormous numbers of almost identical states. The interesting properties of a selectionist system derive from the dynamical magnification of extremely *rare* events – for example, the one bacterium in a hundred million that happens to be resistant to a new drug, or the once-in-a-lifetime combination of unusual ideas into a novel inventive insight. These are not even 'singularities' in the conventional mathematical sense: they simply cannot be represented in the traditional language of functional analysis.

The mathematical concept of a 'fitness landscape' (Perkins, ch.12) is frequently invoked in evolutionary theory, but more often as a metaphor than an analytical instrument. The evolutionary trajectory of a BVSR system obviously depends on the relationship between the probabilistic parameters of its 'search algorithm' and the statistical properties of its fitness function, but this relationship is much easier to explore quantitatively by computer simulation than by abstract mathematical reasoning.[31] The same seems true of game theory – for example, in variations on the theme of the 'prisoner's dilemma' – whose fruitful applications to sociobiology and evolutionary psychology may be expected to diffuse into cultural evolution.

Where formal mathematics may eventually contribute to evolutionary theory is in the analysis of *complexity*.[32] The entities that evolve in most biological systems are themselves, very often, 'complex systems', whose variants are the outward products of internal change. That is to say, they consist of many closely connected components, all in dynamical interaction – biomolecules within organelles, organelles within cells, cells within functional organs, functional organs within organisms, organisms within ecosystems. Generally speaking, these interactions are permanent but haphazard: a specific type of biomolecule, for example, always interacts with or catalyses interactions involving certain other biomolecules, but the pattern of interaction is different from one type to the next.

The question is: how is it possible for such an entity to develop the combination of stability and flexibility required to take part in the selection process at the next level? The answer seems to be that, as the connectivity of the system increases, it passes from a 'frozen' state of complete rigidity, through a relatively broad dynamical regime of internal 'self-organization', until it falls over a sharply-defined 'edge of chaos' into complete disorder. What is surprising is that this occurs even for a system whose internal connections and interactions do not follow a regular plan.

Although this result does not seem to have been proved formally, it is widely-enough confirmed to be taken as a general property of all such systems. It is even conjectured that the BVSR mechanism drives an evolving interactive system towards this transition, since this is where it can adapt most flexibly. Many workers in evolutionary biology now support the hypothesis that *living systems exist in this boundary regime and that natural selection achieves and sustains just such a poised state*. This notion is clearly applicable to cultural entities, despite the moderating effects of artifical selection guided by human intention.

Complexity is another country: they do things differently there. It seems essential to learn an appropriate language for, say, characterizing a system by the diversity of its components and their interactions, for providing a natural definition of the 'function' of a 'part' of a complex system, or for interpreting evolutionary drift towards a phase transition between 'sub-critical' and 'supra-critical' behaviour. This type of analysis is still far from established as a formal theoretical discipline, but it is very instructive in showing that functionally integrated, self-constructing, far-from-equilibrium systems do have their own laws and law-like patterns of behaviour.

5
_____

Evolutionary phenomena in technological change

JOEL MOKYR

5.1 Introduction

We now begin to look directly at the nature of technological change. This is a subject that is now widely studied by economists, typically as an adjunct to the study of economic change in general. The long-standing aspiration that economics should return to its affinities to biology[1] leads those who want to apply evolutionary theory to certain issues in economics to develop detailed economic models where there is a counterpart to each element in that other discipline.[2] The development of evolutionary models outside biology seems to suffer from the same need to shoehorn each concept into analogous concepts in evolution. Such correspondences may indeed be instructive and entertaining, but the usefulness of an evolutionary theory of economic change does not depend crucially on every element in economics being mapped onto a precise correspondence in evolutionary biology. In this chapter I shall argue that this is because the system Darwin was describing is itself only a special case of a much broader set of dynamic theories (Ziman, ch.4) that are often named Darwinian in his honour.

In general, the evolutionary model refers to a system that describes the history of a population composed of evolutionary 'entities'. These typically belong to larger groups of identical or very similar units, which in turn may be usefully classified into even larger families. In the biological world, an entity would be a specimen organism, and the group to which it belongs is the species. The manifested entity is to be distinguished from the underlying structure which constrains its traits but does not wholly define it. Every evolutionary system consists of those elements. In biology, the underlying structure is the genotype, while the manifested entity is the phenotype. In evolutionary epistemology[3] the underlying structure is the knowledge base whose elements determine the traits of cultural entities such as words, artistic forms, ideas, customs, and the like. These elements thus correspond closely to what Dawkins[4] termed 'memes' (§1.2).

Selection occurs because of superfecundity. Any Darwinian system must select, because there are more manifest entities than can be accommodated. The exact way in which this works differs, of course, from context to context. Darwin's famous insight was that each species produces more offspring than can be accommodated, that the reduction in numbers was not random but directed, and that this 'directedness' favoured certain traits. In other, non-biological systems, it seems less obvious why there needs to be selection. I will address this issue below.

Finally, any evolutionary system has certain dynamic properties which connect the present to the past. Inevitably, the past in some way constrains the present, as present characteristics are in large part inherited from the past. By and large, the change of any variable is 'local', that is, it is unlikely for it to change very much from period to period. There are extensive debates on how much 'very much' precisely is, and whether the distribution of possible changes contains, albeit at low probability, changes that can alter the traits of the entity so dramatically that it can produce discontinuities, saltations, and 'punctuating' events (§2.6; §4.5). Such debates are conducted both on a theoretical and an empirical level, and their conclusions are likely to differ depending on the context and the nature of the entities under investigation.

As we have seen (Ziman, ch.4), experiments with highly simplified computer simulations, such as Conway's 'game of life',[5] show clearly that the standard Darwinian model of evolutionary biology is a very special case of a large and diverse set of possible models describing dynamic systems of this nature. What I am proposing to do in this chapter is to define sets of technological information or knowledge and characterize the way in which they behave historically. In doing so, however, I cannot avoid making direct comparisons with biological evolution, not as a stereotype to which all evolutionary change must conform but because it has been studied in such depth that it provides a convenient vocabulary of concepts and processes for which there are often technological analogies.

5.2   Techniques and evolution

As I have argued elsewhere,[6] the unit of analysis (§3.5) that makes sense for students of the economic history of technology is not the artefact[7] but the 'technique' – or, what is often the same thing, the 'routine'.[8] What is a technique? Essentially it is a set of instructions on how to do something that involves production. The set of *feasible* techniques, which I shall call λ, is based on a much larger set, the set of *useful knowledge*, which I shall refer to as Ω. The relation between λ and Ω is somewhat reminiscent of the relation between the phenotype and the genotype: the genotype constrains what the phenotype can be but does not solely determine it. We do not know exactly how knowledge produces techniques, any more than we know exactly how biological genes produce phenotypical organisms.[9] Yet the analogy with 'knowledge' illustrates the limitations of the biological paradigm in explaining cultural phenomena such as technology, where there can be feedback from the manifest entity to the underlying structure.

The mapping from the set of useful knowledge Ω to the set of feasible techniques λ must be one of the central notions in any evolutionary model of technology. It includes the entire relationship between scientific knowledge and its application. But Ω contains a great deal more than science. The mapping is just as real when techniques are based on custom, superstition, or false theoretical concepts, but lead to practices that are in use. The phenotype itself exhibits 'traits' that determine the likelihood that the technique will be selected. The knowledge base of a given technique can be broad or narrow. In the extreme case of a 'singleton' technique – for example, the use of Cinchona bark, containing quinine as a cure for malaria – the only knowledge in Ω that supports it is that it actually works.

Much as in the ontogeny of a living being, there is a distinction between the technique, which is a set of instructions, and what it produces. Just as genes code for certain proteins which then become building blocks for certain traits and organs, the instructions on how to make a pair of shoes or artificial sweeteners differ conceptually from the product or the outcome itself. Selection picks, in the final analysis, on outcomes, not genes, although as long as there is a one-to-one mapping from one to other – or something close to it (but see Jablonka, ch.3) – this should make little difference. If nobody uses the technique anywhere, it can be thought of as extinct. Yet, unlike the biological case, extinction is not irreversible, unless the knowledge underlying it is lost as well and cannot be rediscovered. Such cases are rare in the history of technology. The prevalence of a technique is thus measured by the number of its occurrences and each occurrence is equivalent to an 'entity' living 'one life'.

In some sense a technique is a 'vehicle'[10] for the information underlying it. But this is where the analogy can be misleading. In biology, genetic information cannot exist without a living vehicle. Once the last specimen of an animal species dies, the genes are just as extinct as their carrier. Information, however, can survive outside the technique and evolve quite separately (Stankiewicz, ch.17). Moreover, whereas techniques are 'vehicles' for the information they use, they themselves cannot exist without vehicles such as firms or households. When a technique is used, this happens because some other entity like a firm or a household has consciously chosen to use it, selecting it from a much larger set of potential techniques. To complicate matters further, these vehicles are themselves often the subject of selection. This produces coevolutionary processes, and 'hierarchies' of selection that are still highly controversial amongst evolutionary biologists.

As noted, one of the most challenging issues in the history of technical knowledge is the relationship between the underlying information and the technique itself. One could formulate this in terms of engineering problems and roads to their solution.[11] But it does not seem to be true in general that operational principles have to be understood before design can proceed. Historically speaking, most techniques worked, at least till about 1850, without their designers or users having the slightest idea of their operational principles (Martin, ch.8; Turnbull, ch.9). Trial and error, serendipity, and even totally false principles (David, ch.10) can lead to techniques that work sufficiently well to survive the selection process.

Typically, knowledge provides tools to solve problems, while techniques embody their solutions. Other mechanisms can be imagined. We could distinguish between a complex technique that needs an almost complete understanding of the underlying knowledge for it to work, such as electronics, and a singleton technique that is based on no more knowledge than 'if you follow these instructions, such and such works'. Even today, many feasible techniques, ranging from psychiatry to macroeconomic forecasting, are based on little deep understanding of the basic processes at work.

The advantage of using the technique as the unit of selection in an evolutionary analysis of the economic history of technology is that it focuses on a meaningful and important historical entity while satisfying the basic characteristics of the evolutionary system. First, new items of knowledge (changes in Ω) are introduced through a highly stochastic process of mutation and selective retention.[12] Such mutations may or may not be expressed in the phenotype, that is, be mapped onto λ; indeed, most are not. If they are not and yet are retained in Ω, they can be 'activated' (that is, expressed) at a later time as part of adaptation to a changing environment such as the emergence of complementary knowledge. This is well exemplified by the early work of Cayley in aerodynamics.[13]

Technological innovations thus differ from biological mutations in not being purely random (§1.2). The search for new techniques is clearly motivated, in some sense, by needs and opportunities. Yet it is a long way from being a deterministic process. We do not get all the innovations we desire because of the limitations of Ω – which is a complicated way of saying that we do not know enough. But more is involved. Even if the necessary knowledge exists in Ω, the technological problem has to be well-formulated, and it must occur to somebody to look for the solution. The existence of knowledge in Ω thus does not guarantee that the desired technique will emerge. Why it does or does not remains one of the central questions in the history of technology.

What is more, there is superfecundity in the system, in that there are usually far more apparently feasible techniques than we need. There is more than one way to skin a cat, but for each cat we can only use one way, and if there are fewer cats than ways to skin them, selection will be required. Moreover, when a new technique is proposed, users have to choose between the old and the new ways of doing things. This produces clear-cut selection in the model. From a shoemaker choosing between different materials for the soles of shoes to a committee of engineers choosing between alternatives of machinery deployed in a modern manufacturing plant, there is a process of selection amongst existing techniques. Yet new variants have to be forthcoming if selection is to continue to play a role. In that regard, the history of the evolution of production technology provides a nice fit to the Darwinian model.

Selection mechanisms also apply to general knowledge, but follow quite different principles. This is because the notion of superfecundity does not apply in its simple version. In principle it would be possible to cram the minds, libraries and hard disks of the world with an ever-increasing body of potentially useful knowledge and to draw from this store when circumstances require it. In practice, however, a great deal of knowledge is forgotten, accidentally lost or is superseded by new discoveries. Yet obsolete knowledge, such as phlogiston theory, could be accessed and revived if we chose to. In short, technological change involves a double-layered process of selection. Selection occurs within Ω, as new knowledge is accepted or rejected, and also within λ, as actual techniques are chosen for use.

How do innovations arise? Despite some obvious dissimilarities, the way in which innovations are produced shares some important features with the way that variant organisms are generated. For one thing, the vast majority of all human knowledge is like 'junk' DNA, in that it does not contribute directly to production. Most scientific (let alone other forms of) knowledge has no applications and does not affect production technology right away. Thus most additions to Ω, like most mutations, are predominantly 'neutral' (§2.2) and do not affect the selection criterion one way or another, although they may likely become useful when the environment changes and calls for adaptation.[14] The activation of such previously inert knowledge may be the evolutionary equivalent of what economists think of as 'induced innovation'. General knowledge, too, is being created at a rate much faster than technological knowledge, but if it finds no application in production it is not considered a part of technological evolution. Some biologists argue that 'neutral' gene changes with no obvious phenotypical effect should not properly be regarded as evolution at all.[15] In that sense the expansion of technology is highly fortuitous: new knowledge that seems to serve no obvious purpose is nonetheless created and retained, and is drawn on when needed. As in genetic drift, the 'survival of the fittest' is replaced by the 'survival of the luckiest'.[16]

At times a truly favourable mutation occurs that does not draw much upon previous knowledge, and is immediately 'selected for'. Nature, like techno-logically creative societies, does not abide by the maxim 'if it ain't broke don't fix it'.[17] Examples of technological mutations of this kind are the discoveries of the smallpox vaccination by Jenner in 1798 and of X-rays by Röntgen in 1895. Neither drew upon any obvious prior knowledge, and yet, despite Stebbins's account of evolutionary progress,[18] they did not remain dormant for long. From the outset these inventions were viewed as 'fit' mutations, and selected over previously existing techniques. In that regard, at least, they were more like Richard Goldschmidt's 'hopeful monsters' – sudden large phenotypical changes possessing higher fitness.

Another area where technological evolution escapes the restrictions of evo-lution in living beings is the process of *recombination* (§2.2) – that is, the appearance of entities combining existing knowledge in new forms.[19] In nature, sexual reproduction can produce substantial adaptive gains, but these are inherently local. Human knowledge and its uses, on the other hand, are not so constrained. Although cultural and technological entities have identifiable genealogies, 'even the most distant cultural lineages can borrow from each other with ease'.[20] The biological requirement that a new organism should have no more than two parents, with highly compatible gametes, is a very special restriction on the Darwinian process in living beings.

Innovation can also occur through 'hybrids' – for example, a small internal combustion engine on board a hot-air balloon – which represent combinations of vehicles rather than of the knowledge embodied in them. Thus, the rules of evolution apply not only to the techniques but to the vehicles themselves. Both natural selection and technological innovation occur in hierarchies rather than exclusively at the smallest level of analysis.

Another similarity is the by now well-understood phenomenon of *exaptation*[21] (§2.5). The basic idea is that a technique originally selected for one trait owes its later success and survival to another trait which it happens also to possess. One might venture that, in the history of technology, exaptation is probably more common than in natural history. Many of the most dramatic inventions of the modern age were originally selected for purposes that were quite different from what eventually turned out to be their most enduring trait. The gramophone, for example, was originally intended by Edison to serve as a dictaphone.[22] Moreover, traits often change together in 'pleiotropic' fashion, as distinct 'bundles'. When aspirin was introduced it represented a 'package' that simultaneously reduced fever, alleviated pain, and, as was later discovered, also prevented heart disease. Unfortunately, it is more usual to find positive traits bundled up with negative ones, which is when economists speak of externalities or 'side-effects'.

5.3   Selection units and replicators

To repeat, the main reason why non-biological Darwinian systems should not be seen as analogous to biological systems is that the latter represent a special case, and a rather limited one at that, of all possible Darwinian systems. For one thing, the supposed main function of the entities upon which biological evolution operates, be they organisms or genes, is to reproduce themselves. Yet this seems to be a uniquely biological point of view, which does not seem appropriate in a general model in which entities are chosen consciously according to some external criterion. In principle, technological innovation is driven by the contribution of a novel technique to some abstract 'objective function' that society tries to maximize. This criterion is then transformed into 'vicarious' criteria[23] (§4.1) operated by social agents such as firms maximizing profit, workers minimizing physical effort, or the like. Technological choices are even made by a central agencies – for example, by the US Federal Communications Commission setting standards for digital HDTV.

A potentially difficult problem is the classification of the entities on which selection operates. Cultural evolution has no objective method of classification comparable to biological systematics (§2.6) or to measures of 'genetic distance' in terms of allele frequencies. Intuition suggests broad taxonomic rules for cultural entities such as languages,[24] although these are arbitrary and often lead to confusion. Thus we may class together all entities that display certain general traits, such as all engines, computer programs, or headache medications. We can classify techniques by their purpose (headache medications) or the physical principles on which they are based (engines) or even morphological characteristics of the process (textiles).

It can be argued that selectionism leads to relatively distinct classes of entities in the sense that we do not find a continuum of intermediate forms between automobiles and motorcycles (although some intermediate forms, such as three-wheeled cars and motor-cycles with side-cars, have been tried with varying degrees of success) any more than we do between cats and dogs.[25] Selection typically favours a particularly advantageous set of traits over possible alternative combinations in its immediate environment, so that the evolutionary process settles down on a 'peak' in the 'fitness landscape' (Perkins, ch.12). In general, the intermediate points between such peaks have lower fitness,[26] so that a creature that is half-way between two existing species is less likely to have evolved. It is worth recalling, however, that the distinctiveness of some biological 'species' can be questioned and that higher biological taxa are often found to have been defined quite arbitrarily.[27]

The problem of replication is more serious. Techniques, to use Dawkins's term, are *replicators*: copies are 'made' of a technique over a certain period and these are reproduced in the next period. This is precisely what gives evolutionary systems their dynamic combination of continuity with change. The process of stochastic variation and selection operates on the principle that units normally reproduce themselves almost exactly. Techniques, however, do not replicate themselves the way organisms do. There are no precise cultural analogies to the natural phenomena of birth, reproduction, and death. Nothing in the world of production technology resembles the way that genetic information is transmitted from parents to offspring during meiosis or in the course of epigenesis (Jablonka, ch.3). Instead, we have to define, somewhat arbitrarily, what we mean by the 'life' of a non-living entity.[28] There are cases in which this seems easy, but the concept of a generation is seldom perfectly clear (§18.9) – although the same can be said of some forms of life, such as plants that are regularly propagated by cloning. Nevertheless,[29] this need not be a bar to the selection and acceptance of a trait over time and the evolution of a population of variant entities.

Moreover, while we can think of techniques as undergoing phylogenetic development, the development of knowledge into a precise set of instructions is quite different from the natural process of ontogeny – that is, the development of a phenotype from its genotype through interaction with the environment. In biology the act of transmission of knowledge is usually considered to occur instantaneously when an organism is born (but see Jablonka, ch.3).

Again, at the end of its reproductive career, the specific genome of an organism is irretrievably and abruptly lost. It is like a machine that is ready to run from the moment it is purchased and has to be completely junked at the very end of its life – 'one-horse-shay' depreciation, as economists term it. In technology, on the other hand, an entity keeps on acquiring knowledge throughout its life, whilst the depreciation of knowledge is often continuous: memories fade, skills blunt, artefacts gradually wear out. Yet the general fact of depreciation remains, and for that reason the technological DNA, too, needs to be passed on over time.

5.4   Vehicles and interactors

Replicators need vehicles.[30] As we have seen, a technique is in some sense an intangible vehicle for the useful knowledge underlying it. Although techniques, unlike biological organisms, are not concrete entities, they themselves require carriers or vehicles. What kind of vehicles exist in the history of technology? The answer is less sharp and well-defined than in biology, because technological memes are more protean than genes. But such vehicles must exist, for if they did not the wheel would have had to be reinvented time after time. The main classes of vehicles are as follows.

Artefacts. Because artefacts exist for long periods and are used over and over again, they embody to a large extent the information in them. Thus the technique of hitting a nail into a wooden board is to some extent embodied in the shape of the hammer. Indeed, some theories of economic growth are based on so-called 'vintage models' which define the technology which they have embodied at a certain date. Yet there are different ways of handling a hammer and exact knowledge is required of where to hold the stem, how hard to hit a nail, how to dip the nail in oil to prevent the wood from splitting, and so on. Whilst the artefact imposes a considerable constraint on the range of the technique, it is possible for changes in other knowledge (or other artefacts) to alter the way existing artefacts are used. The notion that, once the artefact is formed, its 'genotype' cannot be altered is only valid if one defines the genotype as pertaining to the particular 'period' in which the artefact is used.

While an artefact embodies knowledge, it rarely defines the whole technique in which it is employed (Fleck, ch.18). A piano embodies the knowledge of how to make sound, so seeing one is enough to suggest how to produce some sound, but more knowledge is require to produce the *Hammerklavier Sonata*. People can learn from artefacts by means of reverse engineering: once we see a wheelbarrow or a cotton gin we can understand how it is used and reproduce it. The relative permanence of some artefacts also makes it possible for knowledge to survive and be resurrected even when that other vehicle, human memory, no longer carries it. It also greatly facilitates, as we shall see, the process of innovation.

Storage devices. Technological information can be stored in a non-performing device such as a textbook, manual, encyclopedia, or an intangible oral tradition. A cookbook is the paradigmatic storage device, just as a cooking recipe is a paradigmatic example of a technique. A 'normal configuration'[31] – for example the convention that almost all cars have four wheels, a round steering wheel, and an acceleration pedal lower than the brake pedal – is also an implicit storage device.

Firms. One solution to the wear-and-tear of knowledge is to entrust it to an infinitely lived organization whose 'routines' include practical techniques.[32] One explanation of the existence of the firm is precisely to perpetuate technological entities through the production processes it performs (Nelson, ch.6; Stankiewicz, ch.17; Fleck, ch.18; Fairtlough, ch.19). Members of the firm – in the past, often members of the same family[33] – jointly carry the technique. Each time the firm produces the goods it specializes in, it acts as the vehicle of that technique.

Human memory. Much of the knowledge transmitted from period to period is simply embedded in the brains of people, who remember how the technique was used the last time around. Thus, the farmer who grows wheat year after year can be said to have access to knowledge and to map this on to a set of feasible techniques by doing the same thing over and over again (Macfarlane & Harrison, ch.7; Martin, ch.8; Turnbull, ch.9).

Direct communication. Techniques can be transmitted by training individuals, who pass them on in turn to others. A farmer who 'knows' how to grow wheat and does so year after year in identical fashion, can also replicate the technique by teaching his son or apprentice how to do so. If the son then emulates the father precisely, the technique can be said to have been replicated. The transmission of 'tacit knowledge'[34] by on-the-job training, or emulation by neighbours, clearly comes under this heading (§8.4).

But instead of thinking of a technique simply as a 'vehicle', it is sometimes useful (§18.11) to think of it as an 'interactor', that is, first and foremost as an entity that interacts in some way with its environment.[35] This is clearly true of biological organisms. The interaction then is governed by traits that ultimately determine the chances that the unit has for survival and/or reproduction. Indeed, the instructions in the technique lead to the production of goods or services, which are meaningful only in the context of interaction. This feeds into the selection process, since a technique that does not produce a workable product at an acceptable price will not survive.

5.5   Selection and teleology

The outline of an evolutionary theory of technological change is now becoming somewhat clearer. To summarize: because the 'entity' that evolves – the technique – is a procedure or a routine, the main actors in the history of technology – human beings, their organizations and artefacts – play somewhat different roles. They are the vehicles that carry each entity from 'period' to 'period'. But evolutionary biologists debate whether the unit on which selection operates is the gene or the phenotypical organism – the 'information' or the vehicle that carries it.[36] In the history of technology it could similarly be maintained (Fleck, ch.18) that the actual unit on which selection operates is the artefact, the person, or firm rather than the technique as such. Indeed, Darwinian models can also be applied to the evolution of firms, people, or artefacts. It all depends on the question being asked.[37]

Whether or not the notion of a hierarchy of natural selection makes sense in biology, it will be readily recognized that in technological production there must be more than one selection process going on at the same time. Thus:

1   The market selects on the outcomes – that is, which products will be produced, how, and by whom. There is superfecundity here because each society is capable of producing many more products, in many more ways, than it actually does.
2   The vehicles themselves select, and here the selection is conscious: they pick and choose from different techniques in 1. Thus firms select on the actual alternative ways of producing the goods.
3   Another level of superfecundity and selection is at the vehicle level: there are a lot more firms that can produce a good, or a lot more engineers that can learn of some technique, than actually do.
4   Finally, as we have seen, there is some measure of selection at the level of Ω.

Selection in cultural systems thus occurs simultaneously at all levels. Differences of opinion about what are the precise units of selection (§3.5) are inappropriate outside biology. In economics there are no 'selfish memes': there are only objective functions of economic agents operating through markets or other aggregators.

There is another basic way in which technological and biological Darwinian systems differ. In the ultra-Darwinian view,[38] the final purpose of existence is existence itself. A successful replicator is one that has an infinitely lived germ line. In this sense its *raison d'être* is wholly recursive. In a more economistic view,[39] living beings have a purpose which we could define heuristically as 'well-being'. In this view, although reproduction is important, in the final analysis it is epiphenomenal to survival and well-being.

In any case, whatever the 'purpose of life' may be, techniques exist for an unequivocal, deeper purpose – namely to increase the utility of human agents. Each technique, when it is applied, serves an 'ultimate' purpose which, while obviously intertwined and correlated with its fitness (§15.5), can be treated separately. In other words, we can regard a technique in terms of what it was supposed to do – produce goods and services – as well as in its success in reproducing itself. Ultimately any selector will have to be judged by its success in satisfying human needs, and the survival of each entity is correlated with that criterion. But the correlation is less than perfect: at times techniques are selected that do not satisfy the objective function of human need as efficiently as others.

It must be a central axiom of any evolutionary view of technology that, despite this difference in fundamental objective, Darwinian logic can be carried over to systems that do not follow recursive objective functions but serve distinct purposes. In this regard, the assessment of 'success' becomes less relative, and terms like 'progress' acquire a meaning they lack in natural history.[40] For instance, the invention of the safety bicycle in 1885 and its rapid adoption[41] was a success not only in that the number of this species increased relative to penny-farthing cycles, but also in that it satisfied in a superior fashion the ulterior social objective function of people needing to transport themselves safely and efficiently. The ability of an innovator to market an invention, and to persuade others of its qualities, contributes to its chance of reproduction, even though it may not be socially beneficial. Fitness, in an economic framework, thus has two distinct dimensions:[42] the ability of the entity to reproduce itself and its ability to make people materially better off. The extent to which those two coincide is exactly the stuff of which the economic history of technology is made.

5.6   Innovation and adaptation

In an evolutionary framework, change occurs through blind variation and selective retention in Ω. The appearance of new techniques is constrained by changes in Ω. Economists regard changes in knowledge as the outcome of search processes rationally driven by incentives. Evolutionary theorists, by contrast, regard them as a spontaneous, largely autonomous processes, in which knowledge begets more knowledge. The main reason why knowledge is constrained in its expansion is that any evolutionary system has inertial forces which resist change. While this kind of resistance is of course an important source of direction, it almost guarantees that change will rarely be precisely what the system needs.

Do technological entities violate the rules of evolutionary biology (but see Jablonka, ch.3) by transmitting 'acquired characteristics' to new generations? In living beings all genotypic change occurs at conception, but this is not the case in other systems. Thus the 'memome' can 'acquire' characteristics during the lifetime of the vehicle, and pass them on – hence the long literature on the 'Lamarckian' nature of cultural evolution. In some respects, this debate is purely semantic, in that the 'lifetime of the vehicle' is not a well-defined concept.

But technological evolution is indeed 'Lamarckian' in that there is feedback from λ to Ω. The underlying structure of knowledge is influenced by the techniques in use. The strict Weissmanian constraints on the evolution of living beings preclude this: there can be no feedback from phenotype to genotype so that the flow is unidirectionally from Ω to λ. Moreover, in living beings the change in Ω is always purely random, and directionality is imparted only by selection on the members of λ. In all knowledge systems, this randomness in changes in Ω is abandoned, since there can be conscious search for new knowledge which imparts an *a priori* directionality on innovative change. However, so long as it is sufficiently imprecise (§4.1), this pre-selection does not preclude the operation of an *ex post* selection mechanism on Ω as well as on λ. In both of these respects, then, technological evolution differs from natural selection in living beings.

The framework developed here is capable of distinguishing between innovation and adaptation. A change in environment (say, due to changes in the availability of complements or substitutes) will cause selection to favour those techniques that stem from other parts in Ω. This would be pure substitution, as the system simply switches to another part of the existing knowledge base. Yet it may also induce a search process that expands into a different part of Ω. For instance, the appearance of a new disease will focus research on isolating the pathogen and uncovering its mode of transmission – although there is never any guarantee that the desired knowledge will be forthcoming. In other cases, extensions of Ω often serve no purpose at all and seem to be uncorrelated with any new 'need' of the system. In that sense, the continuing growth of Ω is spontaneous and unpredictable. In Campbell's terminology, it is effectively 'blind', with ideas 'occurring to someone' in unpredictable fashion. While these mutations are not really 'random errors', they occur as if the result of some kind of stochastic process, subject to the usual constraints of path dependence.[43]

5.7   Summary: information and selection

Technological information, as emphasized above, is embodied in techniques – coherent bodies of knowledge which instruct 'agents' how to engage in production. I have taken the view, thoughout, that the technique is the 'unit' on which selection occurs. Techniques require information that underlies or generates them. As noted, this information can be carried in a person, an artefact, or an organization such as a firm (Stankiewicz, ch.17; Fleck, ch.18). It can also be preserved in storage devices. While an item of information is thus somewhat analogous to a gene, it does not in any obvious way replicate itself and it can remain dormant for centuries. When knowledge 'replicates' by being copied onto another brain, it is not necessarily an exact copy, since it is not obvious that another individual's brain will interpret the words or diagrams the same way. The only observable phenotype is the technique itself.

Needless to say, some of these selection mechanisms appear to be rather trivial logical exercises, a bit like Daniel Dennett's 'Library of Mendel'.[44] Yet they highlight the fact that understanding what we observe begins with asking why there are things we do not observe. Non-existent forms of life may be absent either because there was negative selection or because the necessary mutations never occurred, although they could have.[45] The difference is due either to selection or to constraints. Selection means that they appeared but were selected against. But if they did not appear, is that because they could not have or simply because that particular mutation never presented itself for selection? In the world of technological selection, non-observed techniques in any given period could be either totally impossible, possible but never occurred to anyone, occurred but were beyond practical reach, and so on. These are all in some sense selection mechanisms. What we actually observe is a tiny sliver of what could have been. Theodosius Dobzhansky famously observed that nothing in nature makes much sense except in the light of the theory of natural selection.[46] This, surely, remains true for the history of technology as well.

6

Selection criteria and selection processes in cultural evolution theories

RICHARD NELSON

6.1   Different perspectives on technological evolution

This chapter develops three themes. The first is that the body of writing that proposes that technological change proceeds through an evolutionary process should be understood as part of a much broader body of recent literature concerned with a variety of different aspects of cultural evolution. Second, the scholars who propose that technology evolves come from several different disciplinary camps and, partly because of this, one can discern several distinct views on the nature of the 'selection processes' involved in technological evolution. Third, technology is at once a body of understanding and a body of practice and, to come to grips with how technology evolves, one needs to recognize both of these aspects and to see how they coevolve.

6.2   Technological change as one aspect of cultural evolution

While the focus in this volume is on the proposition that technology evolves, it seems important to recognize that the body of analysis considered here is a member of a broader class of recent theorizing proposing that human culture, more generally, evolves. Specific bodies of evolutionary writing have grown up around not only technology but also science, business organization and practice, the law, and institutions more generally. The following remarks are drawn from a more comprehensive survey, to which I refer the reader for a detailed bibliography.[1]

Scholars theorizing that human culture evolves come from several different intellectual starting places. One is ethology and sociobiology, fields which see animal behaviour, and human behaviour by inclusion, as as much a product of

Darwinian biological evolution as the structure of the human eye or the shape of the hand. At one end of the spectrum of scholars there are those such as Lumsden and Wilson[2] who see patterns of human culture as closely constrained by human biology, and as being 'selected upon' by the contribution of those patterns to human survival, literally. At the other end,[3] there are those who see a much looser biological 'leash', and often back off from the notion that literal inclusive fitness considerations are what is driving the evolution of human culture.

A related, but somewhat different, intellectual base is evolutionary epistemology which, in the eyes of some of its founders, views the growth of distinctly human knowledge as a natural, if very human, extension of biological evolution. Donald Campbell, of course, played a prominent role in developing this line of argument.[4]

Other explorers in this area, while recognizing the basic biological equipment humans need to have before culture can come into existence, have emphasized the role of shared understandings and symbols, and cross-individual and cross-generational cumulative learning, which in their view makes the evolution of science, or technology, or understanding of human history, or the law, a whole new ball game. More generally, historians, economists, and other social scientists, have long used evolutionary language to describe the processes of change they were studying, without proposing any connections with biological evolution. In recent years a number of social scientists have tightened their language and put forth a cleanly articulated evolutionary theory of cultural change.

The notion that science evolves has a long tradition, with recent writing[5] drawing somewhat conflicting themes from Popper on the one hand, and Kuhn on the other.[6] The argument that the law evolves[7] also has a long tradition, but is less coherent. Alfred Chandler's great work on business history[8] puts forth an evolutionary theory of business organization.

A significant body of evolutionary economics has grown up in recent years, much of it following along the lines mapped out by Nelson and Winter,[9] but with several alternative strands being developed also. The Nelson–Winter strand is focused on the key role of technological advance in driving economic growth. The key actors in that theory are for-profit business firms, competing with each other for market dominance. However, in this theory there also is a technological community among which technology is shared.[10] Both of these themes will be discussed at greater length later in this chapter.

There are significant matters that divide scholars of cultural evolution into different camps, and these will be considered in the following section in the context of discussion of evolutionary theorizing about technological advance.

However, all of them share certain perspectives. The central one, of course, is that the processes of change moulding the aspect of human culture under consideration involve mechanisms that 'select on' an extant variety, and forces that sustain the character of what is selected, while at the same time there also are mechanisms that introduce new departures to the evolutionary system.

There also seems to be broad adherence to the position that cultural evolution often takes place at several different levels. Thus, the use of a new business practice, or the use of a new material to make parts of products, may be spreading at the same time within a particular business unit, across different business units belonging to the same firm, across the firms in the industry, and across industries.

While there is less discussion of this matter, an important feature of many aspects of human culture, and most certainly technology, is that what is involved is at once a body of practice and a body of understanding. Some scholars who propose that technology evolves focus on one of these aspects, while other scholars focus on the other. Yet it can be argued that what makes the evolution of human practice, and especially technology, different from the evolution of animal behaviour as studied by ethologists is exactly that extant human practice generally is supported by a rather elaborate body of reasons, or rationalizations.

The above proposition – that extant human practice tends to be supported or rationalized by a body of understanding or belief – holds for many aspects of human culture, from religion, to business practice, to technology. However, it can be argued that what makes certain bodies of human understanding (for example, science and technology) distinct from others (for example religion, or certain areas of theory about business practice, or about how the economy works) is that in many cases (certainly not all) actual practice provides a frequent and sharp testing of the adequacy of the understanding.

One striking feature of the evolution of technology compared with other aspects of human culture is the rapid pace of change. This obviously is not the case with all areas of technology. Consider, for example, the case of the use of the wheel (Macfarlane & Harrison, ch.7) and the manufacture of swords (Martin, ch.8) in Japan. But, on the other hand, compare the pace of change in semiconductor technology with that of accounting practice.

Behind the rapid advance of technology, compared with other areas of human culture, I would propose that there are two major causes, which I have begun to bring out above. One is that the understanding part of technology often provides relatively strong guidance regarding how to improve practice (Mokyr, ch.5; Stankiewicz, ch.17). Put another way, the 'new departure' generating mechanism in technological evolution is more likely to come up with new variants that are significant improvements over what previously existed than is the case, say, with respect to new business practice.

The argument here certainly is not that technologists can clearly see exactly the nature of the new departures that will solve a perceived problem, or make a desired improvement. Walter Vincenti's wonderful study[11] of competition among various ideas as to how to reduce the drag on aircraft caused by fixed landing gears (Vincenti, ch.13) is a good antidote to that kind of thinking. Indeed a hallmark of evolutionary theories of technological change is exactly that significant advances cannot be planned *ex ante*.

However, compare what we know about how new technology comes into existence with the absence of 'understandings' behind the use of a 'safe period' as a method of contraception (David, ch.10). Or think about what is revealed by the literature on 'business fads'. Eric Abrahamson's[12] discussion of the ideas that led to the rise, and for a while the fashionability, of 'quality circles' is a fine example. Or, along a somewhat different vein, reflect on the studies describing how American automobile companies have struggled to try to understand the factors behind Japanese prowess in that industry. Both of these examples reveal the companies shifting from one theory to another, from one tried reform to another, in a manner that highlights the weakness of human understanding in this practical arena.

The second feature of technological evolution that should be highlighted is that selection criteria and mechanisms often are sharp, steady, and rapid. Studies like those of Vincenti show that it often takes a considerable period of time before a consensus judgement is made regarding the merits of competing technologies. However the sharpness and speed of selection criteria and mechanisms revealed in the technological histories with which I am familiar is striking compared with the vague, shifting, and slow, processes of selection involved in many areas of business practice, or the law.

6.3   Differing views on selection criteria and mechanisms

These are contentious issues. One can identify at least four different intellectual camps of scholars currently working to develop and support the theory that 'technology evolves'. First, there are the exponents of *evolutionary epistemology*. A diverse group in terms of their home discipline, some come from the philosophy of science, some from psychology, some from biology, physics, or other fields of natural science. Many had careers studying other things before turning to the study of how technology and other aspects of human knowledge evolve. They tend to be unified by having been strongly stimulated and influenced in their thinking about this topic by the work of Donald Campbell.

Although there is considerable intellectual overlap, the *historians of technology* comprise a different group. While the evolutionary epistemologists often seem to start with a general evolutionary theory of the development of knowledge, and turn to technology as an interesting area of application, the historians of technology tend to start out with an interest in technology and technical change, and as they study the process come to an evolutionary theory. Many of them discover Campbell, but along the way rather than at the beginning.

Contemporary *sociologists* studying technology comprise a quite different camp. While the interests of sociologists in technological advance goes back a considerable time, and so too does the point of view that social structure influences the path of technological evolution, the 1980s and 1990s saw the sharp articulation of a 'social construction' theory of technological advance.

Economists studying technological advance, who have come to see the process as an evolutionary one, tend to form a fourth camp. By and large (there are exceptions) the economists positing that technology 'evolves' are operating somewhat at odds with the canons of mainline economics. Nonetheless, they tend to be thought of by members of the other three camps as 'thinking like economists'.

These differences show up strikingly in the treatment by different camps of the processes involved in selection or winnowing of technological alternatives. In this chapter we will consider later how it matters which aspects of technology are seen as evolving: practice, understanding, or both together. It seems useful to note here, however, that the evolutionary epistemologists tend to have in mind that the evolutionary process is mostly about getting a better fit between the understanding and the reality (which, however, cannot be completely experienced), while the economists tend to focus on how well a technology fits user needs, with the technological historians and the sociologists sometimes going back and forth between these aspects.

The discussion here will also repress the issue that selection would appear to proceed at several different levels. The following will focus largely on how the different camps see selection at the community or user level.

Partly because of a commitment, following Campbell, to the proposition that technological variation occurs through a process that is to a considerable extent blind, evolutionary epistemologists necessarily place heavy weight on selection as a key force shaping technological change. Similarly, the hallmark of historians of technology who propose an evolutionary theory is that multiple possible solutions tend to exist for most important technological problems, and that which works best can only be determined through actual competitive comparison.

Theorists in these overlapping camps do not display, in my view at least, any particular dogma regarding exactly the nature of selection criteria, or even the selection mechanisms involved. However, in much of this writing the assumption would appear to be that there are (in some sense) natural 'technological' criteria of merit, and that a 'technological community' tends ultimately to converge on what is the best alternative. Partly this orientation reflects a focus on the understanding aspect of technology, and partly perhaps that the scholars in question are very interested in the technological communities involved, in their own right. In any case, while often presented cautiously, the flavour of this kind of theorizing is that technology 'gets better' over time.

The sociologists of technological evolution tend, sometimes gleefully and scornfully, to attack this point of view. The first part of the theoretical argument – that selection is a key mechanism at work – is not at issue. But the nature of selection mechanisms is. The sociologists tend to agree with the evolutionary epistemologists and historians of technology that some sort of a 'community' is involved, but see the community as much broader than simply that of 'technologists'. And the notion of natural compelling technological criteria is emphatically denied. Rather, the selection process is seen partly as the battling of competing interests, and partly a matter of more or less effective campaigns to capture hearts and minds. Criteria of merit are established in these processes, which is a very different thing from proposing that these processes assess competing alternatives against a pre-existing and 'natural' set of performance criteria.

Social constructionists take issue both with the historians who propose natural technological standards of merit, and with the economists who propose that there is a market 'out there' that ultimately is doing the selecting. One way of thinking about the latter difference is that the economists, more than any of the other camps, have a set of users in their theory of technological evolution that is distinct from the community of technologists and other internal actors that is involved in advancing the technology.

Evolutionary theories of technological advance in economics represent the coming together of two strands. One of these stems from the writings of Schumpeter, and kindred contemporary economists, who see competition in many industries as largely involving the introduction to the market of new products by different firms, with the fate of the firms depending to a good degree on how well their new product sells in competition with the products of other firms. The other strand represents the growing awareness, of at least some economists, that technological advance, in fact, does seem to proceed through a process in which there is variation, systematic selection on that variation, and then further introduction of new variety.

Together, these elements have led to a theory that is mostly concerned with the practice aspect of technology, and which presumes that users and the 'market' select on new technologies. Technologies that are better than others, given the criteria of the market, win out because customers buy the products embodying them, and the firms that employ them do well relative to firms that do not. The better technology expands in use, both because the firm or firms that employ it expand relative to other firms, and because these other firms are induced to adopt it.

As noted earlier, this body of evolutionary economic theorizing has been part of attempts by evolutionary economists to make sense of the broad patterns of economic growth that have been experienced. It also has been part of a revival of 'Schumpeterian' notions about the nature of competition in high tech industries, with competition focused on the introduction of better products, and not simply 'price'. These economic theories are 'evolutionary' theories in the sense that innovation and ex-post selection are driving the dynamics. But the market, which is doing the selection, is in a sense 'out there' and determining which technological alternatives survive. It is distinct from the technological community that is engaged in advancing the technology.

To sharpen the contrast, one could say that all scholars of the evolution of technology recognize that ultimately a new technology has to be accepted by those who will use it. The technological historians tend to see technologists making their judgements as agents for the users. The social constructionists see user needs as being defined by a political process, which often has some characteristics of a bandwagon. Economists, at least in their first cuts at an analysis, see the users as knowing what they want and ultimately enforcing their choices

Where do we come out on all this? There clearly are quite different theories of selection here. If I were forced to come down strictly for one position or another, I would strongly lean towards the presumption that the selection environment often is reasonably sharp and steady, and that, in most if not all arenas of technological change, selection criteria reflect both technological effectiveness and user needs (§13.1; §15.5). But the issue is not which of these theories is right, and which ones are wrong, in general. Rather, the different theories are applicable in different arenas. Bicycles differ from aircraft. Both differ from electric razors and pharmaceuticals. Indeed, as the examples indicate, in many cases one can find elements of several selection criteria and mechanisms at work at once.

6.4   Technology as both practice and understanding

What is more, it makes a big difference whether the focus is on the practice or the understanding aspect of technology (Fleck, ch.18). A theory of technological change as an evolutionary process that focuses strictly on practice naturally points its analysis of selection towards how well technology meets user needs. Students of technological change may differ on how and by whom the criteria for superiority are determined in this dimension, but the nature of the debate is clear, as are ways to resolve, or combine, the arguments empirically.

To the extent that technology is seen as not simply a body of practice but also a body of understanding, the nature of the evaluation and selection processes become more complicated. While the criteria for selection on the former aspect may well be 'fit' with user needs, the criteria for the latter would appear to be 'ability to explain observed relevant facts and enable problems to be solved and progress made' (Constant, ch.16). The selection processes and those that control them, as well as the criteria, may be very different. For practice, the process ultimately is under the control of users, or their agents; for understanding, control rests with the community of technologists. There may, or may not, be some overlap.

If the user community is diverse, or there are many different kinds of uses, selection may preserve a wide variety of artefacts and techniques. On the other hand, the body of understanding that is the result of selection may be quite unified.

If both aspects are recognized, it would seem that technological advance needs to be understood as a coevolutionary process (Constant, ch.16; Fleck, ch.18; Constant, ch.20). The manners in which the two aspects evolve, and how they relate to each other, would all seem essential aspects of the story. And these processes would appear to differ significantly from field to field.

One central variable is the strength of technological understanding at any time. Another is the knowledge of the technological community regarding user needs, or its ability to control or define them. If both are very strong, techno-logical advance can almost be planned. This seems to be the presumption behind at least some of technology policy, public and private. However, empirical scholars of technological change can cite many examples of disaster when these conditions were, incorrectly, assumed to hold. In general, they do not. Attempts to develop new and better artefacts and practical techniques almost always involve significant elements of technological or user-response uncertainty, or usually both.

Another key set of variables relates to the sources of new understanding, and the relationships between the processes through which these are won, and the processes that create new practice (Mokyr, ch.5; Stankiewicz, ch.17). Until recently, and even in many fields at the present time, the principal source of understanding is experience with practice (Vincenti, ch.13). Thus a hallmark of much of 'inventing' until recently was that those who used or operated a technology were the most likely sources of improvements in that technology. Understanding was largely a body of lore shared by those who practised a trade.

Nowadays, of course, most technologies are understood, in part at least, 'scientifically'. In some fields of technology, advances in relevant scientific understanding come largely from autonomous developments in science, but in most fields this is not the case. Most fields of technology are now marked by the presence of applied sciences or engineering disciplines whose business it is to advance understanding relevant to practice. These disciplines often draw on more basic sciences but are bodies of understanding in their own right.

And, while conventional wisdom often presumes that advances in understanding come before, and enable, advances in practice, scholars of technological advance know that often it is the other way around, or that the process is strongly interactive. The development of thermodynamics as a scientific field, and the development of the theory of how semiconductors worked, by William Shockley after he and his colleagues got a working transistor,[13] are only two classic examples of the understanding following the practice.

In the engineering disciplines and applied sciences, the intimate relationship with practice suggests that one can sharpen up somewhat the selection criteria for developments that are advertised as new or improved understanding. Do they enable problems in prevailing practice to be solved, or advances to be made, that were not possible, or were more difficult, without that proposed contribution to understanding? This kind of operative selection criterion obviously provides a two-way linkage between the evolution of practice and understanding (Constant, ch.16).

The writings that propose that technological change proceeds through an evolutionary process differ dramatically on how they treat the relationships between the evolution of practice and the evolution of understanding, if they recognize the difference and the relationships at all. Unlike the differences in the treatments of selection processes sketched in the preceding section, the differences here have not so much to do with disciplinary background, as with the fact that scholars of technological advance are only beginning to struggle with these relationships.

II
INNOVATION AS A CULTURAL PRACTICE

7

Technological evolution and involution; a preliminary comparison of Europe and Japan

ALAN MACFARLANE AND SARAH HARRISON

7.1 The industrious revolution

This chapter presents in a rough and preliminary way a few brief accounts of the disappearance of what would seem to us to be useful techniques. It considers the difference between an evolution towards the replacement of human labour by non-human power in Western Europe, the basis of the later 'industrial revolution', and the situation in historical Japan where human power increasingly replaced other forms of power, the so-called 'industrious revolution'.[1] It attempts to show that, even when blind variation is not so blind, and even when retention is very selective indeed, the outcome may still be a situation which, in terms of our post-industrial rationality, is a form of involution rather than evolution.[2]

7.2 The West-European trajectory

Let us start at the European end. Mokyr notes that 'As Lynn White has remarked, medieval Europe was perhaps the first society to build an economy on nonhuman power rather than on the backs of slaves and coolies.'[3] What were these forms of non-human power? The first and most important was animal power. There was the traction power of oxen and then horses drawing the improved ploughs using the better harnesses from the ninth century. Animals were also used in other activities – threshing grain, pulling carts and providing proteins for humans. It has been estimated that by the late eleventh century, '70 percent of all energy consumed by English society came from animals, the rest coming from water mills.'[4]

This takes us to the other two major sources of energy, water and wind. To start with water-driven power; the number of water mills was immense. 'In 1086, Domesday Book listed 5,624 watermills in England south of the Severn river, or roughly 1 for every 50 households. Unlike their Roman ancestors, medieval men and women were surrounded by water-driven machines doing the more arduous work for them.'[5] The water mills were used increasingly for a multitude of tasks. They were used for grinding grain and pumping water, pulping rags for paper, hammering and cutting iron, beating hides, spinning silk. For instance, 'The importance of water-power in relation to the iron industries cannot be over-estimated . . .'[6]

The wheel was also increasingly driven by the wind. 'In power technology, the most important invention of the later Middle Ages was the windmill. The windmill combined the ideas of the water mill and the sail. It, too, may have been imported to Europe by Moslems (from central Asia) but, in spite of its apparent advantages in arid climates, it was not used widely in the Islamic world. The first windmills that can be documented with certainty were in Yorkshire in 1185.'[7] All of these technologies depended heavily on that most basic of inventions, the wheel – for carts, ploughs, windmills and water mills, and later for other inventions such as the clock.

The importance of this development is well laid out by Birdsall.[8] 'Perhaps 80 to 90 percent of the total energy consumed at any one time before the Industrial Revolution was derived from plants, animals and men . . . Once this is understood, we can appreciate the importance of the sailboat, the water mill, and the windmill – three energy converters which made it possible for pre-industrial man to exploit the inanimate energy of the wind and of water streams.' It was this that made European development possible. 'Thanks to the menial services of wind and water, a large intelligentsia could come into existence, and great works of art and scholarship and science and engineering could be created without recourse to slavery . . .'[9]

This 'evolution' towards the increasing use of non-human power, appears, after the event, to be so 'natural' and obvious that it is difficult to realize how extraordinary it is. The exceptional nature of the escape from human labour is only brought out when we look at the 'normal' course of human history, as it is shown in most other agrarian civilizations where such developments did not happen. Whether we look at the ancient civilizations of Mesopotamia, Greece and Rome, the pre-Conquest states of the Incas and Aztecs, the Indian sub-continent or China, we find the same absence of this development of non-human energy. This is a huge subject and here all we can do is to look at limited examples in one civilization.

7.3   The decline in the use of domesticated animals in Japan

It was Isabella Bird, coming from animal-rich Britain in the later nineteenth century who most graphically described the absence of domesticated animals by that period. She was struck by the silence and emptiness of the Japanese countryside. 'As animals are not used for milk, draught, or food, and there are no pasture lands, both the country and the farm-yards have a singular silence and an inanimate look.'[10] There were, for example, very few horses: 'there is little traffic, and very few horses are kept, one, two, or three constituting the live stock of a large village.'[11] Horses were not used for ploughing, nor, even, were they used for carrying. 'Very few horses are kept here. Cows and coolies carry much of the merchandise, and women as well as men carry heavy loads.'[12] So rare were domestic animals even in the later nineteenth century, that they were exhibited like exotic species: '. . . monkey theatres and dog theatres, two mangy sheep and a lean pig attracting wondering crowds, for neither of these animals is known in this region of Japan.'[13]

The situation two centuries earlier suggests that there were more animals, though they were still far from the level in Europe. In 1613, Saris[14] had noted that pigs, goats and even cows could be purchased cheaply. Kaempfer's account at the end of the eighteenth century shows that knowledge of the animals was not lacking.[15] Of pigs, he wrote 'They have but few Swine, which were brought over from China, and are bred by the Country-people in Fisen, not indeed for their own Use, which would be contrary to their superstitious Notions, but to sell them to the Chinese, who come over for trade every year, and are great admirers of Pork, tho' otherwise the doctrine of Pythagoras, about the transmigration of Souls, hath found place likewise in China.' Or again 'Sheep and Goats were kept formerly by the Dutch and Portuguese at Firando, where the kind still subsists. They might be bred in the Country to great advantage, if the natives were permitted to eat the flesh, or knew how to manage and manufacture the Wool.' There were some horses, but not a great number. 'There are Horses in the Country: They are indeed little in the main, but some of them not inferior in shape, swiftness and dexterity to the Persian Breed. They serve for state, for riding, for carriage and ploughing.' 'Oxen and Cows serve only for ploughing and carriage. Of milk and butter they know nothing.' Thus in the most pastoral area of the mountains 'we saw no cattle grazing any where all day long, excepting a few cows and horses for carriage and plowing.'[16] 'They have a sort of large Buffles, of a monstrous size, with bunches on the back, like Camels, which serve for carriage and transport of goods only, in large Cities.'[17] A century later, Thunberg noted that 'Sheep and Goats are not to be found in the whole country; the latter do much mischief to a cultivated land, and wool may easily be dispensed with here, where cotton and silk abound.'[18]

7.4   The declining use of the wheel in Japan

An even more mysterious reluctance lies in the decreasing use of the wheel. The wheelbarrow, for example, was traditionally enormously important in China. 'For adaptability to the worst road conditions no vehicle equals the wheelbarrow, progressing by one wheel and two feet. No vehicle is used more in China, if the carrying pole is excepted, and no wheelbarrow in the world permits so high an efficiency of human power as the Chinese . . . where nearly the whole load is balanced on the axle of a high, massive wheel with broad tire.'[19] Yet in Japan the wheelbarrow was not adopted, unlike most things Chinese. For the medieval period, we are told 'The wheelbarrow seems to have been unknown (whereas it was used in China) and earth was carried either in baskets, or thrown on to a screen made of straw or rushes drawn by hand and slid along the ground.' Towards the end of the nineteenth century, Chamberlain noted that 'Japanese rural economy knows nothing of wagons or wheelbarrows.'[20]

Moving up to larger wheeled devices, Kaempfer had noticed a rough cart being used for moving stone.[21] Yet there is very little evidence elsewhere in his work of much use of wheeled carts. A century after Kaempfer, Thunberg noted the virtual absence of wheeled conveyances. 'No post-coaches, or other kinds of wheelcarriages, are to be found in this country for the service of travellers . . .'[22] The only carts he saw were near the city of Kyoto (Miaco), and they were not only the exception, but showed, as Kaempfer noted, how primitive and undeveloped the technology was. 'This day, I saw several carts driving along the road, which were the first I had seen, and indeed were the only wheel-carriages used in and about the town of Miaco, there being otherwise none in the country. These carts were long and narrow with three wheels, viz the two usual wheels and one before. The wheels were made of an entire piece of wood sawed off a log . . . Nearer the town, and in it, these carts were larger and clumsier, sometimes with two wheels only, and drawn by an ox.' He noted that they tended to break up the roads, and were hence confined to one side of the street, carts being 'only allowed on one side of the road – "which, on that account, seemed much broke up".'[23]

In the middle of the nineteenth century, a number of Western commentators noticed the absence of wheeled devices. Oliphant on the Elgin mission noted in almost identical words to Kaempfer and Thunberg, 'I also observed, for the first time, one or two carts of a very rude construction, and drawn by bullocks; but they are apparently very little used in Japan.'[24] Alcock quoted Veitch: 'There are no carts in this district. Everything is transported from and into the interior by horses and bullocks.'[25] Morse noted that 'I have seen no wheeled vehicles except the jinrikisha and there are very few of these,'[26] though he did note 'A very common sight to encounter in the streets is laborers dragging on a two-wheeled dray a fruit, or flowering, tree, such as camellia.'[27] Wheeled transport for carrying people came very late indeed. 'The rickshaw is thought to have been invented in Tokyo in 1869, though the origins are obscure. Japanese no longer had to rely on transport by water or on foot, and within a few years there were as many as 50,000 rickshaws in Tokyo.'[28]

How then was the immense traffic in goods and people in Japan carried? Apart from a very limited use of horses, and good water transport, the answer is basically on the human back and shoulders. Morse described a few of the techniques. The main method was by poles and racks. 'The farmers have long racks with which they lug grass or grain from the fields. They are longer than a man and are carried high on the back.'[29] For stones and dirt '. . . when a sufficient load is collected a pole is thrust through the loops and two men lug it off on their shoulders, the matting suspended from the pole like a hammock. Wheelbarrows are unknown in Japan and this device provides a good substitute.'[30] Immense loads were carried. 'After leaving the market town we met scores of people struggling along with heavy loads hung from their carrying poles. Such loads! I have tried a number of times, without success, to lift them from the ground, and these people will travel miles with them.'[31] Even when there were carts to lighten the load, people were used to pull or push them rather than animals, as in the West.[32]

7.5   The outcome: intensive rice agriculture in Japan

Rice was grown in irrigated fields in Japan. This meant that an immensely complex system of water control had to be developed. A good deal of the water could be taken to the appropriate terrace by using gradients and an elaborate system of dams and sluices which were copied from China. Very often the water needed to be raised from an irrigation channel to individual fields. An obvious way to do this is to use the current of the stream for power, driving a wheel to raise buckets. This was a method early invented by the Chinese, and it was used in parts of Japan. Morse described 'A curious device for irrigating the rice-fields'. 'On the banks of a swift-running river a water wheel was adjusted and was slowly turned by the current. On the sides of the wheel were fastened square wooden buckets; as they dipped into the stream they became filled with water, and as the wheel rotated the water was spilled from the buckets into a trough which conveyed it into the fields beyond.'[33]

While this device was 'not uncommon in the southern provinces', Morse found that it was 'rare about Tokyo and farther north.'[34] In central and northern areas, instead of letting the water take the strain, a much more labour-intensive, if flexible, system was used. The principle of lifting buckets on a wheel was the same but, instead of power being provided by water, it was produced by human muscle. Morse drew a figure showing a man coming down the road with the wheel and box carried in the usual manner. In the same sketch is a man treading the wheel and raising water from the ditch in the rice-field. 'The box is first fitted into the embankment, the wheel drops into appropriate sockets, a long pole is driven into the mud alongside the wheel, and holding onto this the man keeps his equilibrium and turns the wheel with his feet' to lift the water from the channel to the field.[35] Often more than one person provided power. 'It was interesting to see a tread wheel in which were two strong-looking samurai treading away patiently, supplying power for a certain portion of the machinery . . .'[36] A device which could have saved this effort was known and used elsewhere in the country yet people opted for this method.

The soil in Japan is usually baked hard and it is extremely difficult to turn over. This is one of the prime opportunities to apply non-human power, namely oxen or, even more powerfully, horses, as plough animals. It appears that, in the sixteenth and seventeenth century and perhaps before, animal power was used quite extensively for ploughing. In the early seventeenth century it was noted of the Japanese that 'They plow both with Oxen and Horse as wee doe heere.'[37] Kaempfer had noted that 'On the road hither we saw great numbers of calves, which are nurs'd up for ploughing, the country hereabouts being reckon'd the best in Japan for wheat and barley.'[38] Thunberg described a mixture of hoeing and ploughing. 'In the beginning of April, the farmers began to turn over the ground that was intended for rice. The ground was turned up with a hoe, that was somewhat crooked, with a handle to it and was a foot in length, and of a hand's breadth. Such rice fields as lay low, and quite under the water, were ploughed with an ox or cow . . .'[39]

Alcock provided a picture of Japanese ploughing.[40] Morse describes 'a farmer going to his work carrying a plough on his shoulder. It is dragged by a single bull. The point is tipped with iron and the plough is typical of the region, for there are many types of ploughs in different parts of the country.'[41] He adds that 'in mountain regions bulls are used to drag ploughs, and cows are used in softer ground so that boys can do the work'.[42] It would appear that humans also pulled special ploughs; 'peculiar shovel made of wood tipped with iron. The shovel part was over three feet in length and the handle seven feet long. It is used through the western part of this province (Musashi) and seems to take the place of the plough.'[43]

In fact, what seems to have happened is that the Japanese moved from the plough to the hoe as the population built up in the seventeenth century. We can deduce this from the absence of plough animals by the end of the eighteenth century. Horses were not used for ploughing and were in any case very few in number. 'The small number of horses to be met with in this country, is chiefly for the use of their Princes; some are employed as beasts of burden, and others serve travellers to ride on. Indeed I do not suppose that the sum total of all their horses amounts to the number of those made use of in one single town in Sweden.'[44] As for oxen and cows, 'they seem to have a still smaller number . . . the sole use they make of them is sometimes for drawing carts, and for ploughing such fields as lie almost constantly under water.'

Hayami has provided a useful overview of what happened. 'Instead of a plow drawn by livestock, a hoe or spade using human labor became the main plowing tool. This means that the labor that had been carried out earlier by horse-power now came to be done by man-power.'[45] Whereas there had been considerable numbers of draft animals to the end of the seventeenth century, 'after that, their number obviously declined'.

Numerous further stages then occur in rice cultivation, the constant weeding, the harvesting, and so on, but let us move on to the final stages. Once the grain had been brought back to the house, it has to be prepared for use. An immense amount of energy has to be used to de-husk grains and this was one of the areas where water mills became so useful in Europe and in many parts of the world. Given the numerous rushing streams and heavy rainfall of Japan, we would have expected water mills to be widely used at this stage. The principle of water-driven machinery was widely known. Mills could help not only with rice, but also with the equally tedious and time-consuming work of grinding beans and other crops. Such machines were used in parts of China, as King noted in the early twentieth century: 'At several places on the rapid streams crossed, proto-types of the modern turbine waterwheel were installed, doing duty grinding beans or grain. As with native machinery everywhere in China, these wheels were reduced to the lowest terms and the principle put to work almost unclothed.'[46]

Strangely enough, however, we have come across little evidence of the wide-spread use of water (or wind) mills for grinding grain in Japan. Instead, much more labour-intensive methods were used. One was the quern where, with immense effort of arms and shoulders, grain is ground between two heavy stones. Morse describes how 'the mill for grinding grain is turned by hand, and strong arms are required to turn it'.[47] This would turn grain into flour, but it is no use in taking off the outer, inedible, husk. To do this the Japanese used several methods. One of these is similar to that found all over the world, for instance in Nepal, where a heavy weight is dropped repeatedly on the rice until the husk is pounded off. 'The rice is hulled by a sort of trip-hammer made of wood and weighted with stone. This is worked by a man stepping on the end of the beam, thus raising it and letting it drop. This device has endured in China for two thousand years.' Even in the heart of cities, people were employed for hour after hour to step on and off this kind of tread-mill. 'One may see this rice-pounding going on even in the city of Tokyo. The man is naked and is concealed by a curtain consisting of strands of straw rope, a convenient device, for one may pass through this curtain without delay.' A picture of rice pounders is provided by Regamey.[48]

It is not a particularly difficult task to devise a mechanism to allow water to raise and drop this weight. Thunberg at the end of the eighteenth century saw some water-driven machines of this kind. Writing of rice husking, he observed 'Sometimes it was beaten with blocks which had a conical hole in them. These blocks were placed in two rows, generally four on each side, and raised by water, in the same manner as the wheel of a mill. In their fall they beat the rice so that the grain separated from the chaff.'[49] Alongside this was the foot-driven machine. 'Sometimes, when there was no opportunity for erecting similar water-works, a machine of this kind was worked by a man's foot; who at the same time also stirred the rice with a bamboo.'

7.6   The puzzle of the different trajectory of Europe and Japan

Given the different paths taken by Europe, particularly England, on the one hand, and Asia, particularly Japan, on the other, it is obviously puzzling to know why so little use was made of non-human power, in particular animals and wheels, in Japan. The answer to this question would throw much light on the nature of invention and innovation. Japan is an excellent comparative example. The knowledge was there. Earlier from China and later from the Portuguese and Dutch, the Japanese were perfectly aware of the possibility and mechanisms for using animals, wind and water power. For instance attempts had been made to introduce the water wheel from China in the ninth century, but had completely failed.[50]

They were extremely ingenious and inventive in many ways. Morse commented that 'the Japanese are said to have no inventive faculties, but in my rambles around Tokyo I have noticed many mechanical appliances of a simple nature which our artisans might adopt.'[51] He described, for example, how 'at one place in Kii I saw a curious implement used for weeding in rice-fields. It consisted of a long box without a bottom; inside the box were two shafts running from side to side, these shafts being studded with wooden pins; long arms or handles ran up from the box; and the machine was pushed through the rows of the rice-fields . . . It was invented by a man in the village where we saw it used.'[52] Yet the use of these inventions does not seem to have spread. Or again, Alcock noted early instances of colour lithochrome printing in Japan, which suggested that this was 'only one more instance in which by their own unaided genius and ingenuity they have anticipated by centuries some of the most recent inventions and discoveries of Europe'.[53] So it does not seem to have been either lack of knowledge, technical ability, ingenuity, curiosity or a practical desire to increase production that held the Japanese back.

The Japanese were particularly inventive as farmers. Kaempfer paid a tribute to their abilities. 'The Japanese are as good Husbandmen, as perhaps any People in the World. Nor indeed is it very surprizing, that they have made great improvements in Agriculture, considering not only the extream populousness of the Country, but chiefly that the Natives are denied all commerce and communication with Foreigners, and must necessarily support themselves by their own labour and industry.'[54] He noted more generally that 'Nature did not in vain so liberally bestow upon this nation, bodies fit for hard labour, and minds capable of ingenious inventions.'[55] Subsequent visitors also stressed the ingenuity and inventiveness. Thunberg stressed the practical and hard-working nature of the Japanese. 'In mechanical ingenuity and invention, this nation keeps chiefly to that which is necessary and useful; but in industry it excels most others.'[56] In the mid-nineteenth century, the Elgin mission found '. . . for the investigation of their manufactures and appliances, at once so original and ingenious, proved a never-failing source of interest and amusement'.[57] How then are we to explain the apparent movement towards involution?

7.7   Possible reasons for the absence of domesticated animals in Japan

There are a number of possible explanations for the marked absence of large numbers of domestic animals in Japan. The ecological arguments would stem from the nature of the volcanic soil of Japan. Japan lacked the possibility of pastoralism except in certain areas in the west and north. This argument is then supported by a second, namely that, given the small area of cultivable land, people could not afford to keep animals which would compete with grain production. The opportunity cost of giving up precious land to livestock was too high. It was necessary to use every piece of fertile ground to produce the basic grains on a very densely settled strip.

Many people have observed that raising animals is an expensive option – for instance, to feed grains to chickens may produce meat and eggs but many people in the world cannot afford the grain. The fairly desperate struggle to grow enough rice and other foodstuffs may have made animals a luxury the Japanese could not afford. Indeed, as population built up in the seventeenth century, the cereal rather than animal husbandry option may have become increasingly attractive. Thunberg at the end of the eighteenth century had noted the absence of pasturage and animals. 'Meadows are not to be met with in the whole country; on the contrary, every spot of ground is made use of either for corn-fields, or else for plantations of esculent rooted vegetables.'[58] He implied that it was the low number of grazing animals that led to the absence of pasturage. 'They have few Quadrupeds; for which reason there is no occasion to lay out the land in extensive meadows.'[59] As the agronomist King pointed out when he visited Japan in the early twentieth century,[60] 'By devoting the soil to growing vegetation which man can directly digest they have saved 60 pounds per 100 of absolute waste by the animal . . .' He calculated that '1000 bushels of grain has at least five times as much food value and will support five times as many people as will the meat or milk that can be made from it.' On this reckoning, if the agricultural area of Japan had been based on pastoral agriculture it could only have supported 6 million, rather than the 30 million actual inhabitants in 1800.

A similar theory was put forward by several anthropologists in the 1950s who described how 'Land shortage accounts particularly for the rarity of grazing animals. On arable land, crops grown for direct human consumption are much more efficient than natural vegetation or fodder crops for grazing animals.'[61] There is not enough waste or spare grazing for larger animals. This view is supported by the agricultural economist Boserup, who points out that 'draft animals fed on produced fodder are not an efficient source of energy supply. The mechanical energy supplied by them is probably only some 3–5 per cent of the energy contained in the fodder they consume.'[62]

While all this is undoubtedly a powerful factor, there is clearly also a cultural or religious dimension.

> There is a mixture of ritual prohibition and a feeling of disgust which alone can explain why, even when chickens or cows were kept, they were not eaten by ordinary Japanese. This was an aversion that lasted into the middle of the twentieth century. For instance, an anthropologist describes how in a Japanese village in the 1930s 'horses and cows are kept, but they are used only as beasts of burden. The animals are backed into their stables, where they spend all their time when not working. Milk is considered dirty and is only drunk on doctor's prescription.'[63]

There is evidence that some Japanese interpreted the Buddhist scriptures as putting a ban on consuming the products of four-footed creatures. Hence meat and milk would be banned. That Buddhism in Thailand, China or much of South-east Asia has not lessened the consumption of sheep, goats and other animals suggests that this can only be a partial explanation, but it does not make it an invalid one. There is obviously something more, however, which concerns the classification of what is 'good to eat'. Many were genuinely disgusted at the thought of eating meat or drinking milk; it was not merely a matter of religious prescription.

Of course, Japan is not quite the most extreme case of the avoidance of animals, and it may also have other roots, as Mokyr suggests. He notes that large domesticated animals 'were entirely lacking in pre-Columbian America and Africa, and scarce in most parts of Asia. This scarcity may have had deep historical roots: African and East Asian adults suffer from lactase deficiency and cannot digest large quantities of fresh milk (although they can digest milk in the form of cheese or butter).'[64] Whatever the reason, it is clear that, while knowing about most useful animals from at least the sixteenth century, the Japanese kept few domestic animals. This affected every branch of their life.

7.8   Theories to explain the declining use of the wheel

In accounting for the non-use of the wheel and plough in sub-Saharan Africa, Goody suggests that it was low population density which was a major factor.[65] This can hardly be the explanation in Japan. Again, in trying to explain the fact that the wheel faded out of use in Muslim societies, Bairoch suggested two major reasons.[66] One was religious. 'At least throughout the Middle Ages, there seems to have been a sort of prohibition in the Muslim West outlawing the use of wheeled vehicles, a prohibition of which it would be worth finding a plausible explanation.' Yet this cannot be the major explanation why the wheel had disappeared earlier. 'What makes the search for some sort of explanation all the more intriguing is the fact that carts or wagons and other wheeled vehicles were used in the Middle East for at least a few millenia before the advent of Islam. It appears, however, that the use of the wheel had already disappeared before the Arab conquest.' Instead, it looks as if an alternative had been found, namely the camel. 'The most plausible explanation may be the introduction and eventual predominance of the camel as a beast of burden in Muslim societies. And it should be noted in this connection that, as Bulliet (1975) has clearly demonstrated, the replacement of the cart by the camel in traditional societies, especially in semidesert regions, constituted technological progress and not a step backward.'

If we take this argument to Japan, and substitute the human being for the camel, we move into the most commonly advanced type of argument, which is also used to explain the situation in China, namely that, as the population becomes very dense, the cost of human labour becomes so low that it is cheaper to use people. This, for example, is the explanation given by Fairbank for the drying-up of labour-saving inventions in China. As population built up 'the abundance of muscle power made labor-saving devices less needed. Kang Chao notes that the 77 inventions for use in agriculture (like the bucketed water wheel or *noria* for irrigation) listed in a 1313 handbook were not appreciably added to in later such works.'[67]

The Japanese were not averse to labour-saving devices. 'Everywhere in Japan may be seen the most successful efforts to economise labour.'[68] Alcock was aware that labour was very cheap; nevertheless savings were still worth while. He noted that the fields were carefully trodden down with bare feet, 'to keep the seed from being worked out', which gave '. . . equal evidence of the care of the husbandman and the cheapness of labour. Yet they are always economising it, cheap and plentiful as it may be, and instead of the feet they sometimes use a simple roller made out of a transverse section of a tree.'[69]

Part of the solution lies in a remark of King's. He noted that 'extensive as is the acreage of irrigated rice in China, Korea and Japan, nearly every spear is transplanted; the largest and best crop possible, rather than the least labor and trouble, as is so often the case with us, determining their methods and practices.'[70] With a very dense population and, relatively, too little land, the really scarce resource is not human labour, but land. Hence, one criterion used to decide whether to use human or non-human labour is which is more efficient, not in terms of amount of human effort, but in total yield. If labour is very cheap, it becomes uneconomical to sink money into plant or animals. To build, maintain and operate a water mill to grind grain was more 'expensive' than doing it oneself. To pump water using one's legs, was less 'expensive' and much more flexible than constructing large machines for doing so, which would not easily be moved from tiny field to field.

All this tends to support those famous arguments which suggest that one of the reasons for the development of industry and manufactures in England was the very high cost of human labour. It became economical to invest money to replace labour. Of course it is a circular situation. The low cost of human labour in Japan was one of the reasons why it was not practicable to use other sources of power. This tended to make escape from this form of the 'poverty trap' or 'high level equilibrium' more difficult. The exigencies of rice cultivation, particularly where there was extreme shortage of space – to graze animals, to construct paths and tracks down which wheelbarrows and carts could proceed, to build mills on – did not help. It is no coincidence that it was not in the rice-growing south of China that wheeled carts were to be found. 'It is only in northern China, and then in the more level portions, where there are few or no canals, that carts have been extensively used, but are more difficult to manage on bad roads.'[71]

In conclusion, we may suggest that a set of ecological, social structural and demographic pressures coincided to take the Japanese economy towards a form of Geertz's classic 'agricultural involution',[72] which was the total opposite of the situation in Europe – and which could not be blamed on colonialism. Very roughly, if there was a problem in Europe, one tried to get round it by applying more ingenious machinery. If there was a problem in Japan, one tried to apply more thought, social organization and human labour. One led to the 'industrious revolution' in Japan. The other led to the 'industrial revolution' in Europe. Judged by the standards of the latter, the former was a trap in which alternatives to human labour were increasingly replaced. Only a very small part of the story has been told here, but it illustrates how complex the development of technology can be and the need to avoid certain ethnocentric assumptions based on the West European experience.

8

Stasis in complex artefacts

GERRY MARTIN

8.1   A skilled craft in a sophisticated civilization

Most of this volume is concerned with mechanisms for the generation of new knowledge, particularly the type of knowledge of the world around us which helps us to create new resources. During the long history of *Homo Sapiens* – a timeframe subject to much debate, but certainly in excess of 40,000 years – a great deal of resource-producing innovation has taken place, embodying much new tacit and reliable knowledge, but the rate of innovation has been far from constant. Periods of change have been punctuated by long periods of stasis, so that for most of history the world would seem to be pretty much the same when a person died as at the time when they were born.

This phenomenon, of high rates of innovation followed by stasis extending over many generations, is not confined to the less developed hunter gatherer or early farming societies. It is also exhibited in advanced and sophisticated civilizations, which had learned to beat the Malthusian trap, to produce resources to provision steadily rising populations. Japan provides us with one such example of a sophisticated society, many of whose artefacts – resources – had remained largely unchanged, apart from decorative variation, for many centuries before substantial contact with the Western World (Macfarlane and Harrison, ch.7).

We shall examine here one of the most famous of Japanese artefacts, embodying extraordinarily high levels of craft skill, but which remained virtually unchanged in form, function and manufacturing process for over 700 years: the Japanese sword. We will show how the complexity of the manufacturing process, and its sequential nature, in which each step is crucially dependent on the meticulous and precise completion of the previous stages, tends to produce

90

'lock-in' (§2.5). Variation and innovation become increasingly hazardous and are confined to minor decorative features which, while largely irrelevant from a functional point of view, assume great social importance.

8.2   The Japanese sword

Steel swords have been found in tombs in Japan datable to the fourth or fifth century AD, but it is probable that these were made in China and brought to Japan through Korea. Swords were being made in Japan by the Heian period (AD 794–1185), but the special characteristics which made the Japanese sword such a formidable artefact were not developed until the Kamakura period (AD 1185–1333). The functionality and process of manufacture remained virtually unchanged from this time until AD 1876 – eight years after the Meiji restoration – when the warrior class, the Samurai, were formally abolished. Within a century, the sword had become obsolete, except for the small numbers produced purely as art-objects, and mass-produced swords of much lower quality for army officers.

The Japanese sword is a cutting tool, originally developed for use from horseback with a slashing action, not the thrusting action of many Western swords. The blade is typically 24–30 inches long, and has a handle suitable for two-handed operation. It is slightly curved, with a sharpened edge on the convex side of the curve. The concave side is not intended for cutting, and is blunt. The functional requirements are straightforward: the main one is that the cutting edge be capable of extreme sharpness, neither dulling nor chipping when striking violently armour of cotton, leather or steel, often flesh and bone. (Good reader, at this point I ask for your sympathy and understanding; we are discussing an object constructed with consummate and loving skill, revered, collected and exhibited in the world's greatest museums, but whose sole purpose is to violently cut up living human beings. I cannot start to reconcile these conflicting attributes.)

The second important functional requirement is that the blade remains straight and in one piece in use, even when struck hard by, say, another sword in battle. There can be few occasions more embarrassing than to be left holding a handle alone. There are other slightly subsidiary requirements: the shape of the blade should assist a slicing action; the weight distribution should be such that the sword has plenty of momentum on the one hand, but is balanced and not awkward to use on the other; and the surface is of a composition and finish that does not encourage rusting, as long as care is taken.

The requirements of hardness and toughness do not normally go together, particularly in steels. To remain sharp, an edge must remain very hard –

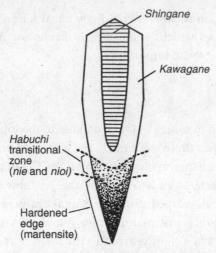

Figure 8.1 Cross-section of a sword forged in the kobuse style with a core of soft steel and a jacket of hard steel. From Kapp, L. & Kapp, H. & Yoshihara, Y. *The Craft of the Japanese Sword*, Kodansha International.

certainly much harder or resistant to deformation than the thing it is intended to cut. But hardness usually goes with brittleness, the opposite of toughness. The strength – literally – of the Japanese sword blade is that it combines these two qualities, hardness and toughness in a most successful manner, using steel in its hardest form for the cutting edge, but using steel in its toughest form to back up the cutting edge and to form the body of the blade. The creation of this combination of properties requires the most extraordinary skill on the part of the swordsmith, and it is the recognition of this level of skill that ensures the blades a special place in our museums, and the accolade of 'Living National Treasure' to the best of their makers in Japan.

8.3   Iron + 0.7% carbon = steel

Of all the metals, it is iron which has had the most profound effect on the course of history. It is abundant, comprising around 5% of the Earth's crust – 200 times more abundant than copper and 3,000 times more abundant than tin, the two constituents of the alloy which gave us the Bronze Age. Pure iron – rarely encountered even today – is quite soft and bendable, and has very few practical applications. The properties of the metal change markedly, however, with the incorporation of quite small amounts of carbon. As small an amount as 0.4% carbon permits the material which we call steel to be capable of heat treatment to a greater hardness than the best bronze tools, with which the Egyptians carved their huge granite statues.

Japanese swords typically have an edge of iron containing 0.6% or 0.7% carbon, which permits, with suitable heat treatment, a hardness sufficient to scratch glass. Higher amounts of carbon are used in a variety of applications in Western industry. The massive tools used for moulding plastic or forming car bodies may have around 1% carbon and the cast iron which used to be the standard material for stoves and grates, manhole covers, railings etc. had up to 3%. For any given application, the percentage of carbon, its distribution in the iron and the heat treatment the iron/carbon material (steel) has undergone, are crucial to the satisfactory functioning of the artefact − in our case, the blade.

The Japanese knew nothing of carbon. Neither did anyone else in the heyday of the sword: it was not identified as a separate material, an element, until the end of the eighteenth century. Nor did they know that they were adding this all important material accidentally to the iron during the process of extraction of the iron from its ore, iron oxide.[1]

The extraction of iron (many other metals are extracted in a similar manner) entails mixing pulverized ore with charcoal, igniting and blowing air (oxygen) through the mixture in a furnace large enough to ensure a very high temperature is attained. The charcoal burns to carbon dioxide and carbon monoxide; the carbon monoxide reacts with the ore at high temperature, removing the oxygen and leaving iron. The temperature is not high enough actually to melt the iron which stays in the fired mass as friable lumps, to be picked out when the furnace has cooled down. These lumps are then reheated and hammered repeatedly to produce a useable and workable material.

During this process, hot iron extracted from the ore comes into contact with hot charcoal − which is essentially carbon − and absorbs it quite readily. However, the amount of carbon absorbed cannot be controlled, and the process is quite haphazard, producing a carbon content range from less than 0.5% to over 1.5%. The swordsmith is highly skilled at sorting the lumps of raw steel from the furnace and judging from appearance and texture their suitability for the hard outer shell of the blade, including the cutting edge, and the softer, more ductile core. To do this, he takes lumps of the raw, friable, steel from the furnace, reheats it to a bright red heat and hammers it into sheets about a quarter of an inch thick. While still hot, the sheets are dropped into water. The rapid cooling leaves them in a very hard and brittle state. They are then broken up with a hammer to form flat but rough-shaped pieces, each with an area somewhat less than 1 square inch.

The finished sword needs steel with 0.6% to 0.7% carbon at the cutting edge, but many reheats are required during the forging processes, and each time the steel is reheated in the forge, a little carbon is burned away. For this reason, the smith starts the whole process with steel containing 1% to 1.5% carbon and the carbon content gradually drops during the long, and many times repeated, heat and hammer, heat and hammer.

It is worth remembering here that the smith has no real knowledge of what he is doing in metallurgical or chemical terms. Only in this century has he been able to think in terms of percentages of carbon. Indeed, he was unable to think in terms of carbon, or some unidentified substance, being present in the iron and dramatically altering its properties at all. In Europe, for instance, in the middle of the eighteenth century, it was widely thought that steel was a particularly pure form of iron, purified by the heating process. Cyril Stanley Smith quotes the *Dictionnaire de Chymie* by P.J. Macquer, published in 1776:

> From what we have said, we may judge that steel is much better purified iron than any other iron impregnated with a larger quantity of the inflammable principle, and hardened by the temper.[2]

And yet, by observing tiny clues in the process and thoughtfully relating them to the properties of the finished article, Japanese swordsmiths produced a remarkable consistency in composition and hardness. The little chunks of hammered and broken raw steel are carefully examined at their fractured edges and those which are heavy and dense, with a fine crystalline structure and a bright silvery colour – which we now know to have between 1% and 1.5% carbon and a few other minor impurities – were then just known to be the right material to start the blade-making process. They were carefully sorted out and stacked in a pile, six or seven layers high and 3 to 5 inch side, heated to a bright yellow colour (around 1,300 °C) in a charcoal forge and hammered to a solid mass.

Iron and steel in this carbon content range (but not with above 1.5% carbon) may be welded one piece to another by heating to a high temperature and then hammering them together. The hammering is done in a manner that elongates the block, and after a long flat bar has been obtained it is folded over on itself and the process repeated, rather as one might do making pastry, but in this case with white hot metal and a heavy hammer. Yoshindo Yoshihara, a modern swordsmith who describes this process in great detail,[3] folds and hammers out the steel about 13 times (producing around 16,000 layers per inch of thickness). The repeated heat–hammer–fold, heat–hammer–fold homogenizes the material, distributing the carbon evenly, while a little is constantly being lost. Eventually the material has reached a stage at which the smith considers it ready to fashion, by again heating and hammering, the rough shape of the blade.

However, this would produce a blade of uniform material; a blade more able to resist shock can be produced by inserting right down its length a core of a softer, more ductile steel. Just as the original broken chunks of raw steel were selected to produce a steel that would give a hard edge, so other chunks are

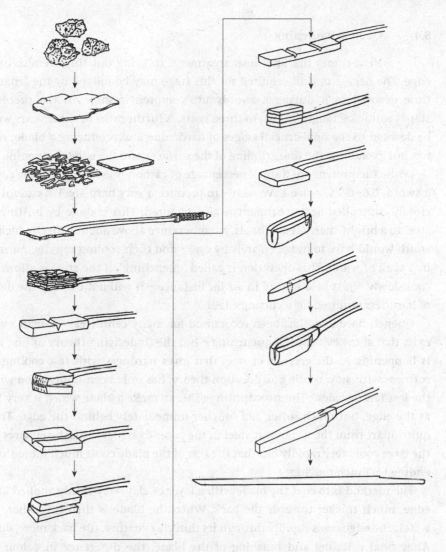

Figure 8.2 Forging sequence. From Kapp, L. & Kapp, H. & Yoshihara, Y. *The Craft of the Japanese Sword*, Kodansha International.

selected – those with a darker, muddy-looking fracture – which have a much lower carbon content. These are destined, after their own heat–hammer–fold processing, to form the softer, ductile core.

This is done by chiselling a deep V-cut along the length of the edge of the steel, performed when the steel is white hot. The softer core material is forged to a shape suitable for insertion into the V and the two are, at very high temperature, inseparably hammered together, the core inside and towards the back of the blade. The composite blade is now reforged into its final shape.

8.4   Quench hardening

Next comes the final heat treatment, to bring out the strength of the edge. The degree of skill required for this stage may be judged by the length of time devoted to it during a swordsmith's apprenticeship. All the preceding stages would be taught in two to three years. A further two to three years would be devoted to the final crucial stages of hardening and producing a blade, ready for, but excluding, the fine grinding of the surfaces and the edge sharpening.

While the presence of a small percentage of carbon is essential – in the case of a sword, 0.6–0.7%, as we have seen – to produce a very hard steel, a careful and closely controlled heat treatment is also required. This is done by heating the steel to a bright cherry red – that is, a temperature above about 720 °C (which the smith would have to judge entirely by eye) – and then cooling rapidly, normally in a tank of water. This operation is called 'quenching'. If the steel is allowed to cool slowly – just held to cool in air for instance – it will not develop the degree of hardness required for a cutting edge.

Quench hardening has been recognised for many centuries – Roman swords exist that show evidence of quenching – but the underlying theory of just what is happening in the crystals of steel that gives hardness with fast cooling and softness with slow cooling (dislocation theory) has only been understood within the last few decades.[4] The swordsmith wishes to make a blade which is very hard at the edge, but much softer and tougher immediately behind the edge. This is quite apart from the use of soft steel in the core. To achieve this, he ensures that the edge cools very rapidly but that the rest of the blade cools much more slowly, ending up much tougher.

The method is to coat the blade with a layer of clay, very thinly applied at the edge, much thicker towards the back. When the blade is thrust, red hot, into water, the edge cools rapidly through its thin clay coating, the back more slowly. After final grinding and finishing of the blade, the difference in colour and texture between the hard edge and the softer rear steel can be clearly seen. Many centuries ago, this dividing line was taken by the client warrior to be good evidence that the blade had been made out of the correct materials and properly hardened. Over the centuries, swordsmiths have learned to slightly contour the clay coating, producing patterned effects in the steel behind the edge. Different schools of swordsmith and different individual smiths would develop different patterns that have become important, though almost entirely decorative, features in the assessment of the sword.

If you take a piece of steel, 2 or 3 feet long and approximately 2 inch × 1 inch section, heat it to a bright red heat and suddenly thrust it into a tank of cold water, you can expect trouble. The whole thing can bend or twist, or cracks can

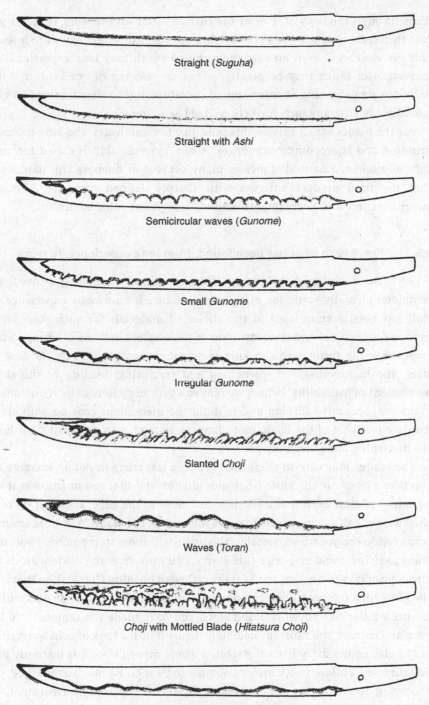

Figure 8.3 Common hamon. From Kapp, L. & Kapp, H. & Yoshihara, Y. *The Craft of the Japanese Sword*, Kodansha International.

form through the body of it or at the surface. After all the work that has gone into the blade up to this stage, the average swordsmith would expect to lose over half his work, and even an extremely skilled smith may lose a quarter. Some correction of faults may be possible – not the closing of cracks, but a little judicious straightening or correction of curvature, but the loss figures just given would apply after any such corrections had been attempted.

For the blades which survive this ordeal by fire and water, the long process of grinding and sharpening commences. Again, extreme skill is called for, using carefully selected natural stones in many degrees of fineness and hardness so that the final product emerges with sharply defined contours, a smooth sweeping curve and a very finely textured, though not polished, finish.

8.5   Keeping to what has been found, from long experience, to work

The swordsmiths' skills evolved over centuries. Having evolved, they remained virtually static for many more centuries. Enormous experience and skill has been accumulated in the choice of materials for each stage of the process – the choice and concentration of the original ore; the cleaning away of impurities; the firing with carefully prepared charcoal to extract the first raw steel; the hammering and quenching and fracturing, leading to the skilled assessment of 'suitability' (which we now call carbon content); the refinement by many-times-repeated forging and folding; the meticulous coating with clay to insulate one part of the blade more than an immediately adjacent part during the hardening and quenching process.

The reader may care to think through this last stage in detail. Imagine that you take a large carving knife from your kitchen and that you must cover it with a coating of clay, as thin as a visiting card close to the edge, stepping up to the thickness of half a dozen cards towards the back of the blade. You must scour the countryside for your own suitable clay (you will have to try many), you must learn to refine it and prepare a thin slurry. You must heat the blade with its thin envelope of clay to a bright red heat over a bed of burning charcoal, so that it has the same high temperature right along the length and on each side – all without disturbing the clay anywhere. And when you have made the temperature both even and correct, you must immediately thrust it into a tank of cold water. If any of the clay comes away from the steel, a whole month's work is instantly lost – and there are children with hungry mouths in your house next to the forge.

Having found a clay that works, in spite of this violent treatment, you treasure it. You lay hands on enough to last you through your career. You take great care to make sure there are no oversized particles to give little drag lines as you apply it with a spatula to the blade, no minute specks of organic matter

Figure 8.4 Yoshindo gives the finished blade a rough polishing. From Kapp, L. & Kapp, H. & Yoshihara, Y. *The Craft of the Japanese Sword*, Kodansha International.

which could burn with the heat and lift the layer. You will develop extreme caution in the surface finish of the steel to which you apply the slurry – not a hint of grease, not too smooth, a nice even oxide coating, but not a scale which could become detached.

The Japanese sword has been used as an example of this phenomenon, but there are many others. It particularly affects processes in which slight changes in one ill-understood variable produce large changes in the product. This occurs frequently, for instance in early glass manufacture, in which the absence of a small percentage of calcium oxide (accidentally incorporated in Roman glass by shells in the silica sand deposits used as the raw material – very similar to the accidental incorporation of carbon in iron) leads to an unstable, water-sensitive glass.

This type of complex sequential process thus leads almost inevitably to lock-in and evolutionary stasis. Even when using the best and most careful techniques a high failure rate occurs. Innovation is extremely risky and unacceptably expensive. The only way to achieve a success rate that can be lived with is to repeat each stage as exactly as possible.

So when you have got it right, you adopt one golden rule: **don't change anything**. Use exactly the same materials – even though you don't know in any rigorous way what they are. Prepare them and process them in exactly the same way each time. Suppress variation, suppress innovation, teach your apprentices to stick rigidly to the rules. The evolutionary model explains technological conservatism as well as 'progress'.

9

Gothic tales of spandrels, hooks and monsters: complexity, multiplicity and association in the explanation of technological change

DAVID TURNBULL

9.1   Explaining technological change

The general field of historical change across the range of the natural and social science is characterized by heterogeneity. There are widely divergent theories of change and no agreements about what is to count as an explanation. This is no bad thing. Rather it is to be taken as symptomatic of contemporary theorizing and understanding in that phase of modernism where universalizing and totalizing theories vie for supremacy but none shows any likelihood of predominating, and where at the same time a multiplicity of alternatives are flourishing. It is a good thing in the sense that it reflects the underlying complexity of the phenomena of historical change and our own role both in it and in explaining it.

We live in an age of complexity and multiplicity both in our technoscientific knowledge production and in our understanding of it. What is also symptomatic but less healthy is the isolation of areas of explanations of historical change from one another. The study of social, political and cultural change – that which we broadly take as history – has shown little interest as a discipline in the role of technology in historical change and even less in the explanation of technological change itself. Likewise biologists concerned with historical change and the processes of life have paid little attention to historiographical debates, though they have on occasion ventured into areas that overlap with the sociology of scientific knowledge (SSK) – the social history of technoscience and architecture. This separation of disciplines and modes of explanation is unhealthy. Cross-disciplinary discussions are vital to what is an essentially pluralist exercise.

Evolutionary theory may well have something to contribute to the under-standing of historical change in all arenas but its role is limited because it cannot be the sole explanation in any of them.

In this chapter I argue that, although Darwinian adaptationism is very seductive to theorists of technological change, it is of relatively little value in explaining the highly contingent nature of our artefacts and technological systems. On the other hand, I suggest, somewhat tentatively, that there are two principal areas where evolutionary theory can provide possible links between explanations of the natural and the social. I hope to reveal something of contemporary constructivist theories of technoscientific knowledge production, including my own views on knowledge assemblage, by looking at the example of Gothic cathedral building. This discussion shows that theories of symbiosis and associationism, and of epigenetic inheritance, can complement and even provide a positive heuristic for non-deterministic accounts of technological change.

9.2   Two contrasting stories

I start with the two contrasting stories about evolutionary change told by Gould and Lewontin[1] on the one hand and by Dennett[2] on the other in the 'spandrel debate'. I use the term 'story' quite deliberately to emphasize the narratological: all history, as I see it, is ultimately story telling. This is not the place to enter into the technical pros and cons of a debate where the arguments on both sides seem about even in their strengths and weaknesses. I just want to take a little from each, since they both are so keen on the injection of architectural and engineering examples into biology.

Gould and Lewontin have a double target: they want to demolish the Panglossian version of adaptationism and they want to argue that a pluralist approach is desirable because, among other things, there are architectural or structural necessities that are not adaptations but which nonetheless play a role in the development of life forms. Their example is 'Spandrels – the tapering triangular spaces formed by the intersection of two rounded arches at right angles are necessary architectural by products of mounting a dome on rounded arches.'[3] They also consider another example of 'architectural constraint' – the spaces created by the meeting of fan vaulting across a vault.

Gould and Lewontin argue that to suppose that spaces like spandrels are primary, or anything other than a by-product of architectural necessity, is to suppose like Voltaire's Dr Pangloss that 'things cannot be other than they are', and 'everything is made for the best purpose'. Further, 'evolutionary biologists in their tendency to focus exclusively on immediate adaptation to local con-

Figure 9.1 Spandrels at San Marco. From Gould, S.J. & Lewontin, R. 1979.
The spandrels of San Marco and the Panglossian paradigm: A critique
of the adaptationist paradigm, *Proceedings of the Royal Society of London*,
Vol. B205, 581–98, pp. 147–65.

ditions, do tend to ignore architectural constraints and perform just such an
inversion of explanation.'[4] The adaptationist programme or the Panglossian
paradigm 'regards natural selection as so powerful and the constraints upon it
so few that the direct production of adaptation through its operation becomes

the primary cause of nearly all organic form, function and behaviour.'[5] Adaptationist programmes, according to Gould and Lewontin, proceed by atomising an organism into traits and explain everything thereafter as interaction and trade-offs between traits, while ignoring the organism as a whole. This holistic approach which they characterise as continental is exemplified by the concept of a *bauplan* (§2.5). A bauplan is literally a groundplan or floor plan. The significance of their invocation of the notion of a plan is something I will come back to but I take them to mean something like structural form. The bauplan of an organism produces architectural constraints or necessities like spandrels.

Dennett, by contrast, wants to claim that adaptation by natural selection is the universal algorithm that explains all the forms of life. So what Gould and Lewontin describe as architectural constraint or structural necessity he sees as an 'obligatory design opportunity' and hence an adaptation.[6] But what he claims as an adaptation is the specific design, that is, the smooth minimal energy surface that Gould and Lewontin focus on in San Marco's. He thereby overlooks their main argument which is that the bauplan necessitates or constrains certain outcomes, independent of adaptation. He is partly correct. It doesn't make much sense to describe spandrels as constraints or necessities, since they have no structural role. Structurally they are accidental and indeed they provide a design opportunity. And yet they are not as a consequence adaptations in any even quasi-biological sense, since they play no functional role. How then can they be best described?

In the name of the kind of pluralism that Gould and Lewontin cite Darwin as favouring,[7] I suggest that there is heuristic value in considering *associationism*, or *symbiosis* (§2.5) as it is more commonly called. On this view evolution is in part to be explained by 'the merging of organisms into new collectives'.[8] Quite apart from the biological arguments about this theory, what is striking is its seemingly mundane emphasis on association. Associations are contingent not necessary, yet have profound consequences, as in the case of the cell, which according to Margulis is the result of an association between the nucleus and the mitochondria.

Thus, a spandrel can be seen as a contingent consequence of the association of arches and domes – an association that in providing a design opportunity becomes contingently linked in a chain of further associations forming a network. Equally significantly for me, as a sociologist of scientific knowledge who takes a lot from network theorists like John Law and Bruno Latour,[9] associations are fundamental to understanding sociotechnical systems and assemblages such as cathedrals. This associationist view thus provides a possible middle ground between Gould and Lewontin and Dennett in the 'spandrels debate'.

What then of Gould and Lewontin's other criticism of adaptationism? Dr Pangloss is something of a straw man since there are no adaptationists, even those as extreme as Dennett, who claim that all traits are optimally adapted. Some, as he allows, are simply accidents. But concealed within the Panglossian argument there is a relevant point which has its equivalent in historical and technological change

This point comes in various forms, that can be captured by Staudenmaier's contrast between *rational* and *contingent* explanations.[10] Rational explanations – those that Popper called historicist – display an outcome as the inevitable consequence of the interactions of a determinate set of forces in the manner of a physical explanation. Various forms of determinism including technological determinism come under this heading. Contingent explanations, by contrast, come closer to being descriptions in that they attempt to explain historical outcomes by a detailed and fully contextualized description of all the circumstances leading up to the event, or artefact.

The significant difference between the two approaches is the difference between a Panglossian paradigm and the constructivism of the sociology of science and technology. For Dr Pangloss, the ultra rationalist, the world could not be other than it is. For the sociologist of scientific knowledge the essential point is that it could be 'other than it is'. Indeed, successful contingent explanations are to be distinguished from simply descriptions because they aim to provide sufficient understanding to change the world *à la* Feuerbach.

Perhaps, therefore, Dennett avoids falling into the determinist version of Panglossia: but is he not just a little tarred with the 'best possible world' brush in selectionist terms? He is such an arch-adaptationist that he insists that 'Biology is not just like engineering it *is* engineering. It is the study of the functional mechanisms, their design construction and operation.'[11]

But then you have to ask 'is engineering biology?'. For Gould obviously not, since elsewhere he argues that 'Biological evolution is a bad analogue for cultural change because the two systems are so different'.[12] For this, he gives three fundamental reasons: 1. Cultural evolution is faster by orders of magnitude, and timing is of the essence in evolution. 2. Cultural evolution is direct and Lamarckian. 3 The topologies are different; biological evolution is irreversibly divergent and cultural evolution can be convergent.

Gould allows that they are both processes of historical change that may have general structural rules in common. This may be so, and for Dennett there is no problem; biology is engineering and both are driven by an adaptationist algorithm. Dennett, then, is an extreme functionalist and hence falls to some extent into the Panglossian category. But Gould slips too easily into a nature/culture dichotomy of the kind that Latour has done so much to sensitize us to. His first point about time may be a matter of scale, which I will leave to the experts. His third point seems possible to challenge by raising the convergent possibility of symbiotic mechanisms. His second point is the standard and most important problem for evolutionary theories of technological change and relates to the questions of performance, practice and tradition that I will come to in the cathedral case.

9.3   The social construction of science and technology

For the past 30 years or so – more or less since Kuhn – sociologists of scientific knowledge have focused on what scientists actually *do*, as opposed to the more traditional historical or philosophical approach which had a more representationalist orientation, emphasizing theory and observation and the relationships between ideas. As Joseph Rouse has put it:

> One might say that the traditional philosophical model of the local site of research is the observatory. Scientists look out at the world or bring parts of it inside to observe them. Whatever manipulative activities they perform either are directed at their instruments or are attempts to reproduce phenomena in a location and setting where they will be observable. The new empiricism leads us instead to take seriously the laboratory. Scientists produce phenomena: many of the things they study are not 'natural' events but are very much the result of artifice.[13]

This switch to the laboratory and practice has not only resulted in a dissolution of the science/technology distinction but has also led to a recognition of the local, complex, messiness of technoscientific practice. This anthropological perspective portrays knowledge as being produced by particular people with particular skills at particular sites, working with their own understandings of theory and method. So knowledge is, in origin, local and heterogeneous and the more sociological approaches suggest ways in which processes of negotiation, trust, standardization and assembly account for the way knowledge is enabled to move and achieve universality, objectivity and efficacy.

My own particular analyses of comparative knowledge traditions show that, to move knowledge from the local site and moment of its production and application to other places and times, knowledge producers deploy a variety of social strategies and technical devices for creating the equivalences and connections between otherwise heterogeneous and isolated knowledges.[14] The standardization and homogenization required for knowledge to be accumulated and rendered truthlike is achieved through social methods of organizing the production, transmission and utilization of knowledge. In addition to social strategies, the linking of the heterogeneous components of a knowledge tradition is achieved with technical devices which may include maps, templates, diagrams, and drawings.[15]

The anthropological focus on local practice, and the sociological focus on association and assembly have also to be combined with a constructivist focus on the ways in which knowledge and society coproduce one another. In effect the kinds of knowledge spaces in which people, practices and sites are linked serve to produce knowing subjects and relationships between them, and vice versa.[16] This produces the kind of complex and interactive sociocognitive ordering I mentioned at the beginning. But I also think that there are a plethora of other foci and approaches, all of which have some virtue whether they are economic, political, psychological, evolutionary or whatever. There are a variety of strategies for finding a way through the labyrinth; none can any longer be especially privileged. I find myself currently easiest with the history as narrative performance approach, so now to a story within the story – the construction of the Gothic cathedrals, Chartres in particular.

9.4   Thinking with cathedrals

Cathedrals are 'good to think with'. They are frequently invoked in trying to explain something else like scientific theories or design. This is especially true of Chartres. It is also somewhat problematic as an object of design, since it has largely been understood in representational terms rather than in performative ones. The very fact of its existence sets up a misplaced concreteness leading to its being analysed with whiggish, even Panglossian, retrovision.

Consequently, there has been a tendency to overlook the collective dynamic of the design process, which is also suppressed in the other trope that is so 'good to think with' in design discourse – the organistic evolutionary metaphor. Indeed, Gothic cathedrals have proved irresistible to theorists of the evolution of design.[17] Arguably, the emphasis on the representational and the evolutionary has served to suppress the complex heterogeneity of the collective of human actors and historical contingencies while also favouring the neo-Darwinian metaphor over the associational and symbiotic.[18]

First, let me reconsider the cathedral construction process so as to bring out the suppressed heterogeneity and make some suggestions about what this can tell us about the design process and incidentally about the nature of techno-scientific knowledge production.[19] Two points about linking spandrels and Chartres are the limited role of theory, geometry, and structural analysis, and the performative rather than representational use of drawings and plans. These relate to the central question (Jablonka, ch.3) of the nature of information and its transmission.

Much of our thinking about and understanding of design has been distorted by the undue influence of the concept of a plan. Lucy Suchman has pointed out that 'the view, that purposeful action is determined by plans, is deeply rooted in the Western human sciences as the correct model of the rational actor.'[20] Similarly, drawings are held to be the *sine qua non* of design, and Gould can unreflectively invoke the concept of a bauplan without concern for what it takes for granted.

Nowhere are the presupposition of plans and drawings more evident than in the question of how the great Gothic cathedrals like Chartres were designed and built. The question is contentious and problematic in the first instance because, in the case of Chartres, there are no plans and no known architect. There has been a tendency on the part of historians to make a mystery out of the building of Chartres; creating a great divide between then and now, by attributing either secret knowledge to the masons or special building skills now lost. Alternatively historians have denied any divide and any historicity by asserting that 'they must have had plans'.

There is of course the evolutionary school perhaps best represented by the greatest architectural Panglossian of them all, Viollett le Duc, who, in his evolutionary interpretation of the overall progress of the Gothic style, talks of medieval architecture as a whole being an organism 'which develops and progresses as nature does in the creation of beings; starting from a very simple principle which it then modifies, which it perfects, which it makes more complicated but without ever destroying the original essence'[21] – the original essence being a sort of 'idea plan'.

Thanks to the detailed analysis of the structure of Chartres by the Australian architectural historian John James, such explanatory strategies now stand in need of revision. Previously we had been presented with accounts of a glorious example of Gothic architecture, a harmonious unity embodying light, spirituality, innovation and complexity; following James we still have the acme of gothic expression, soaring height and sumptuous glass walls, and we still have the innovative flying buttresses and the complexity of pinnacles and spires; but where previously there was unity, coherence and unfolding of original essences we now have 'an ad hoc mess'.[22]

As is obvious once you break with the 'Gothic style' mindset, Chartres has a bewildering variety of spires, pinnacles, buttresses, fliers, roofs, doors and windows. Rather less obvious, but more important in structural terms, the bays and the axes in the nave and transepts are completely irregular, the only regularity being in the interior elevation. James's brick-by-brick, moulding-by-

Figure 9.2 Viollet le Duc's 'Cathédrale Idéale'. From Steadman, P. 1979. *The Evolution of Designs: Biological Analogy in Architecture and the Applied Arts* (Cambridge: Cambridge University Press).

moulding, analysis reveals thirteen major design and structural changes in the unusually rapid process of rebuilding between 1194 and 1230, undertaken following a disastrous fire. James argues that there were nine different contractors or master masons and that there was no overall designer; just a succession of builders. 'Chartres was the ad hoc accumulation of the work of many men.'[23] A similar ad hoc quilt-like character has been attributed to English cathedrals.[24]

James's claim that there was no one architect remains contentious and beyond the scope of this chapter. However, his identification of major design changes are very significant, as are the other factors which he points out – the lack of plans, the lack of a common measure and the key role of the template.

The question of the apparent lack of plans is crucial.[25] It is important to distinguish between what we now call blueprints or architectural drawings and site drawings. Site drawings, either in the form of full scale drawings on plaster or on the floor, or more detailed formal or informal sketches, had been in use at least since Roman times. However, apart from the plans of the monastery of St Gall AD 840, the earliest evidence of the use of detailed architectural drawings in the medieval period is 1225.[26] 'Remote control' of the design and construction process through drawings was not technically possible until the sixteenth century,[27] and even now is not fully realized or realizable. As the architect Peter Brooke commented on completing a 39-storey skyscraper in Melbourne in 1990: 'The principles of the shape and structure are well known, but the detailed solutions always vary.' Actually building it is rather like a gigantic experiment and you are never really sure what it will end up looking like.'[28]

The balance of evidence suggests that architectural drawings to control the design process were developed during the great cathedral-building period and that therefore a large proportion of the Gothic cathedrals were built without them. How was this possible? The achievement of the order of complexity and structural innovation involved in the construction of the cathedrals required a high degree of precision in the production of the stones and it also required the coordination of large numbers and types of workers. These factors created production and organizational difficulties which turn crucially on a fundamental problem – communication, that is, the transmission of information.

Knowledge, skills and instructions had to move between many participants. For a given building there had to be clear communication between the ecclesiastical client and the master mason, and between the master and other masons on and off site, since some stones were cut at the quarry. Masons had also to communicate with other teams of masons and other workers, principally carpenters, who were responsible for the invisible but nonetheless essential scaffolding and form work as well as the roof and all the heavy lifting equip-

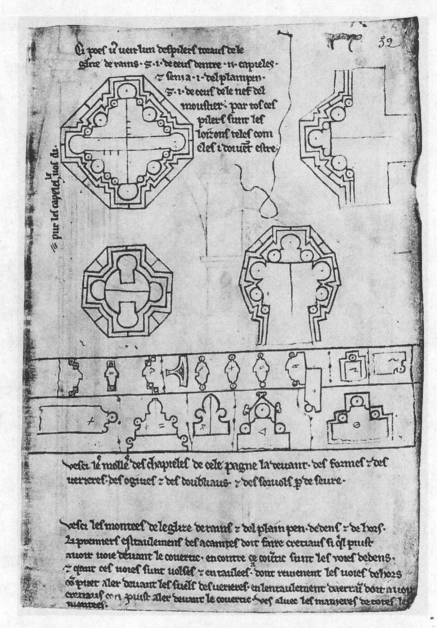

Figure 9.3(a) Template by Villard de Honnecourt. From Coldstream, Nicola, 1991.
Masons and Sculptors (London: British Museum Press). © Bibliotheque Nationale.

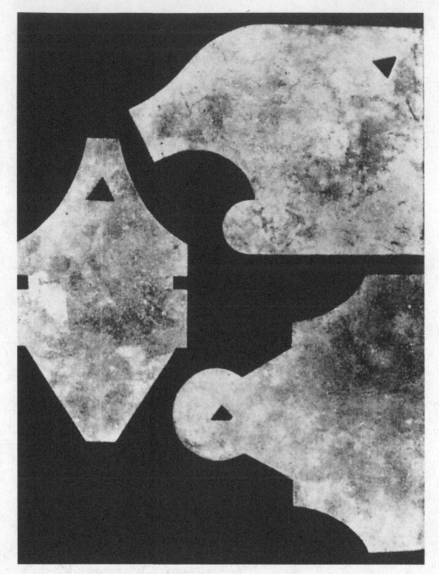

Figure 9.3(b) Mason's templates. From Coldstream, Nicola, 1991. *Masons and Sculptors* (London: British Museum Press). © Bath and Portland Stone.

ment. In addition, knowledge and skills had to be transmitted between sites and across successive generations of masons.

9.5   The power of the template

The key device is the *template*, a pattern or mould that permits both the accurate cutting and replication of shaped stone and the transmission of knowledge between workers.[29] The power of the template lies not only in the way in which it facilitates accurate mass production, but also in the fact that simple geometrical rules of thumb will often suffice for the template itself to be accurately reproduced as often as required.[30] The template helps to make possible the unified organization of large numbers of men with varied training and skill over considerable periods of time:

> On them were encapsulated every design decision that had to be passed down to the men doing the carving in shop and quarry. Through them the work of all the masons on the site was controlled and coordinated. With them dozens, and in some cases hundreds, of men were guided to a common purpose. They were the 'primary instruments' of the trade.[31]

Though templates were essential representational devices, other more performative elements were also essential: talk, tradition, and trial and error. Masonry as a knowledge tradition, while not having any secrets, did differ in two important ways from other organized trades whose skills and practices were handed on through guilds and apprenticeships (cf. Martin, ch.8), both of which made masonry, literally, an 'ad hoc' profession.

Firstly, they seldom practised their profession in one place; the job did not come to them; they went to the job.[32] This exposure to new sites and the work of others was a constant spur to innovation. Secondly, the construction site was essentially an experimental laboratory in which they were able to try to see whether an innovation was successful. Close observation of the drying mortar enabled the builders to detect areas of stress in the fabric and to take appropriate remedial measures through the placement of buttresses, pinnacles or reinforcement.[33]

9.6   The power of talk

This then leaves the question: who designed Chartres and the other medieval churches, given that there were no architects in the modern sense? Or to put it another way: in the absence of plans how did the scholastic and the craft traditions work together, how was it possible for them to form an association or collective? The patron scholar couldn't build and the master mason couldn't read, so how did the spiritual and intellectual values of the day become integrated in the building?[34] Lon Shelby's deceptively simple but profoundly important answer is *talk*.

'Frequent – sometimes daily – consultations between the master mason and the patron or his representative were the normal routine in medieval building.' It was 'this symbiotic relationship' which provides the essential clue to who

Figure 9.4 'Mastermason and client talking'. From Coldstream, Nicola, 1991. *Masons and Sculptors* (London: British Museum Press).
© British Library.

designed the medieval churches.[35] It was this dynamic interaction that enabled the radical innovations as well as the constant design changes, but it was also the use of constructive geometry in the drawing of the templates that allowed individual masons to make their own design decisions.[36]

Talk, tradition and templates provided for a *distributed* design process. The design as well as the construction of the cathedrals was literally the result of 'the ad hoc accumulation of the work of many men'. It should be noted that the geometry in question is not the abstract Euclidean variety but the common-or-garden rule-of-thumb variety. The builders were guided neither by theory nor by structural mechanics.

Chartres cathedral is an example of distributed design where the achievement of homogeneity in a welter of heterogeneity serves to create new knowledge spaces or associations in which people, skills, and technical devices interact to coproduce cognitive and social order. The essential factors in this complex multiplicity are the social strategies and technical devices which allow for different associations between the heterogeneous components. These factors are not typically allowed for in an evolutionary account and they make a significant difference.

9.7   Theory and practice

Robert Mark, who has revolutionized the structural analysis of great architectural achievements of the past, is driven to conclude that 'building technology follows an evolutionary pattern' which he claims had three components: 1. practical experience with earlier building, that is, tradition; 2. the experimental approach to construction which I mentioned earlier, but which Mark claims we can no longer use because we have separated the theoretical design process from construction; 3. masonry buildings are very forgiving – a fact that is very rich in possibilities for the exploration of the material constraints, fudge factors and bricolage which are common to both biological and technological change[37] (cf. David, ch.10).

The second point is crucial; but the relationship between theory, design and construction is a social variable which is not as historically fixed as Mark implies. It has varied historically and its apparent fixity derives from our unexamined assumptions about the nature of knowledge and its transmission. The split between theory and practice, so often claimed as an essential characteristic of the process of the division of labour was, for example, disavowed by Vitruvius, the Roman architect and military engineer who defined architecture as a discipline consisting of both practice (*fabrica*) and reason (*ratiocinatio*) – though his *De Architectura*, which was one of the very few available throughout the

Middle Ages, was unlikely to have been read by, or known to, any masons.[38] Nevertheless, in antiquity it had been commonplace to deprecate handiwork and separate it from theory and mathematics.[39] This was again usual by the fifteenth century, as can be seen in the work of Alberti, the first Renaissance architectural writer, who separated design from construction, and it has remained so till recently.

The master masons of the Gothic period, by contrast, represent what Mark and others have seen as a unique blending of theory and practice, which vanished with the advent of the architect. However, contemporary bridge-building shows a countervailing trend. The most recent advances, in which cable-stayed bridges are being built across unprecedented spans are not due to any breakthrough in materials or design. Rather, 'The construction industry . . . has been able to extend the boundaries of bridge technology by getting its designers, builders and computer specialists to work together more closely',[40] thus recognizing the necessity of moving people together to achieve a 'resynthesis of existing knowledges and techniques'.[41] Indeed it is now commonplace to use the so-called 'design and construct' technique in construction, where the overall concept is prepared in advance and the detailed design is done as the project proceeds.[42]

9.8   Analysing artefacts in use

In calling the process evolutionary, Mark meant presumably that cathedral construction changed over time through a process of trial and error. He, therefore, would appear to agree with Gould that it is a Lamarckian form of evolution. But if my remarks about the historical variation in the way knowledge is assembled are taken as significant, then neither the Darwinian nor Lamarckian form apply here.

It is possible to argue, however, that the sociological perspective necessarily entails seeing the whole as a *sociotechnical system*, which is bound to tell against the evolutionary account. Indeed if you change the scale and level of analysis, artefacts do appear to follow a kind of sequential development, though of a more open-ended contingent kind than either Viollett le Duc or Gould had in mind. Steadman argues, in his classic analysis of evolution and design, that the evolutionary account gains its strength from the simple fact that, especially in craft and vernacular traditions, one object is very often copied in its design from another, farm wagons and ships for example.[43] The most florid version of this thesis is perhaps George Kubler's *The Shape of Time* where he claims: 'Everything now made is either a replica or a variant of something made a little time ago and so on back to the morning of human time.'

Steadman's own account appears more acceptable in its contextualist approach:

> The explanations of artefact sequences made according to a situational logic will be related to the cultural and social circumstances in which the demand for the artefact is created and given meaning; to the constraints imposed on design by technological and material means available at each historical juncture; and to the body of knowledge scientific or otherwise, by which the designer is informed and on the basis of which his design 'hypotheses' are made and tested. Changes in form, and the emergence of 'types', will be the result of processes which represent responses to problems, and which must be referred to purposes.[44]

Like Martin's example (ch.8) of the Japanese sword, such sequences of artefacts seem very persuasive, and seem even more so when 'contextualized' as Steadman suggests. However, as several chapters in this book show, we are now learning (Jablonka ch.3; Fleck ch.18) ways in which to talk about artefacts that are less representationalist, less abstracted from their use. Artefacts are not simply objects with or without context. Like history itself, they are performative, and better thought of as closely coupled with activities (§18.6). It is not just a question of putting theory and practice together, or looking at problems and purposes; artefacts must be analysed in *use*.

But, in order to make the break with our representationalist preconceptions complete, we also need to revise our understanding of the kind of information that is transmitted in processes of technological change. The standard evolutionary approach is to conceive of DNA as a set of encoded instructions. But the example of the cathedrals shows that the information that is transmitted is not necessarily formulaic or a 'recipe' that evolves over time, and that the process of reproduction is accompanied by changes in the design process, and vice versa.

Jablonka's account of epigenetic inheritance systems (ch.3) illustrates an evolutionary model in which the rules themselves can also evolve. This is the key to the possibility of a broadly based evolutionary account of cultural and technological change. With such a model we can break with the limitations of genetic information as a code and with technological information as a plan. If we can develop the 'epigenetic model', and allow for the performative coproduction of design and artefact, then telling the monster-making story of the complex multiplicity we and our machines create will be much more persuasive than evolutionary accounts have been hitherto.

10

Path dependence and varieties of learning in the evolution of technological practice

PAUL A. DAVID

10.1 Introduction: varieties of learning in the economics of technology

A large part of modern discourse about technological innovation and other modes of cultural change is concerned with one or another among the varieties of learning that are characteristic of human populations. The dynamic, cognitive processes through which people interact with, interpret and alter their technological control over their external environments are variegated in their complexity and degree of sophistication. Nevertheless, it is possible to view all of them as extensions, refinements and elaborations of the primitive practice of improving productive performance through sequential experience-based modifications – so-called 'cut-and-try', or 'trial-and-error' methods. Such elementary forms of adaptive 'recursive practice' (Constant, ch.16) have been familiarly described by economists and students of the economic history of technology under the rubrics of 'learning by doing' (Fleck, ch.18) and 'learning by using'.

This chapter ventures inside the 'black box' of popular behavioural models of learning, in order to identify some of the psychological and epistemic mechanisms that may be thought to mediate the cognitive transformations which drive purposive technological adaptation. My concern is to uncover the connection, or lack of connection, between 'learning' of different kinds, and specifically between cognitive progress towards understanding complex and imperfectly observed physical processes, and the evolution of human control over such processes under conditions of uncertainty. An aspect of novelty in this consists simply in examining a framework of analysis that brings together two conceptualizations of learning in the sphere of technology that, curiously, have undergone rather separate development in the economic literature.

More often than not, however, the conceptualization of innovation as a learning process is rather loosely applied, so that the precise nature of the learning that takes place, as well as how it occurs, and who or what it is that 'learns', are matters that often remain largely unexamined. At the level of individual workers in a production system, learning might be supposed to include the enhancement of the blacksmith's motor-skills and other 'tacit' forms of competence, as well as an understanding of how to reprogramme a numerically controlled machine tool. Likewise, learning might encompass not only organizational routines that were the products of conscious, rational design efforts on the part of management, but others which were 'emergent properties' of the interactions among work-groups and larger organizations such as multi-plant management teams. Thus, economists studying modern business organizations today speak of the development of firm-level 'capabilities' and 'competences' as the outcome of learning that does not yield forms of knowledge which any single individual can comprehend or access.

Consequently, as has been noted, the usage of the term 'learning' in the economics literature[1] today carries at very least two quite distinct basic connotations. One one hand, it may simply refer to the existence of a systematic empirical relationship between some measure of productive experience and the improvement registered by an index of productive performance in the activity in question. Employed as a popular management tool, the 'learning curve' or 'industrial progress curve' tells us that a fall in real unit costs of production can be expected to accompany the accumulation of 'experience' gauged, say, by the rising cumulative volume of the output – from an individual worker, a team, a business organization, or, indeed, an entire industry. In contrast with this, the second of the principal ways in which 'learning' is conceptualized in modern economics refers to processes that are explicitly cognitive, in that they concern the transformation of the beliefs that the individual holds regarding the material world and the actions of other agents, and posits that such transformations result from changes in the information set of the person in question.

Economists quite evidently took over the first of these ways of treating learning from the school of behavioural psychology that still flourished in the US during the 1950s and early 1960s: the behaviourist tradition viewed learning simply as a transformation mechanism, hidden inside the black box that stood between observable stimuli and observable responses. Its effect was that of modifying the responses of agents (or experimental subjects) to external stimuli – whether in the form of observable events or symbolic representations of ideas. That was the model immanent in the pioneering contributions made to the economics of learning by doing by Kenneth Arrow and others at the RAND corporation in the late 1950s and early 1960s, and it shaped the formulation of

Arrow's seminal contribution to the theory of endogenous technological change, and the many empirical studies of industrial learning curves and related phenomena that it inspired.[2]

The second, cognitive conceptualization is one whose elaboration in economics has been shaped mainly by theories of human inference under conditions of uncertainty. Here 'progress' typically is viewed as the systematic alteration of the probabilities that agents assign to different possible states of the world, in such a way as to bring beliefs into closer and closer correspondence with the 'true state'. In theoretical and applied decision analysis, as well as in the work of game theorists on games of incomplete information, a central idea is that learning by individual agents and collectivities can be usefully modelled as a dynamic process involving Bayesian inferences from observable outcomes (Constant, ch.16).

Not only did Arrow pioneer in modelling the economic implications of the existence of 'industrial progress curves', but, during the same era and in the same applications context, he also essayed a path-breaking information-theoretic formulation of the notion of cognitive progress achieved through organized research and development. In 1955 he proposed to treat technological development as 'a special case of learning . . .' which he characterized in the following terms:

> In technical language we may say that in development the *a priori* probability distribution of the true state of nature (the unknown performance characteristics of possible models) is relatively flat to begin with. On the other hand, the successive *a posteriori* distributions after more and more studies have been conducted are more and more sharply peaked or concentrated in a more limited range, and therefore we have better and better information for deciding what the next step shall be.[3]

Given the historical conjunction of these two conceptualizations of learning in Arrow's seminal work during the 1950s, the subsequent bifurcation in the economics literature's treatment of the subject is all the more remarkable. Whatever the explanations for it that may be found by the historians of economic thought and sociologists of economic knowledge, perhaps it is time to consider whether there is something further to be learned about the evolutionary character of technological innovation by trying explicitly to see what relationships may obtain between the varieties of learning when they are coupled within a single, dynamic framework of analysis.

10.2   A concrete historical application

The mode of analysis on which I report here is that of exploring by means of computer simulation a specific recursive model of intentional adaptive behaviour under uncertainty, guided by Bayesian inference. For the foundations of what follows I am drawing freely upon previous joint work with Warren Sanderson of the State University of New York (SUNY), Stony Brook, the background for which along with the technical details have been set out fully elsewhere.[4] Our investigation of the possibilities of learning by doing, and its relationship of observable performance changes with transformations in the belief states of the individuals involved, was rather particularistic as it had been motivated by a concrete problem in the history of biomedical technology − the use of so-called 'safe-period' methods for the control of human fertility. As will be seen, this mode of contraception can be viewed as a special form of technological practice, one involving the management of a probabilistic production process by biologically unsophisticated persons. Nevertheless, however far removed it may at first appearance be from the usual matters on which students of technological change have focused, the way in which Sanderson and I chose to study this particular matter turned out to provide a more generic framework for analysis, and raised several issues of more general interest about the interpretation of evidence about experience-based learning.

The historical background is this, in a nutshell: in the 1930s, the modern 'rhythm method' was introduced on the basis of confirmation of the periodicity of ovulation in human females by laparyscopic observations. Before this there had been a variety of theories about the timing of susceptibility to impregnation within the intermenstruum. Most of these theories, and the prescribed practices based upon them, were physiologically incorrect, since they had been built upon the maintained hypothesis that menstruation was an oestrous in the higher primates − an erroneous belief in the medical tradition of the West that traces back at least to the gynaecological writings of Soranus of Ephesus (AD 98−138). That fundamental confusion notwithstanding, belief in the existence of a 'safe period' and in the potential for or even efficacy of contraceptive action based upon it, persisted for century after century. Books published in England and the US late in the nineteenth century were still offering medical advice on contraception that repeated the mistaken belief, and so instructed couples to confine sexual intercourse to the supposedly 'sterile mid-month' period. Not all the 'medically informed' advice published on the subject of the safe period was quite so bad, however; some of the more cautious prescriptions called for abstinence from coitus throughout an interval extending from the onset of the menses until the eighteenth day of the menstrual cycle. But, when considered in toto, the information that emanated directly and indirectly from physicians as to the whereabouts of the 'safe period' had an exquisitely treacherous cast to it. A seemingly modest reduction of 'caution' and a consequent shortening of the interval of abstinence by 5 or 6 days would shift the would-be contracepting couple from a quite reliably safe regimen into one that maximized exposure to the risk of an unwanted pregnancy.

This story poses a two-fold puzzle. How could the mistaken supposition that ovulation coincided with menstruation in human females (as it actually does in the case of other mammalian species, including the lower primates) remain the reigning medical theory when, seemingly, it was so vulnerable to disconfirmation by the readily observable experiences of the people to whom it was a matter of immediate interest? Moreover, it appears no less paradoxical that a general belief in the safe period as a method of fertility control should have become so widely diffused geographically, and have persisted in Western societies for so considerable a span of time despite the treacherous nature of the practices to which it led. Contemplation of these puzzles prompted the thought that perhaps belief in the efficacy of a safe-period method survived because the process of implementing it behaviourally, in a trial-and-error fashion, actually had discernable anti-natal consequences; especially if those trying it for themselves did not adhere rigidly to the 'medically informed' *a priori* beliefs as to the location within the intermenstruum of the period of maximum susceptibility to conception.

10.3   Inquiry by means of computer simulation: the 'Bayesian adaptive rhythm' (BAR) model

Could an informal policy of searching for the elusive safe period, implemented by an individual woman throughout the course of her reproductive life-span, constitute an effectual means of controlling her fertility? Would it be possible to adapt behaviour so as to avert a significant number of conceptions, by trial-and-error variation of the timing of unprotected coitus within the intermenstruum? Would successful 'learning' (in that behavioural sense) on the part of isolated experimenters require them to make significant cognitive progress towards discovering 'the truth' about the rhythm of ovulation, or might it be compatible with the persistence of substantial confusion, and the co-existence in such a population of a variety of contradictory beliefs? These were the questions that Sanderson and I undertook to answer with the aid of a computational model.

The model in question is a dynamic stochastic process that has a very specific structure. It implements a Bayesian strategy (Constant, ch.16) for averting conceptions by selecting regimes of sexual intercourse that minimize the subjective expectation of pregnancy, through sequential interaction with a dynamic algorithm that provides a probabilistic represention of the actual physiological cycles of ovulation and menstruation, and the reproductive processes of conception, gestation, and parturition. The expectation of conception that is being minimized is not the actual biologically grounded probability, but instead the subjective expectation held by the rational actor. It derives from an evaluation of the monthly probabilities of a conception under each of the alternative time-patterns of ovulation, conditional upon a choice of the timing regime for unprotected coitus, and given the *a priori* probabilities that the woman currently assigns to those alternative conceivable 'states of the world'. On the basis of such evaluations for each of the possible arrangements of a fixed numbers of days of coitus in the month, the selection of the *ex ante* 'best anti-natal' regime is made 'myopically', for each successive monthly 'trial'. But, month by month, the *a priori* probabilities are 'up-dated' – transformed into *a posteriori* probabilities by the calculation of the likelihoods of the observed outcome of the previous trial and the appropriate application of Bayes's Rule.

In sum, the Bayesian adaptive rhythm (BAR) model simulates a generic method of averting pregnancies by sequential experimentation that is constrained to always minimize the subjective probability of the unwanted outcome, given the parametrically set monthly frequency of coitus. The underlying specification of a Bayesian strategy means that the conditional likelihoods of the observed outcomes drive the revision of beliefs about the location of the 'safe period', by generating cognitive innovations (new information) which may or may not induce a modification of the next regime to be followed. This, of course, gives rise to the possibility that there will be a course of behavioural innovations, or 'learning' determined by the trial outcomes. The BAR model thus belongs within a very broad class of statistical learning models that have been investigated in the field of mathematical psychology.[5]

But, unlike the majority of such models, the structure of the learning algorithm in BAR is non-Markovian: the probabilities of the possible alternative allocations of the fixed number coition-days within a particular month are not strictly 'state dependent'. That is to say, they cannot be fully described simply by knowing the immediate past state of the system – the date allocation which had been selected for the previous month. Being driven by a Bayesian inferential engine that is linked to a sequential selection process that conditions the probabilistic observable events, our model of learning behaviour is recursive: the probabilities of a response at every moment in time are functions of the preceding sequence of response probabilities and of the intervening consequences of each probabilistic trial, for it is those that have been compounded into the distribution of subjective probability weights currently assigned to the admissable 'states of nature'. Learning in this kind of model thus evolves in an historically contingent, path-dependent manner.[6]

An 'adaptive' or functionally successful learning process based upon BAR can be defined as one in which the distribution of individual fertility rates, or completed lifetime fertility, fell significantly below what would otherwise be generated by a regime of randomly timed coitus. By parametric variation of the frequency of coitus, holding that constant throughout the entire reproductive lifetime of the subject(s), it is then possible also to investigate how this dimension of the regime of intercourse might affect the possibilities of trial-and-error learning. Clearly at the extrema of this parameter's range – coitus on every day within the intermenstruum, or on none of the days – nothing whatsoever can be 'learned'. The interesting question is whether within the intervening range some significantly adaptive learning effects appear and, if so, where in that range these are most pronounced.

10.4   The historical frame and the computational bounds upon inferential learning

The ubiquity of the sort of myopically constrained, sequential trial-and-error mode of learning represented by the BAR model scarcely needs to be pointed out. Throughout much of human history, and, indeed, until very recent times in a number of important fields of human endeavour – for example, the control of disease, and other biological processes – the quest for useful knowledge has been heavily dependent upon trial-and-error learning. Situations continually arise in the course of everyday practice where systematic, open-ended, long-term 'experimentation' is not feasible, affordable, obviously beneficial, or even desirable. The creation of learning opportunities through field implementation of production systems, and the sequential incremental modification of such systems, has been found to constitute an important, and for some industries the dominant, source of cumulative productivity improvements. More generally, this kind of 'by-product' learning contributes immeasurably to the extension of our technological knowledge, in both codified and tacit forms, and to other manifestations of cultural progress, tangible and intangible.

It is my view that such conditions (Turnbull, ch.9) have characterized 'learning' environments for extended epochs in the evolution of technology; that although it came to persist in fewer and fewer areas of the mechanical technologies down to the modern era, it has continued until comparatively recently to characterize the problems of 'engineering' in agriculture and animal husbandry, and the practice of clinical medicine involving human beings. Even now, when it has become possible to go beyond controlled laboratory experimentation with 'models' whose outcomes are isolated from the immediate functioning of the learning organization. Indeed, when it is possible to experiment with computer-based 'virtual models', it cannot be said that the mode of adaptation envisaged in the BAR model has ceased to be economically relevant.

In highly sophisticated engineering regimes (Vincenti, ch. 13), engineers still have to 'learn from their failures'. In principle they ought to be able to figure out which subsystem(s) in a complex artefact will be most likely to be stressed to the point that a 'low tolerance' specification would be likely to fail. In many instances they can do so, either from first principles or by instrumentation that will allow the operation of each component (subsystem) under field conditions to be monitored in real time. But this may be too costly, or not technically feasible. At a certain point they have to 'do' – that is, build the best full-scale system they know how, and put it into operation – learning what they can from a 'post mortem' examination of the physical evidence if it fails. But, in the absence of the appropriate remote monitoring equipment there will be significant residual uncertainty as to exactly what it was that 'went wrong'; all that they may have to go on is that the dam burst or the rocket blew up. How can they, as Edward Constant argues, nevertheless make 'design progress' by adopting a recursive Bayesian procedure?

The first question to be addressed in that regard is whether a learning algorithm of this sort is really practically feasible when pitted against a stochastic process of any complexity. Thus, one of the interesting issues illuminated by working with the BAR model pertains to the bounds placed upon human learning and rational optimizing behaviour by limitations on the capacity for real-time processing of information. The computational task of evaluating the likelihoods of all the relevant combinatorial possibilities for regimes of intercourse timing turns out to be staggeringly large – once one considers even an approximation to a biologically realistic, stochastic model of the reproductive process in which a conception might ensue from a single act of unprotected coitus held on any one of 6 days in a randomly chosen month. Faced with this implementation problem, Sanderson and I ventured an intuitively appealing way to overcome it, namely, by supposing that our hypothesized learning agent would have had first to simplify her *a priori* theory of reproductive physiology. This means re-specifying her 'model of the biology' to the point that a computationally feasible Bayesian learning algorithm could be implemented.

It turns out that in this case there is an appealing, albeit a rather drastic, simplification that will suffice to reduce to quite trivial dimensions the computations required for strict Bayesian information processing and the monthly re-optimization of the allocation of coition days within the intermenstruum.

One has simply to presuppose that all the relevant monthly 'rhythms' of susceptibility have the same, very simple form: a common conditional probability profile, in which there is but a single day of peak susceptibility in the monthly cycle, surrounded by days on which the hazard of a conception is negligibly low. There then can be only as many 'admissible' states of nature to be considered as there are days of the intermenstruum, each 'admissible rhythm' being uniquely identified with the sole day on which the peak level of suceptibility is located.

By means of this bit of *a prioristic* modelling finesse, the woman's assignment of subjective probabilities to the admissible states of nature is made equivalent to her subjectively assessing the relative 'dangerousness' of allowing unprotected intercourse on the corresponding days of her month. Given such a subjective rank-ordering of the days in her 'month' (i.e. of the alternative hypothesized times of peak susceptibility), her procedure for mimimization of the subjective monthly expectation of conception is then readily accomplished without need of any further calculation at all. She should allocate the (paramentrically fixed) number of days of coition by starting from the 'least dangerous' day and moving upward through the ranking, as far as is necessary until all have been assigned.

It is then evident that if she begins in a state of complete confusion, or total skepticism regarding pre-existing beliefs as to the location of the period of peak susceptibility, she initially will hold every date as being equally 'dangerous' (or equally 'safe'). Starting in that way, with initially flat priors, her initial allocation would correspond to a randomly timed pattern of coition. But thereafter, depending upon the outcome of that trial, the inferential procedure would differentiate between the initially selected sets of coition days and non-coition days. In the event of a conception (which we assume to be immediately detected) the 'relative dangerousness' rating of the previous month's coition-days would be raised; whereas, in the event of no conception, she would raise her evaluation of the 'relative dangerousness' of the days on which coition had been omitted. The cognitive demands of the routine for implementing BAR in this bounded form can thus be shown to be quite undemanding, requiring no memory beyond recall of last month's regime of intercourse along with the corresponding subjective rating of the relative dangerousness of the days of the month; such calculations as are required are no more intellectually exacting than the repeated mechanical application of an invariant rule for revising upwards or downwards the former relative dangerousness rating of the days in the fashion just described.

A more general point suggested by the foregoing considerations is that arriving at a feasible learning routine under severe computational constraints might very well mean embracing 'a model of reality' that is so inexact as to exclude the 'true state of the world' from among the set of 'admissible states of nature' that are to be evaluated by means of Bayesian inference. Under such conditions there is nothing to guarantee that the revision of priors will tend to converge in the way described by Arrow in 1955, progressively piling up the *a posteriori* probabilities assigned to the 'true' states of nature, or something reliably close to those. In other words, being subjected to fundamental computational limitations, the evolution of the inferential aspect of the process may well be characterized by the frustration of significant cognitive progress. But whether that entails the accompanying frustration of behavioural adaptation is another matter, on which the simulation results obtained with BAR prove enlightening.

10.5   Simulation results

Only two of the many interesting sets of results obtained in experiments with the BAR algorithm need be summarized here. The first is that reliance upon BAR turns out to be behaviourally adaptive; indeed, when practised consistently it is a surprisingly effectual means of family limitation. On average it is possible for the learning agents in our model to avert a significantly large number of births by following the indicated routine over the course of a 264-month reproductive life-span, compared with the expected levels of completed fertility that they would experience had the same number of days of intercourse per month been allocated randomly. This transversal measure of learning reveals that the expected performance improvement is particularly large when the frequency of coitus is confined to the lowish range between 3 and 6 times per month; but there still are appreciable gains in terms of births averted in regimes where coital frequency is maintained at 8 times per month. At the lower coital frequency levels it is also found that the mean subjective expectation of conception undergoes statistically significant monotonic reductions during the first half of the reproductive span, so that one may speak of behaviour exhibiting the dynamic signature of a classic learning curve.

The second set of findings confirms the suspected limitations of the algorithm's ability to generate corresponding cognitive progress. Our learning agents as a group remain very confused about the timing of the 'fertile' and 'safe' periods within the intermenstruum. Even among those maintaining the behaviourally adaptive lower levels of coital frequency, much less than half manage to attain even a modestly significant degree of correlation between the *a posteriori* distribution of beliefs about the rhythm of susceptibility and 'the true rhythm' generated by the objective biological process they are trying to control.

Thus, there is an evident disjunction between the performance of this algorithm in the behavioural and the cognitive domains. A trial-and-error process that is grounded on a mistaken view of the underlying biological reality, under some conditions, can use a simple inferential procedure as the basis for evolving particular behavioural routines that are efficacious. This is so despite the inability of the Bayesian inferential engine to reject the mistaken *a priori* model specification (concerning the biologically general profile of susceptibility); or to make significant headway in dispelling the initial confusions of belief concerning the location of the average woman's 'fertile' and 'safe' periods within the intermenstruum.

The persisting 'cognitive uncertainty' arises in the hypothesized learning situation because we have supposed that the decision-agent, the representative woman, is motivated to draw inferences from her own experience only insofar as these direct her to select an effectively anti-natal time-pattern of intercourse; she is not functioning as a researcher seeking to discover the timing of ovulation, for that is taken to be a derivative aim, one that is of interest only to the extent that it helps her to make month-by-month regime choices that avoid pregnancy. Consequently, should she happen upon a set of beliefs (i.e. 'priors') which enabled her successfully to avert any further births, her beliefs about the safe period – however far they might be from 'scientific validity' – will never progress further towards correspondence with the underlying reality. For these rational adaptive agents, the operative 'truth' towards which they are myopically driven is simply the pragmatic test: 'whatever works'.

How, then, does cognitively bounded BAR nonetheless manage strongly to curtail fertility? Evidently the learning algorithm must be working in a way that does not entail a tight coupling at the micro-level between 'correct knowledge' and 'correct action'. The key to the explanation of this puzzle lies in an often overlooked generic feature of Bayesian adaptive routines. They all have embedded within them the following primitive but powerful rule: 'Stay with a winner.'

Staying with a winner – that is, in the present example, not abandoning a regime that has just avoided generating a conception – is the behavioural consequence of the fact that, in response to contraceptive successes, the woman's beliefs undergo revision in a particular direction; she raises her subjective evaluation of the relative dangerousness of those days on which coitus did not take place during the preceding month. This causes the same regime to be repeated whenever it has yielded a success in the previous trial. Regime selections that initially met with a run of successes must therefore tend to become 'locked in'. Correspondingly, agents whose initial regime choices issued in unwanted outcomes will eventually be pushed by the revision of their beliefs to discard such regimes in favour of others which come to appear as less 'dangerous'.

The tendency, therefore, must be for an ensemble of agents who are BAR-users to be jiggled and jostled toward the subset of 'relatively safe' time-distributions for unprotected coitus. These regimes, being 'functionally adaptive' with reference to the individual agents' goals, also will possess the property that their selection causes the agent's behaviour to stabilize – thereby incidentally fulfilling the behavioural psychologist's minimalist criterion for the 'completion of learning'. As a larger and larger proportion of the ensemble becomes thus stabilized in relatively safe regimes, the average monthly probability of a conception must be reduced.

At the macro-level, this would look like a process of continuous learning on the part of a 'representative (contracepting) agent'. In evolutionary terminology, the collection of different regimes that were in use in each month by the ensemble of agents would constitute a set of different 'allomemes'. The BAR routines followed by the agents could be viewed as a mechanism which drives the 'regime population' to evolve in the (desired) anti-natal direction, by differentially selecting and retaining for use those regimes that were relatively adaptive, and so would 'survive' further modification.

Our simulation studies revealed that the BAR method works best at low average rates of coital frequency. The reason for this can now be seen to be because the woman is then more likely, purely by chance, to start out by avoiding unprotected sexual intercourse on those days on which the probability of a conception is relatively high. One may calculate that under random allocation with 6 susceptible days out of a possible 24 on which intercourse could take place, a monthly coital frequency of 4 would give the woman a 23% chance of becoming 'locked in' to a perfectly 'anti-natal' regime by the subsequent BAR-driven revisions of her initially diffuse prior beliefs. This falls away to 14% when the coital frequency is 6, and to a mere 2% when it is 8 times per month. Thus, the hidden 'stay-with-a-winner' rule would tend to come into operation sooner when coital frequency was held below the levels observed in modern populations equipped with reliable methods of contraception.[7]

10.6   Some extrapolations

These findings would seem to have a wider validity, pertaining to many other contexts involving a reciprocal and mutually reinforcing interplay between cognitive evolution and behavioural adaptation. The usefulness of the iterative learning model transcends the particular application that originally prompted and shaped its construction. For example, the same abstract model might also be interpreted as representing an organizational 'software' management problem, in which a cadre of more expert workers are being assigned under conditions of uncertainty to some among a larger number of tasks, in an effort to minimize organization failure by properly staffing a suspected number of 'critical' activities. Alternatively, the problem could be reformulated as one involving the trial-and-error design of a multi-component production system, for which the engineers are constrained by budget limitations to install relatively more costly high-tolerance components in only some of the total number of subsystems.

The merit of the Bayesian strategy is that it provides a familiar and appealingly rational basis on which to develop a computationally tractable algorithm describing path-dependent stochastic learning behaviours that coevolve with a structure of beliefs about the natural world. It is likely, however, that qualitatively similar results can be obtained with many other algorithms that are statistically less well founded, but which might be justified as arising from psychologically more plausible heuristic procedures for processing information acquired sequentially, in the manner of trial-and-error experimentation.

What these results show, in general, is that behavioural adaptation and cognitive evolution need not be closely coupled for 'learning', in the sense of improved practice, to occur. In most instances of 'learning by doing', the 'feedback' from 'experience' to inferred 'understanding' is severely constrained. The 'doers' have limited facilities for accurately observing and recording process outcomes, or for hypothesizing about the structure of the process they are trying to control. Advances in knowledge that are empirically grounded upon inferences from trial-and-error in a myopic control process cannot be a big help when they are restricted in both the number of trials they can undertake, and the states of the world they can imagine as worth considering.

In any case, cognitive evolution within a Bayesian decision-theoretic framework is clearly an instance of a path-dependent epistemological process. Coupled to myopic optimizing behaviours, it will generate perceptible hysteresis effects in actions and observed consequences. At each stage in the process those acting on their *a priori* views have made themselves, in effect, creatures of a potentially tenacious past – a past that will have been distilled down and incorporated into their prevailing subjective evaluations of alternative hypotheses about the nature of the world with which they have to contend. In this way, historical experiences are able to lay down a (boundedly) rational basis for behaviours that come to resemble habits. Indeed, the beliefs on which they are based are themselves so 'habit-like' that they grow more and more difficult to cast off quickly, even after they have revealed themselves to be responsible for the recurrence of unwanted consequences. In economic terms, this is 'sunk cost hysteresis' in the realm of cognitive evolution.

Even when they are not petrified into habitual beliefs, 'heuristics' such as 'stay with a winner' show up in this model as a means of overcoming the cognitive limitations of human agents caught up in evolutionary change. Limitations on computational capability, as well as on information collection, storage and retrieval, may preclude attempting to base decisions upon representations of the world that reflect the full range of its perceived complexities. Those limitations, indeed, may be at the root of perceptual distortions of the sort that have been identified by behavioural psychologists as 'framing biases', 'salient aspect' comparisons in multi-dimensional choice situations, and suchlike, not to mention the manifold difficulties that ordinary people display in grasping elementary concepts of probability, and in applying such notions to everyday decision problems.[8]

This is close to the spirit of Herbert Simon's[9] original observations about rationality being bounded by cognitive constraints on decision analysis. In our simulation model, by grossly simplifying the representation that the decision agents give to the time-pattern of fecundability within the intermenstruum, the updating of priors in strict Bayesian fashion is rendered equivalent to an easily implemented heuristic procedure. One may hypothesize that rational economic agents needing to manage complex technical and social processes (the underlying structure of which is incompletely known to them) will in general be found to be relying upon heuristics and decision-rules that had been formed by framing some model of the world so as to render feasible the necessary 'calculations'. Implementable rules of this kind may far outlive the conditions under which they were formed, especially when they become embedded in the ritualized operating procedures of social organizations.

10.7   Instructionist versus selectionist mechanisms in evolution

The results of carrying through the 'thought experiment' represented by BAR also expose the semantic confusions engendered by questions about the nature of the 'units of selection' and the 'level of selection'. Suppose that we only had an empirical 'learning curve' to go on. Would we know whether this was due to (a) the incremental acquisition of knowledge and/or skill by most members of the population, or (b) discrete changes from dysfunctional to functional behaviour by a progressively larger fraction of individuals in such a population? Each of these processes could be thought of as evolutionary, and selectionist at some level. Yet the 'units' being selected are different in each case, and the implications of 'learning' are thus quite different. Even the average data produced by the simulation runs do not enable the analyst to differentiate between cases (a) and (b). The ambiguity on this score is associated with another characteristic problem in applying the evolutionary metaphor to technological innovation – the difficulty of giving an unequivocal definition of 'memes' (§1.2, §3.2, §18.5). Are memes to be seen as the individual's belief structures (e.g. about the timing of ovulation) or the associated behavioural regimes (e.g. the patterns of inter-course)? This evidently is another semantic issue on which confusion is endemic in the literature about cultural evolution.

In either case, however, in the illustration at hand it is clear that the variations in the 'memes' over which selection takes place will not be 'blind'. Such evolution as may occur within the framework of BAR is driven by a conscious cognitive process. Under the constraints of the information currently available to them, agents interpret 'the data' to optimize or re-optimize their behaviour. They envisage mentally the likely outcomes of various possible actions and act accordingly. In other words, the processes simulated in the BAR model are 'evolutionary' in that whatever can exist at each point in time remains strictly bounded by that which existed previously. They are equally 'selectionist', in that they depend on a mechanism involving 'variation and selective retention'.[10]

But they are not strictly 'Darwinian'. The variants from which a selection is to be made are not presented 'blindly', as if with disregard for their putative outcomes. They introduce – into a system that already operates a process of 'adaptation by selection' – further elements of 'adaptation by coding' and 'adaptation by revision' (Perkins, ch.12). Or, as one might put it, a certain amount of 'Lamarckian instructionism' has been injected into the model, although this is not the natural instructionism (inheritance of acquired characteristics) en-visaged by Lamarck, but rational design based loosely on Bayesian reasoning.

What is the effect of this admixture of cognitive intentionality on the performance of the model? Is this mode of 'learning by doing' qualitatively superior to a strictly Darwinian process in adapting to what is required of it in practice? The computer algorithm is actually very flexible. Suppose then that we employ the same set-up to simulate purely Darwinian evolution by starting with a random set of alternative choices and substituting the simple rule 'win, stay – lose, start again' (WSLSA) in place of the Bayesian formula at each monthly moment of selection. What happens to the learning agent's average perform-ance? In fact, remarkably little. Considered as a practical procedure for reducing expected completed fertility, WSLSA is only a little less efficacious than BAR at low coital rates, and is even slightly better at higher rates. To put it another way: in our simulation runs, 96% of population-level performance was accounted for by selection mechanisms acting on blindly random variants, even though these were operating at the heart of what is ostensibly a non-biological Lamarckian evolutionary system.

Thus, a process that appears Lamarckian (intentional, planned, designed), as the BAR algorithm does, may be working effectively in the behavioural domain because it is largely 'selectionist' in the Darwinian sense. Is it then so obviously important to insist upon the qualitative difference between the two species of evolutionary theory? In social science invocations of evolutionary metaphors, the effort spent in carefully distinguishing non-biological processes that are Lamarckian from those that are properly Darwinian may be misplaced, and reflective of analogical rather than analytical reasoning. What appears outwardly to be a thoroughly intentional instructionist mechanism has turned out here to be effectual only to the extent that it mimics Darwinian selectionism. There evidently are contexts where the instructionist–selectionist distinction is operationally less important than would be suggested by the amount of paper and ink that has been devoted to it in the literature of evolutionary epistemology. It is a distinction that is likely to be worth insisting upon when cognitive evolution must underpin behavioural adaptation. But that is something to be demonstrated in future research, rather than presumed.

III
INVENTION AS A PROCESS

11

Invention and evolution: the case of Edison's sketches of the telephone

W. BERNARD CARLSON

11.1　The evolutionary role of the inventor

The good news of this book is that technological artefacts do evolve. The bad news is that we don't know much about the processes by which this evolution takes place. If we restrict ourselves to assuming that the evolution of the artefacts takes place in the market-place, then it is easy to argue that technological evolution is caused by something that looks like adaptation and natural selection: the products that are adopted and survive are those that customers think offer the best combination of functions at the best price. Commerce, like nature, is red in tooth and claw.

Yet the problem with assuming that technological artefacts only evolve in the market is that it overlooks the role of the individual inventor or designer. Many, if not most, inventions start with an idea that some individual had. The idea may be modified, tested, and changed through a myriad of social interactions (Fleck, ch.18; Fairtlough, ch.19), but inventions have to start in a single mind. Individual inventors and designers may create an artefact with a target market in mind (Nelson, ch.6; Vincenti, ch.13; Miller, ch.15) but, as Michael Gorman and I have learned in developing a cognitive approach to invention, this target market in the inventor's mind may be very different from the actual market-place. In fact, the real challenge for the historical analyst studying individual inventors is to pay attention to differences between what the inventor says or thinks he is doing versus what actually happens in the market-place.

The purpose of this chapter is to inquire if it makes sense to think about the production of technological artefacts by individual inventors in terms of evolution. Do some inventors work in an incremental, step-by-step process that looks like biological evolution, and if so, what is the nature of that process? If there are similarities between biological and technological evolution at the level of the individual inventor, should we go so far as to argue for an evolutionary model or should we settle for simply using evolution as a metaphor to describe the gradual nature of technological change?

I believe that it is possible to answer these questions by drawing on a large body of evidence that my students and I have been studying. In 1877, Thomas Edison was asked by the Western Union Telegraph Company to develop a telephone which would be an improvement on the one just invented by Alexander Graham Bell. In taking up this task, Edison produced over 500 sketches which show in detail how he went about inventing or, perhaps more accurately, designing a better telephone. As I will discuss below, my students and I have organized these sketches into 'maps' which allow us to reconstruct the course of his thinking and experimenting. I believe that these maps provide us with meaningful answers about how artefacts evolve on the benchtop or in the minds of individuals.

11.2   Demystifying the process of invention

Before plunging into the maps and investigating how Edison's work on the telephone might be construed as an evolutionary process, we need to address a tension embedded in the main title of this chapter. For some people, invention and evolution are polar opposites. The invention of new technology is typically viewed as a sudden, discontinuous, and revolutionary process through which an inventor stumbles onto a new machine or process by accident, not by planned effort. Occasionally (cf. David, ch.10), an inventor may develop a new creation by trial and error, but we still tend to assume that serendipity plays a large part in this search for a better mousetrap. It is easy to assume that invention, which springs from the inventor's imagination and personality, is a magical process that cannot be analysed.

Biological evolution, in contrast, is commonly viewed as a process character-ized by small, continuous changes; new species are not supposed to burst suddenly into being. Evolution, moreover, is a process which can be analysed (Jablonka & Ziman, ch.2). Through the mechanisms of variation and selection, species change in order to adapt to their environments. Seen as a revolutionary, discontinuous process, one might conclude that technological invention could not possibly be interpreted using the concepts and terms associated with biological evolution.

I would argue, however, that revolution and discontinuity do not describe how inventors actually work. An examination of their notebooks, artefacts and letters reveals that inventors work methodically and purposefully. Yes, they do have Eureka moments in which they perceive things in radically new ways, but these moments are preceded and followed by much hard work. Successful inventors such as Edison knew only too well that 'luck favoured the prepared mind', and they prepared their minds for breakthroughs by collecting information, conducting experiments, sketching and tinkering. In the course of working purposefully, they often had breakthrough insights though, at the time, the breakthrough was frequently not appreciated; it is simply one more good idea in a steady stream of good ideas. For instance, Edison's associates recalled that they did not realize in October 1879 that the carbon filament incandescent lamp they had been working on was the breakthrough; it was only looking back several months later that they realized that late October was when they had turned the corner. Likewise, once a breakthrough had occurred, an inventor often faced a great deal of work in figuring out how to utilize the insight, and it might take years of additional creative work to convert the insight into a successful product.

While it is exceedingly difficult to fathom the Eureka moment, the purposeful work performed by inventors can be analysed. We can demystify much of what goes on in the mind of inventors. In particular, invention can be seen as a process by which an inventor represents a device in his mind, on paper, or as test model and then proceeds to manipulate the device through a series of transformations. These transformations can be readily understood by looking at the sketches an inventor produces and so, to appreciate how the invention process can be demystified, let us consider an example of how Edison used sketching in 1875 to analyse a telephone invented by Philipp Reis.

11.3   A historical interlude: Edison, acoustic telegraphy and the Reis telephone

To understand Edison's sketches of the Reis telephone, we need to examine Edison's situation in the 1870s and who Reis was. Although Alexander Graham Bell is credited with the invention of the telephone in 1876, he was not the only inventor investigating the possibility of a telephone in the early 1870s (cf. §21.3). Both Edison and Elisha Gray were also studying the relationship between sound and electricity. Like Bell, Edison and Gray were fascinated with the problem of multiple message telegraphy. But unlike Bell, who struggled in vain to create a working multiple system, Edison successfully patented a number of multiple message schemes, including several duplex (two-message) and quadruplex (four-message) systems. Edison sold his duplex patents to Western Union who used the patents to block the development of rival telegraph systems, including the postal telegraph scheme proposed by Bell's future father-in-law,

Gardiner Hubbard.[1] Edison also helped create one of these rivals to Western Union by initially selling his quadruplex to Jay Gould and the Atlantic & Pacific Telegraph Company in 1875.[2]

Because multiple telegraphy had proven to be central to his livelihood, Edison carefully monitored developments in this field. In the summer of 1874, he learned that Gray was demonstrating a new harmonic telegraph scheme which might use several acoustic tones to send multiple messages. Anxious to protect his quadruplex patents, Edison began studying acoustics and thinking about how sound and electricity might be integrated in future telegraph systems.[3]

In the course of his thinking about acoustics and telegraphy, Edison became familiar with the electromechanical apparatus devised by Hermann von Helmholtz to reproduce vowel sounds.[4] However, Edison was even more intrigued by an idea proposed in a popular account of electricity. In *The Wonders of Electricity*, J. Baile speculated about the possibility of using electricity to transmit the human voice.[5] Would it not be possible, suggested Baile, for a person's voice to vibrate a plate on one end of an electrical wire, have the vibrations carried along the wire by an electric current, and somehow have these transmitted waves vibrate a plate on the other end? Based on his experience with electricity, Edison sensed that this was possible, but he thought that the real challenge would be to reproduce both loud and soft sounds. To do this, he decided that it would be necessary to vary the current in the telegraph circuit proportional to the volume of the sound. Edison thought that the best way to accomplish this was to vary the resistance in the circuit, and he had already patented a variable resistor which functioned by having a electrode move up and down in a small vial of water (fig. 11.1).[6]

Although Edison thought about how sound and electricity might be used together in a telegraph system as early as 1874, he did nothing with these observations until July 1875. By then, the president of Western Union, William Orton, had become sufficiently concerned about the acoustic telegraph schemes of Gray and Bell for him to propose that Edison investigate the field for Western Union. Orton had let Edison's quadruplex slip through his fingers, and he was now anxious not to have a rival company gain a competitive edge through acoustic telegraphy. Consequently, Orton offered Edison a contract which paid Edison $200 a week for experiments. Edison accepted this contract and, as a starting point, Orton sent him an English translation of a German report of a curious electroacoustic device, a telephone invented by Philipp Reis.[7]

Reis was a school teacher at a private school outside Frankfurt who learned physics by attending lectures at the Physical Society of that city. In the early 1860s, he became interested in understanding the functioning of the human ear and, to do so, he invented a series of instruments 'by which it is possible to make

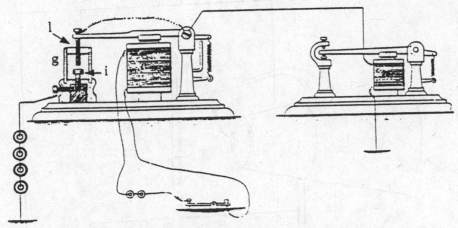

Figure 11.1 The liquid rheostat that Edison patented in 1874 for use in his duplex telegraph is shown attached to the telegraph relay on the left. It consists of a cup g filled with either water or glycerine with contacts l and i. As l moves closer or further away from i, the resistance varies proportional to the distance that the current has to pass through the liquid. From Thomas A. Edison, Relay Magnets, US Patent No. 141,777 (executed 7 March 1873, granted 12 August 1873).

clear and evident the functions of the organs of hearing, but with which also one can reproduce tones of all kinds at any desired distance by means of the galvanic current'.[8] Reis called his instruments 'telephones', and he appears to have employed them to understand how the human ear processes complex sounds such as the voice or musical notes. In demonstrating his telephones before several audiences in Germany, Reis mentioned the possibility of using them to conduct conversations at a distance, but it is unclear how well his instruments were at transmitting speech.[9]

In 1862, Reis demonstrated his telephone for Inspector Wilhelm von Legat of the Royal Prussian Telegraph Corps. Legat published a report of the demonstration in the journal of the Austro-German Telegraph Society, and it was this report that was translated and given to Edison in July 1875.[10] In his paper, Legat explained how complex sounds were the result of the superposition of individual tones and how the Reis telephone could be used to study this phenomenon.

In the Reis telephone, the transmitter consisted of a cone, membrane and sensitive switch (fig. 11.2). As one spoke or sang into the cone, the collodion membrane vibrated. Resting on the membrane was a metal lever, pivoted in such a way that a small motion of the membrane produced a large motion at the opposite end of the lever. At its end opposite the membrane, the lever formed a switch in the telegraph line. Normally, this switch was closed but, when one

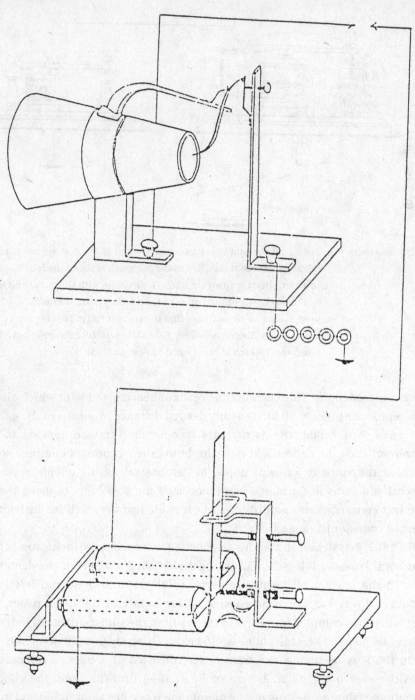

Figure 11.2 Sketch of Reis telephone included with Western Union translation of Legat's report. From *The Speaking Telephone Interferences. United States Patent Office. Evidence on Behalf of Thomas A. Edison*, Vol. 2, pp. 509ff. on reel 11, *Thomas A. Edison Papers, A Selective Microfilm Edition*.

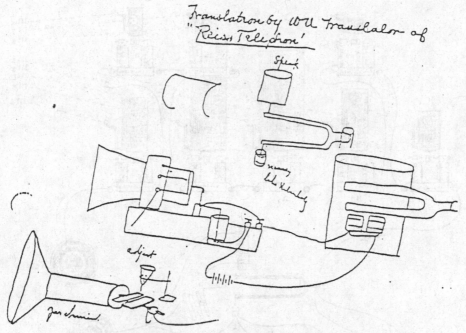

Figure 11.3 After reading the report on the Reis telephone and seeing fig. 11.2, Edison decided that the Reis telephone failed to transmit loud and soft sounds because the make-and-break contact did not vary the voltage in the electrical circuit. To vary the circuit's voltage, Edison decided it was necessary to vary the resistance of the circuit, and to accomplish this he drew two different telephones. In the centre sketch, Edison varied the resistance by having a needle move up and down in a vial of water. Edison borrowed this idea from an earlier patent, shown in fig. 11.1. In the bottom sketch, Edison varied the resistance by having drops of water fall between two knife-edges. In both cases, we see how Edison quickly transformed the Reis telephone into a variable resistance transmitter. From Edison microfilm, reel 11, frame 675.

spoke into the cone, the vibrations interrupted the current, creating a series of pulses. These pulses were then sent to the Reis receiver or analyser. Here the intermittent current energized and de-energized an electromagnet which in turn attracted and repelled a metal reed. Properly adjusted, the reed reproduced the basic tone of the voice or musical note sounded at the transmitter.

11.4   Edison's transformative sketches of the Reis telephone

Upon reading the translation of Legat's report, Edison quickly sized up Reis's telephone and sketched several alternatives (fig. 11.3). These sketches are

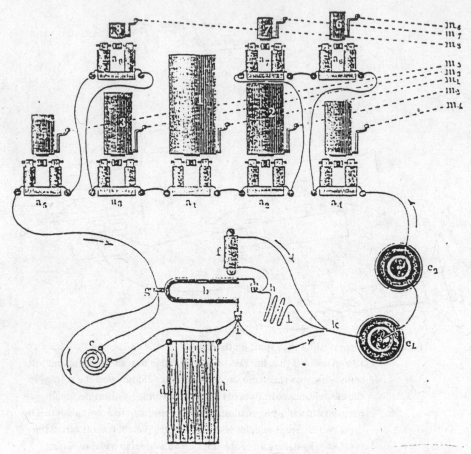

Figure 11.4 Apparatus used by Hermann von Helmholtz to produce artificial vowel sounds. In this apparatus, Helmholtz had electromagnetic f vibrate tuning fork b at a regular frequency. In receivers a1–a8 are smaller tuning forks of the seven harmonic upper partials of main fork. By controlling the resonant cavities above each receiver's fork, Helmholtz was able to include or exclude different tones and hence combine the tones necessary to create a vowel sound. We believe Edison borrowed the mercury cups marked h or i and one of the receivers marked a. From Helmholtz, Hermann F.L., 1875. *On the Sensations of Tone*, trans. A.J. Ellis (London), p. 606.

his 'notes' on the Legat report.[11] To us, these sketches may seem surprising because Edison used very few words to assess the Reis telephone, but he probably found it easier to think in terms of pictures and not words.

In the top sketch, Edison compared the Reis device with Helmholtz's apparatus for reproducing vowel sounds (fig. 11.4). Instead of using a delicate (and probably temperamental) pivoted lever as Reis had, Edison proposed substituting a tuning fork and a needle-in-a-cup switch which used 'mercury like Helmholtz'. Drawing on his earlier thinking about sound waves and electricity, however, Edison decided that, if one wanted to produce both loud and soft sounds, one did not want to interrupt the current; rather, one wanted to vary the current in the circuit proportional to the sound.

Accordingly, Edison re-represented the Reis telephone as a variable current device. In the centre sketch, he replaced Reis's switch with the liquid variable resistor he had patented in 1873 (fig. 11.1). Because the liquid resistor required one electrode to move vertically in a vial of water, Edison bent Reis's lever 90° at the pivot, but he still proportioned the lever so that a small motion at the membrane produced a large motion in the vial. For this second transmitter, Edison added a receiver modelled after the receivers in the Helmholtz apparatus.

In the bottom sketch, Edison did away with the vial and introduced a high-resistance liquid directly into Reis's switch. In this design, Edison positioned a funnel with a drip wick over the switch. Drops of water would fall from the funnel between the switch's two contacts and create a path of high resistance for the current. Because of electrolysis, the passage of current would cause the drop to evaporate, thus making it necessary for the funnel to create a new drop. Finally, because it would be difficult to have a drop fall precisely between two points (as in Reis's original switch), Edison stretched the points of the electrodes into lines and hence sketched two small plates or knife-edges as the electrodes in this transmitter.

In describing these three sketches, we can see that Edison transformed the Reis telephone by making a series of specific 'moves'. On one level, Edison made several substitutions. Deciding that the switch in the Reis transmitter should be changed, Edison tried using a combination of a tuning fork and mercury cup, then a liquid rheostat, and finally a novel arrangement using knife-edges and drops of a high-resistance liquid. In the centre sketch, Edison dropped the Reis receiver and substituted a Helmholtz receiver. Significantly, with the exception of the droplet scheme, Edison substituted devices which already existed; in the first and second sketch, he borrowed items from the Helmholtz apparatus while in the third sketch he carried over a device from one of his own patents.

In studying a large number of sketches made by Edison, Bell and Gray, Gorman and I have found that inventors frequently make these sort of substitutions. The inventor takes something that he has seen in another machine and places it in a new design. Frequently, the components used in the substitution are relatively stable, in the sense that the inventor carries them over from one machine to next without making significant changes to them. Further, as one surveys a number of sketches from one inventor, one notices that an inventor favours some components, and uses them repeatedly. For instance, Edison frequently used a special telegraph device, a polar relay, in many of his inventions, ranging from his multiple telegraph schemes to his motion picture machines.[12] Because these stable components or building blocks seem to play a prominent role in the invention process (Stankiewicz, ch.17), we have chosen to give them a specific name, mechanical representations, and we are trying to identify and trace them through the sketches from each inventor.

In the process of making substitutions, it is important to note not only *what* an inventor substitutes (the mechanical representations) but also *where* he chooses to make the substitution. In these three sketches, Edison wanted to convert the Reis transmitter from an intermittent current device to a variable current device and he chose to do so by focusing his attention on a particular part of the Reis transmitter, the sensitive switch.

To highlight what the inventor chooses to problematize in a particular invention, we refer to the place or zone he or she studies as a 'slot'.[13] In comparing Bell, Gray and Edison, we see that they focus on different aspects of their telephone-like devices, thus making it possible to compare their activities in terms of their choices of slots. In using the word 'slot', we are construing the process of substitution in terms of the imagery of inserting an object (like a square peg) into a space (like a square hole). However, for inventors, the mechanical representation does not always fit precisely in the slot, and a poor fit between object and analytical space may lead an inventor to try substituting new mechanical representations and identifying new slots in his or her design.

While substitution is one kind of 'move' or operation that an inventor can perform in a sketch, a second kind of 'move' involves spatial manipulation. An example of this sort of manipulation can be seen in Edison's bottom sketch of the Reis transmitter. In this sketch, Edison converted Reis's switch into a liquid rheostat by having droplets fall between the contacts. Because it would be very tricky to get the droplets to fall and rest between two points, Edison replaced the points with two horizontal plates. In doing so, one could say that he created the plates by stretching the two points into two horizontal lines.

Viewed in the isolation of a single sketch, this sort of spatial manipulation may seem far-fetched, but as one scrutinizes many sketches, one finds that an inventor tends to make some spatial manipulations and not others. Moreover, in comparing inventors, one finds that different inventors manipulate their ideas in different ways. For instance, Bell generated several different multiple telegraph devices by rotating and positioning the reed in different ways in relation to the pole of the electromagnet. Although Bell frequently rotated devices as he moved from one sketch to another, Edison appears never to have recast a device by rotating one component relative to another. Conversely, we have yet to see Bell stretch a point into a line.

We know as much as we do about these three sketches because Edison discussed them in his testimony during the telephone litigation in 1880. He used the sketches in the testimony to establish that he had possessed the idea of using variable resistance to convert sound waves into electric current waves in 1875, months before Bell filed the patent application for his telephone. However, in the testimony, Edison did not enter this page of sketches as a regular exhibit because the sketches were not witnessed or dated at the time they were created. Indeed, Edison's chief assistant, Charles Batchelor, testified that he had seen his chief make these sketches in the summer of 1875 but they had not been explained to him and so he did not sign and witness them.[14]

Rather than serving to communicate Edison's interpretation of the Reis telephone to his assistants in 1875, these sketches were part of the process of generating ideas, and Edison himself testified to this:

> I have spoken as if these devices were made. I do not mean to say they were made . . . My sketches were rough ideas of how to carry out that which was necessary in my mind, to turn the Reiss [sic] transmitter into an articulating transmitter. They were notes for future use in experimentation.[15]

In calling these sketches 'rough ideas' and 'notes for future use in experimentation', Edison signals that these sketches were not handed over to a machinist or experimenter to be built and tested. Instead, these sketches embody something more general in his thinking about a possible invention. In sketching these variations, Edison transformed the Reis telephone (which was not very useful or interesting to him) into an idea or vision he could use in future experiments. Edison was trying to capture on paper his vision of how sound waves could be converted into electric current waves. Because aspects of this vision were hazy and uncertain, he could only attempt to capture it by representing and re-representing it, using devices with which he was familiar. As a telegrapher, manufacturer and inventor skilled in manipulating objects on the benchtop, Edison had found that the best way to track a hazy vision was to sketch it; although he was aware of how physicists and engineers were using words (theories) and mathematical statements to represent electrical phenomena, Edison had little use for words and equations. And although this vision seems painfully elusive and fleeting to us (especially if we expect Edison to put it into words), we would nonetheless argue that this sketch embodies a key component of the invention process, a mental model.

As an inventor thinks and sketches, he or she does not simply manipulate mechanical representations using moves or heuristics in some random fashion; instead these manipulations are an attempt to capture, to flesh out, a mental model (§14.6). On one level, a mental model is a goal, as in Edison's desire 'to turn the Reiss [sic] transmitter into an articulating transmitter'. Yet, on another level, a mental model is an incomplete picture or image of how the inventor thinks his new invention should function. Mental models are frequently the 'rough ideas' which the inventor converts into a smooth-running invention only by representing and re-representing his or her mental models in a variety of ways.

For Edison, these sketches from the summer of 1875 represent a powerful mental model that he used through much of his research on the telephone from 1876 to 1878.[16] Many of Edison's sketches for the telephone were based on the idea of variable resistance secured by either the notion of moving or dragging an electrode through a high-resistance medium (as in the centre sketch where the vertical electrode moves through the vial of water) or the idea of securing variable resistance by squeezing a high-resistance medium between electrodes (as in the bottom sketch). Significantly, I would emphasize that while Edison derived this mental model from the Reis telephone, the Reis telephone (as Reis or Legat or anyone else conceived it) did not constitute Edison's mental model. Rather, Edison's mental model was his *transformation* of the Reis telephone, as illustrated in these sketches.

Though seldom articulated by historical actors and elusive to the historical analyst, mental models play a central part in the thoughts and actions of inventors (Solomon, ch.14). More than the 'moves' and mechanical representations, they are a bright thread that runs through the tapestry of an inventor's work. Moreover, as the elemental abstraction which the inventor is trying to realize, one could argue that the mental model is, in evolutionary terms, the meme (§1.2) that shapes the actual artefacts and is replicated in them. However, to appreciate fully how mental models might be the memes in the invention process, one needs to look not at how an inventor transforms one sketch into another, but at how he or she transforms dozens of sketches in sequence. To capture this sequence, we must turn now to looking at how sketches can be arrayed in the form of a map of the invention process.

11.5   Another historical interlude: Bell, Western Union and Edison's contract

Before delving into the details of mapping Edison's work on the telephone, it is useful to provide some additional historical background. Although

Edison thought about the Reis telephone in 1875, he devoted his energy to developing other forms of an acoustic telegraph and he did not consider the possibility of transmitting speech. Instead, this possibility animated Bell, and he succeeded in transmitting speech in March 1876. After demonstrating it at the Philadelphia Centennial Exhibition that summer, Bell and his principal backer, Hubbard, offered to sell their telephone patent to the Western Union Telegraph Company for $100,000. For several technical and business reasons, Western Union decided not to purchase Bell's patent in the autumn of 1876. Instead, the company again turned to Edison and asked him to develop a series of patents by which Western Union could either block Bell's entry into the electrical communications business or which the company could use to develop its own telephone business at a later date.

As a result of this request, Edison spent much of 1877 working at his new laboratory in Menlo Park, New Jersey, developing an improved telephone, and filing patent applications for various telephone designs, as a result of which he secured ten telephone patents. In this course of this work, Edison rejected Bell's basic idea for the telephone. Bell had patented a magneto telephone in which the power of the speaker's voice generated an electric current. Edison quickly realized that the feeble currents generated in Bell's magneto telephone would severely limit the commercial applications of the telephone, and so he concentrated on developing a variety of telephones which used batteries. Most of Edison's telephones converted sound waves into electric signals by varying the amount of resistance (and hence the current) in the circuit, and the trick was to come up with a way by which one could vary the resistance very precisely and hence pick up all the nuances of the voice.

11.6   Sketches as fossils: taking a palaeontological approach

To develop a variable resistance telephone, Edison drew hundreds of sketches. These sketches show different conceptualizations of how a telephone might work, they depict models Edison wanted his associates to test and build, and they record the results of experiments. These sketches have survived largely because they were included as evidence in a telephone patent case in 1882. Although they were accompanied by testimony given by Edison and his associates, the testimony generally does not explain their contents. One should not be surprised about this because in the five years between when the sketches were made and the testimony was taken, Edison and his associates invented two other major inventions, the phonograph and the incandescent lamp.

On several levels, these sketches are a nightmare to study. First, they are accompanied by very little text or supporting documentation in the form of correspondence or diaries. Consequently, my students and I had to develop a high degree of skill in deciphering and interpreting them. Fortunately, however, most of the sketches are dated. Second, the sketches are not systematic in any way – we do not necessarily have, say, a conceptual sketch followed by a drawing showing an experiment, followed by another drawing showing how the results were incorporated into a patent. We have samples of each sort of drawing but few cases where we have anything that looks like a meaningful sequence.

Given these problems, we adopted a palaeontological approach. We decided to treat each sketch as a fossil; our task was to try to arrange them in a pattern that would reveal clues about how Edison thought and worked. As good palaeontologists, we began by looking for family resemblances. On one level, we looked for similarities in the devices making up the telephones, the mechanical representations. But on another level, we sought deeper, structural resemblances – devices that shared an underlying principle of operation, a common mental model or meme. As we found telephones sharing an underlying principle of operation, we began to group them chronologically in horizontal rows on a map, letting them constitute different 'lines' of research.

11.7   Making maps to find patterns in the fossil record

As we grouped the sketches into the lines of research, my students and I became interested in how Edison moved from one sketch to another. Over the next several years, we painstakingly reconstructed the hypothetical manipulations and transformations which Edison may have used to generate new telephone designs. This effort resulted in eighteen huge maps, which depict every sketch from 1877 and how each sketch is connected to other sketches. We are still mining these maps in order to create a catalogue of all the moves or heuristics Edison employed as an inventor.

Gradually, however, my students and I came to see this set of eighteen maps as being too detailed, and that they did not permit us to discern general patterns in Edison's work. To press the cartographic metaphor, our first set of maps had a scale of 1:10 when what we really wanted was 1:100 or 1:1000. We were interested not in just how Edison generated each new telephone design; we also wanted to know how he orchestrated the overall research process. Consequently, one student, David Tjader, and I generated one large map which reduced each sketch to a single numbered box, taking care to preserve the lines of investigation. And I boiled this map down once more, showing the number of sketches Edison produced in each line of investigation each month (fig. 11.5). With this map in hand, it became possible to look for general patterns in Edison's efforts to develop a telephone.

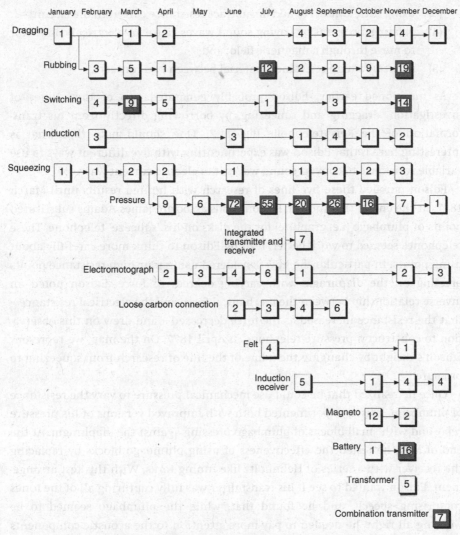

Figure 11.5 Overview of Edison's development of the telephone, 1877.

11.8   A narrative overview of Edison's work on the telephone

Turning to this map, one can see that in early 1877 Edison started his work by pursuing five basic lines of investigation. He developed variable resistance telephone transmitters by:

(1) *dragging* an electrical contact through a high-resistance medium (often a liquid);

(2) *rubbing* an electrode along a high-resistance material (such as coil of platinum wire);

(3)   using little *switches* to add or subtract resistance coils into the circuit;

(4)   *inducing* a current by having sound waves cause an electrical conductor to move through a magnetic field; and

(5)   *squeezing* a high-resistance material between two contacts.

As mentioned earlier, Edison probably generated two of these lines of investigation, dragging and squeezing, by borrowing directly from his transformation of the Reis telephone (fig. 11.2). One should note that what is interesting here is that Edison was experimenting with five different ways to use variable resistance to convert sound waves into electric current waves.

Edison pursued these five lines of research with limited results until March 1877. At the end of that month, Edison or his associate James Adams substituted points of plumbago (i.e. graphite) for the disks on his 'squeeze' telephone. These telephones seemed to work better, leading Edison to think more carefully about using points. In particular, he now considered using four high-resistance points pressing on the diaphragm with varying degrees of force. Edison noted an inverse relationship between the mechanical force and the electrical resistance – that the resistance increased as the force decreased – and drew on this observation to construct a pressure telephone in April 1877. On the map, we represent Edison's insight by changing the name of the line of research from squeezing to pressure.

Once he realized that he could use mechanical pressure to vary the resistance of plumbago, Edison experimented both with improved versions of his pressure relay and with small blocks of plumbago pressing against the diaphragm. At the end of May, he tested the effectiveness of using plumbago blocks by replacing the receiver with a series of Helmholtz-like tuning forks. With this test arrangement, Edison wanted to see if his transmitter was fully capturing all of the tones comprising speech, and he found that, while the plumbago seemed to be working all right, he needed to pay more attention to the acoustic components of his telephone: the diaphragm and mouthpiece or resonant cavity.

Consequently, during the summer of 1877, Edison and his associates simultaneously focused their research on one kind of telephone (pressure) but at the same time broadened their investigation by undertaking intensive studies of the acoustic components of the pressure telephone. While the map in fig. 11.5 shows that Edison greatly increased the number of sketches he drew of the pressure telephone, fig. 11.6 reveals how this research was spread across a number of new lines of investigation.

Not convinced that pure plumbago was necessarily the best material to use, Edison began by having his associates test the relationship of mechanical pressure and electrical resistance for hundreds of mixtures of plumbago and

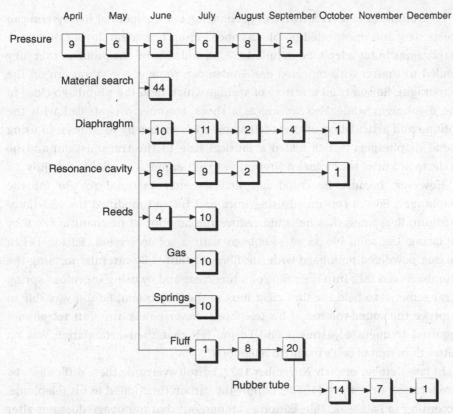

Figure 11.6 Edison's lines of research on pressure telephones, April–December 1877.

other substances. At the same time, Edison and his associates made a series of changes to the diaphragm (such as replacing the metal membrane with a mica one) and they tested several kinds of resonant cavities. Because sounds such as 's' (sibilants) were not being transmitted (they failed to vibrate the diaphragm sufficiently), Edison tried adding a vibrating reed to his telephone. Mounted on the top of the resonant cavity, the reed vibrated in response to sibilants, opened and closed an electrical circuit, and hence added additional current at the appropriate frequency to the signal. Eventually, however, he decided that the slight improvement from using the reed did not offset the extra circuitry that it required and so he abandoned this line of investigation. Perhaps to help think about how mechanical pressure was being converted into electric signals in these telephones, Edison made a few sketches in which he substituted gas pipes for the electrical wires. In these designs, sound waves impinging on the transmitter diaphragm caused waves to propagate through the gas line which in turn caused a diaphragm to vibrate at the receiver.

As is often the case (Vincenti, ch.13), changing one component in a system can create new and unexpected problems; once Edison began employing thin mica diaphragms in his telephone in June 1877, he had to solve the problem that they tended to shatter with repeated use. To dampen some of the vibrations on the diaphragm, Edison tried a variety of springs which held the plumbago close to the diaphragm while also serving as a shock absorber. Dissatisfied with the volume and articulation of these telephones, Edison briefly went back to using metal diaphragms (which added a musical ring to the transmission) and he undertook a brief search for an alternative to either metal or mica in late July.

However, because he could not find a better material to use for the diaphragm, Edison continued using mica and instead modified the resistance medium. Reasoning that he could reduce the amount of mechanical force by replacing the solid blocks of plumbago with a softer version, Edison began mixing powdered plumbago with silk fibres or fluff.[17] By carefully forming the plumbago and fluff into the shape of a tiny cigar and by using ingenious spring arrangements to hold the fluff cigar next to the diaphragm, Edison was able to improve the sound-volume of his telephone. Nevertheless, the fluff telephones required frequent adjustment and Edison felt that their articulation was no better than that of Bell's magneto telephone.[18]

In late October or early November 1877, Edison overcame these difficulties by developing a new mental model of how the carbon functioned in his telephone. According to Batchelor, 'Mr. Edison . . . found out that plumbago does not alter its resistance by pressure as we at first thought; but the increased pressure made better contact.'[19] On the basis of this new mental model, Edison now decided to employ a hard button made of lampblack. Because a button of pure lampblack would have a high resistance, Edison lowered its resistance by mixing in ground rubber and he reduced the force on the button by placing a small rubber tube between the button and the diaphragm.

While Edison and his associates devoted most of their attention from April to December 1877 to perfecting a pressure telephone, they also pursued a few other lines of research (fig. 11.5). Beginning in April, Edison began developing his own distinctive receiver, drawing on an earlier invention, the 'electromotograph'. Intent on using his pressure transmitter combined with his electromotograph receiver, Edison experimented in August with how these two devices could be combined into a single instrument, somewhat like the handsets we use today. In August and September, Edison sketched a number of magneto transmitters and induction receivers which closely resembled Bell's basic design; in doing so, he may have been looking for ways he could patent improvements on Bell's design, and these patents would have then become a hindrance to the Bell Telephone Company. And finally, Edison toyed with a few additional ways of securing variable resistance in his transmitter using a loose carbon connection, felt, and transformers. Of these new approaches, only the designs in which the diaphragm vibrated a baffle between the electrodes in a chemical battery (and hence varied the current generated by the battery) resulted in a patent filed by Edison.

11.9   So what do these maps tell us about invention and evolution?

Although figs. 11.5 and 11.6 may look daunting, if you have followed the narrative thus far, you really have all the information you need to begin thinking about the evolutionary mechanisms in Edison's creative work. So what can we learn from this map?

One thing that jumps out from these maps is that Edison did not work through one line of investigation and then move to another; rather, he was always juggling several lines of research. In the early months of 1877, he investigated several different classes of telephone (dragging, rubbing, switching, induction and squeezing); in the summer months he broke his research on the carbon pressure telephone into several different areas; and in the autumn he supplemented work on the rubber tube telephone with studies of induction receivers, battery telephones and a few alternative lines. In pursuing simultaneous lines of research Edison was like other creative individuals in that his creative work constituted a network of enterprises (Fairtlough, ch.19); Edison was not simply investigating one kind of telephone but rather a network of possibilities.

While not illustrated on the maps shown here, Edison pursued several simultaneous lines because these generated lots of new devices and mechanical representations which could be transferred from one line to another. (To see these transfers, one would need to return to the original series of eighteen maps.) These transfers were often like the grafts that plant breeders make, and for Edison they often resulted in improved performance in the telephone under study at any particular moment.

But more than this, Edison's pursuit of multiple lines of research, as shown in fig. 11.5, can be interpreted in terms of variation and selection. To see this, we need to redraw fig. 11.5 once more, this time dropping the boxes which depict the monthly totals so that we can focus on when the lines of research begin and end. In addition, if we are going to think about selection, we must ask 'Selection for what?'.

Here we must recall that Western Union contracted Edison to develop both a production model telephone that it could install on the company's lines and patents which could be used to block Bell. Consequently, in fig. 11.7 I have added

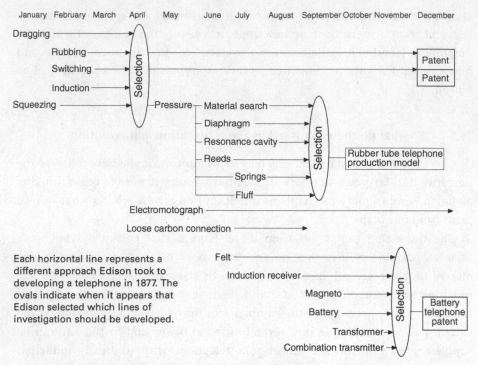

Figure 11.7 Variation and selection in Edison's work on the telephone in 1877.

boxes for the US patents that Edison filed during 1877 for the telephone as well as a box indicating when Edison had a production version of his telephone. By comparing the shaded boxes on fig. 11.5 with the patent boxes on fig. 11.7, one can see a correlation between sketch production on certain lines and when patent applications were filed. For instance, the flurry of sketches for rubbing and switching telephones in November 1877 corresponds to a patent Edison filed in early December.

Figure 11.7 suggests that Edison's work on the telephone can be viewed as variation and selection. During certain periods, Edison varied his lines of research, and then at particular moments, he appears to have selected one line for further development. In late April 1877, he chose to drop four lines in favour of the pressure line, and in December, after playing with several miscellaneous lines, he chose only to submit patent applications for his battery telephone. In September, with the pressure telephone, the selection process was somewhat different. Rather than singling out one line of research, Edison combined the most promising results from the six lines to create the rubber tube production version.

It is clear that, if there is a selection mechanism operating in invention, it may operate in several different ways (cf. Mokyr, ch.5; Nelson, ch.6; Miller; ch.15).

On one hand it may be a 'singling out' process as in the case of April 1877 or it may an 'amalgamation' process as it was for Edison in September 1877. Similarly, the selection process is affected by different goals. In April and September, Edison was guided by performance: which type of telephone worked best; what features should be brought together for a practical model? In contrast, in December, Edison selected only his battery telephone designs since he thought he could get these patented.

And finally, one should note that the timing of selection can be affected by both an inventor's judgement and outside events. It is likely that the April and December selections were based on Edison's own shrewd assessment of the situation – that it was time to shift from a divergent research strategy to a convergent one. In September, however, Edison's selection process was driven by outside events; planning to step up pressure on Bell, Western Union placed an order with Edison for 50 telephones. To meet this order, Edison was obliged to make some selections from the several lines of research he had going forward.

11.10   Edison as breeder

As we scrutinize how Edison varied and selected his lines of research, it suggests to me that we should view invention not as *natural* selection but rather as *artificial* selection and Edison in the role of *breeder*. If we think of each line of investigation as a 'strain', then one could argue that Edison varied each sketch in different ways in order to learn as much as he could about how that particular class of telephones functioned. He then periodically 'bred in' devices developed in one line onto another, and several of these 'hybrids' paid off in terms of being better performing telephones.

Now one should note that in describing Edison as breeder I would deliberately avoid the idea that he was attempting to optimize one particular characteristic in each line of investigation, which is something that a real breeder would perhaps try to do (cf. Miller, ch.15). Indeed, I think that Edison tended to play out each line of investigation just to see where it would go and what it might teach him about how the telephone might or might not work. The individual lines of research were valuable because they provided ideas and devices which could be periodically 'recombined' (§2.2; §12.6; §15.2). And just as recombination in the living world often leads to hybrids with superior performance, so Edison seemed to realize that the breakthroughs came by juxtaposing devices and ideas from two different lines. Of course, why Edison chose to introduce a device from one line onto another line at a particular moment is very difficult to answer, and these decisions may be the real points where genius or serendipity come into play.

Another important feature in thinking about invention as artifical selection is that Edison was not simply a breeder in the old-fashioned sense but actually more of a genetic engineer. Unlike the traditional breeder who must work with the basic biochemical make-up of the plant or animal species, Edison was able to change substantially the make-up of a particular telephone. If he wanted he could constitute a telephone in a particular line of research using different basic building blocks – the telephone could use the same principle as others in the line but be made up of a different combination of mechanical representations. Like modern geneticists, he really could 'engineer' the species he was breeding.

In conclusion, I suspect there are points where this comparision between invention and artificial selection breaks down, and I think we need to consider those problems as well. However, overall, this approach underlines the value of borrowing all sorts of ideas from evolutionary biology to understand techno-logical change. In trying to understand how technological artefacts evolve, we should not limit ourselves simply to thinking about natural selection; if we do, we fail to acknowledge the powerful and crucial role played by individual inventors and designers.

12

The evolution of adaptive form

DAVID PERKINS

12.1   The evolution of adaptive form

The evolution of adaptive form is all around us. In the biological realm, our own bodies give us evidence from the opposable thumb to the useless appendix. In the social realm, the languages we use are descendants and transformations of other languages from other times, our customs and governmental structures adjustments of earlier ones. In the technological realm, the word processors we write on are superpens, and pens are super-quills. All such forms have a history of development. One form leads to another and another, becoming more complex or simpler in ways adaptive to circumstances.

A number of tempting analogies align the worlds of biological evolution and human invention (Ziman, ch.1). These include[1] adaptation to environmental change, satisficing fitness, convergence, punctuated stasis, emergence, extinctions, arms races, self-organization, irreversibility, and 'progress'. Of course, to speak of all such processes as evolution begs an important question. Does Darwin's concept of evolution, generalized beyond biology, provide a good account of the full range of these phenomena? At first thought, the prospects seem dim. Human invention clearly and profoundly involves the intentional ideation of inventors (Carlson, ch.11). Yet Darwinian evolution (Jablonka & Ziman, ch.2) is an emphatically non-intentional process, a huge history of accident – the work of a 'blind watchmaker' as it has been aptly described by Richard Dawkins.[2]

Nonetheless, the Darwinian concept of evolution may extend beyond biology better than at first it seems. As Campbell pointed out in a classic paper,[3] accident figures far more in human discovery than is ordinarily recognized. Processes of blind variation and selective retention are central to discovery, and the more fundamental the discovery the more prominent their role. A process of 'continual breakout from the bounds of what was already known, a breakout for which blind variation provides the only mechanism available' is needed.[4] In other words, the minor twiddles by which theories and inventions improve may not owe much to a Darwinian mechanism, but the major transformations do. Similarly, according to Margaret Boden, creativity requires escape from prevailing rules: 'A merely novel idea is one which can be described and/or produced by the same set of generative rules as are other, familiar ideas. A genuinely original, or creative, idea is one which cannot.'[5]

In contrast, Langley *et al.* offer a far less accident-prone account of scientific inquiry.[6] They examine a range of scientific discoveries and through artificial intelligence techniques demonstrate that these could have been accomplished through relatively methodical processes that minimize haphazard search. They argue that similar conclusions might apply to discoveries beyond the scope of their investigations. Perhaps a Lamarckian conception, eschewed in biology, could offer an enlightening account of some non-biological evolutionary processes. Although protogiraffes that stretch their necks toward the higher leaves may not thereby yield progeny with longer necks, prototype fruit-picking machines that have to stretch to reach the fruit may find their blueprints revised so that the next generation incorporates longer booms.

All this issues an invitation to look deeply at the logic of evolution-like processes in general. What is Darwinian about them, what Lamarckian, and what neither? Such an inquiry should lead to a better judgement about whether speaking of diverse biological and non-biological processes as evolutionary marks a deep unity or, on the contrary, means no more than saying a robin's egg is blue as the sky, a matter of superficial similarity.

12.2   The challenge of adaptive form

If evolution of some sort is the solution, what is the problem? Why does arriving at adaptive forms, from opposable thumbs to doorknobs to theories of relativity, require a lot of fuss and bother of one sort or another? What makes adaptive forms hard to discover?

Contemporary cognitive science and complexity theory (Ziman, ch.4; Miller, ch.15) have taken over from evolutionary theory the concept of a 'fitness landscape'[7] (§2.4; §4.4; §5.3). The general idea is that all possible forms can be represented as points in an abstract space. In principle, this space has a great many dimensions, but we can think of it as something like an ordinary two-dimensional map. Within such a space, some forms are more 'fit' than others, according to whatever criteria happen to be applicable. The degree of fitness of a particular form is then represented by a 'height' above the corresponding point on the map. We thus construct a 'landscape', where, for example, a high peak would stand for a particularly 'fit' form. Evolutionary change can then be viewed as a process of traversing this landscape in search of such peaks.

In the guise of 'problem space' or 'design space' (Stankiewicz, ch.17) this concept has been used to understand problem-solving and invention.[8] A range of seminal experiments has demonstrated that such cognitively demanding activities as proving theorems in logic, solving certain kinds of puzzles, and playing chess could be modelled as heuristic search processes guided by indicators of 'closeness' to a solution. The same notion has been applied to creativity in the arts and the sciences and to scientific discovery.[9] The term 'fitness landscape' has also diffused from strictly biological contexts into contemporary complexity theory and studies of artificial life.[10] Whether one speaks of a fitness landscape with criteria of fitness, or a problem space with measures of closeness to solution, the import is essentially the same.

Although the notion of fitness derives from biology (§2.4), fitness in this generalized sense need not just mean survival. In the context of solving a problem, more fit means closer, by some measure, to a full solution. In the context of marketing an invention, more fit may mean more effective as a mechanism and more viable in the market-place. In the context of biology, fitness involves not just survival or breeding but sexual selection. Indeed, organisms may incorporate compromises between general viability and sexual selection: shorter tail feathers may serve flying better, longer ones attract a mate better, with an evolutionary compromise converging on tail feathers somewhere in between.[11]

It is also important to recognize that fitness is a moving target. Evolution, both in the biological world and the market-place of inventions, is often a matter of coevolution, with the rules of the fitness game changing as forms compete or develop symbiotic or parasitic relationships. Also, outside forces – a meteor striking the Earth, or a declaration of war – may abruptly change the fitness landscape for biological or technological innovation. A fitness landscape may involve shifting dunes rather than granite ledges, depending on the circumstances.

In the language of fitness landscapes, what makes adaptive forms hard to find? This depends very much on the topography of the fitness landscape. Friendly topographies can offer easy paths to good solutions. For example, the fitness landscape in the case of the protogiraffe's neck or the fruitpicker's boom would seem to be fairly friendly. It's clear that more length would serve well. Darwinian or Lamarckian processes would easily arrive at this solution on behalf of the protogiraffe, and the inventor would do just as well by the fruitpicker.

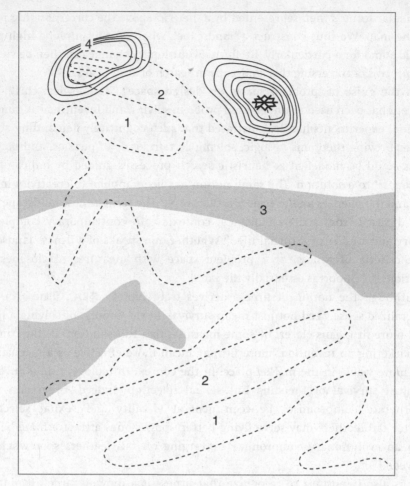

Figure 12.1 Search in a Klondike space. 1. Large space, few solutions (wilderness gap). 2. No clues pointing direction (plateau gap). 3. Solution isolated from where search starts (canyon trap). 4. Area of high promise but not over solution threshold (oasis trap).

All this is true enough provided the protogiraffe's neck and the fruitpicker boom are not too long to start with. Otherwise, the fitness landscape gets complicated. Blood must be pumped all the way up that longer neck (and in fact giraffes have adaptations to deal with this problem). A long boom is unwieldy. It's no longer more fit simply to make the neck or the boom longer. Other things need to happen in support of such adjustments. After a certain point, another entirely different kind of adaptation may make more sense as a way of reaching the higher leaves or fruit.

In previous writings, I have found it useful to characterize fitness landscapes that are unfriendly to easy search as 'Klondike spaces'.[12] The metaphor here is with searching for gold in the Klondike. Gold is where you find it – sparsely distributed with relatively few signs pointing a clear direction. In this respect, Klondike spaces contrast with 'Homing spaces' where a clear gradient points the way to a viable, satisficing (§4.1) outcome.

To formalize the notion of a Klondike fitness landscape, here (fig. 12.1) are its four key characteristics:

Wilderness gaps. Solutions – forms of sufficient fitness – are sparsely distributed in a wilderness of possibilities. In the world of biology, relatively few bioforms are viable. In the world of invention, relatively few 'technoforms' do their jobs sufficiently well. In the world of insight problems, relatively few solutions among solution-like constructs are genuine.

Plateau gaps. In large regions of the landscape, there are few clues pointing a clear direction toward greater fitness. The fitness landscape is relatively flat. It's hard to know where to go next, what to try next. Only in small regions close to a solution can a search process reliably follow a gradient to it.

Canyon traps. Search often occurs within regions almost fully bounded by areas of low fitness or outright impossibility. Canyons are what the complexity literature (Ziman, ch.4), inverting metaphorical gravity, would call the ridges surrounding 'basins of attraction'. Unfortunately, adequate solutions may lie in other regions, representing completely different approaches to the fitness problem. Thus both human inventions and biological adaptations sometimes come from surprising directions.

Oasis traps. Search tends to linger in the neighbourhood of pseudo-solutions or near-solutions that will not quite work. Finding a full solution requires leaving these oases behind. This is the classic problem of unsatisfactory local maxima.

Klondike spaces are typical of problem-solving and search phenomena in insight problems, human invention, and biological evolution. Sudden discovery is one of the most conspicuous phenomena in all three cases: the abrupt insights that occur in response to insight problems;[13] breakthrough discoveries as with Gutenberg's adaptation of the design of a wine press to the printing press or Darwin's discovery of the principles of natural selection;[14] and punctuated equilibrium in evolution (§2.6), where sudden transitions occur in adaptive form.[15] All three, I have argued,[16] can be explained in terms of the Klondike topography of the search spaces involved. A Klondike fitness landscape inherently leads to episodes of sudden discovery, regardless of the particular search process involved.

So what makes adaptive forms hard to find? The suggested answer is that the search for adaptive form – whether by inventors or biological evolution – commonly occurs within fitness landscapes with Klondike characteristics. Why, it might be asked, do such landscapes occur so often? The law of large numbers provides a first-order answer. A friendly Homing landscape is something of a statistical anomaly, in the ideal case offering a clear gradient leading up to a single global maximum. This does not mean that problems in Homing landscapes are trivial, because they may require careful reasoning to track the gradient to the top. But they are problems of a different character. In the wide and somewhat disorderly world, Klondike landscapes rule. Orderly hill-climbing is an important tactic but a local one. Good solutions cannot simply be figured out by some algorithmic or near-algorithmic procedure. They have to be discovered by some kind of far-ranging flexible search process (Miller, ch.15).

12.3   Strategies of search

By itself, the notion of Klondike spaces leaves hanging the story of the evolution of adaptive form. What kinds of search processes can cope with the troublesome Klondike features? Here it's illuminating to compare and contrast three different strategies of search: adaptation by revision, adaptation by selection, and adaptation by planning.

Adaptation by revision

Adaptation by revision is the basic pattern of search by which human beings handle most immediate and accessible needs that require search at all. A longer more descriptive phrase would be adaptation by trial and informed revision. The general pattern goes like this: a form (theory, invention, whatever) undergoes some kind of test or trial. This test or trial yields information about the adequacy of the form that leads to implications for revision of the form. The revision, carried out, yields a better-adapted form. In the case of the fruitpicker, the user observes the difficulty of reaching the higher fruit and translates this into a revised form with a longer boom.

Adaptation by revision has a Lamarckian character. The fruits of experience, so to speak, get translated fairly directly into a better adapted form. Adaptation by revision is commonplace in the human world – for example, in the process of 'recursive practice' (Constant, ch.16). In the biological world, epigenetic inheritance (Jablonka, ch.3), physiological adjustments such as muscle strengthening and some forms of learning are adaptation by revision – not of the genome but of the individual organism within its life-span. At the genetic level, gene repair is a case of adaptation by revision.

The great advantage of adaptation by revision is its directness and efficiency. One limitation is its requirement for embedded intelligence. The processes of evaluation and interpretation toward revision must incorporate considerable knowledge of the situation in question, in effect knowledge of the fitness landscape (Vincenti, ch.13; Miller, ch.15). Of course, to speak of intelligence and knowledge is not to imply that selection by revision need be a conscious process or one that involves elaborate problem-solving. The process of muscle strengthening, for example, entrains mindless mechanisms that embody physiological knowledge in much the way that a fish's fin embodies knowledge of hydrodynamics.

A second limitation is that adaptation by revision in its simpler forms, even if consciously directed, can have trouble with Klondike features of the fitness landscape. It can get trapped by oases (local maxima) and canyons, through its persistent effort to revise directly on the basis of what it 'knows'.

Adaptation by selection

Adaptation by selection follows a Darwinian pattern. A longer but more descriptive name would be adaptation by constrained variation and selection. From an initial form, variations are generated by a specific mechanism. The variations are constrained within some range – breeding roses does not produce elephants, inventors developing microchips do not produce sonnets (or only on the side) – but not constrained in ways that guarantee improvements in fitness over the initial form, otherwise the process would not be blind to position in the fitness landscape. A process of testing ensues, yielding a selection of the more fit variants.

Adaptation by selection occurs widely. Besides the example of Darwinian evolution, it is essentially the mechanism of trial-and-error or reinforcement learning, important both in animal and human cognition. Inventors sometimes engage in near trial-and-error searches too, finding useful forms by selection. One example is the 'draghunts' that Edison conducted for materials for the filament for a light bulb and for other projects as well[17] (Carlson, ch.11). Another is the increasing use of genetic algorithms (§4.3; §15.3) for technological innovation. Still another is mechanized scanning for natural antibiotics, where many thousands of biological samples are processed for signs of antibiotic action. Ivermectin, a cure for river blindness, was one antibiotic discovered by this procedure.[18]

Adaptation by selection has advantages and disadvantages complementary to those of adaptation by revision. On the downside, it is inefficient, investing in many trial variations (§4.1; §5.1). On the upside, the process need not embody much intelligence about the fitness landscape. Thus, it can function adequately in a Klondike fitness landscape, which lacks helpful order or where the helpful order has not yet been detected. Adaptation by selection can escape canyon and oasis traps, provided the process of variation is wide-ranging enough.

Adaptation by coding

To do its work, adaptation by coding depends on combining adaptation by revision with selection. The general idea is that search involves adaptive forms on two levels that may be called the *code* and the *construction*. The code might be a sketch or blueprint, the construction a prototype built from them. The code might be the genome, the construction an organism. The code may determine the construction closely (a blueprint or genetic code) or only loosely constrain it (a sketch or rough idea). The code may specify the construction by specifying its structure (e.g. a blueprint for a building) or a procedure to produce it (e.g. a recipe for a cake), which is the way the genetic code operates.[19] It does not so much specify the structure of organisms directly as cause organismic forms to emerge from complex chemical and physical interactions (Jablonka & Ziman, ch.2; Jablonka, ch.3) by triggering and biasing the direction of natural emergent tendencies.

The aim of the search is to produce a more fit construction through adjustments in the code – that is by 'vicarious' or 'virtual' selection (§4.1; §5.3; §12.3; §13.3; §16.4; §20.5) – rather than through adjustments directly in the construction. This happens when the inventor, rather than just tinkering with the prototype, revises the sketch and from that builds a new prototype. It also happens when natural selection selects more viable organisms and thereby the genes that yield the next generation. As these examples show, the feedback from construction back to code can occur either through adaptation by revision (as when the inventor makes calculated changes in the sketch – Carlson, ch.11) or adaptation by selection (as in natural selection).

Note that adaptation by coding allows a full characterization of what Lamarckian versus Darwinian evolution means in a genic system (Jablonka, ch.3). Adaptation by revision at a single level might be called Lamarckian and adaptation by selection Darwinian, but the full-scale phenomena involve bi-level code-construction relationships. Lamarckian evolution entails feedback from construction back to code via revision, whereas Darwinian evolution entails feedback from construction back to code via selection.

In addition, coding creates the opportunity for independent search at both the code level and the construction level. Inventors tinker with the code itself and the construction itself, besides revising the code on the basis of experience with the construction. In the biological world, gene repair tinkers with the code. Physiological adjustments, learning and epigenetic inheritance tinker with the construction. Finally, in the human world, we find complicated chains of code-construction relationships – for example, evolutionary cycles for 'virtual' arte-facts.[20] For instance, an idea functions as a code for a sketch derived from the idea, which functions as a code for a prototype constructed from the sketch, which functions as a code for a production version constructed from blueprints made from the prototype. Problems at any level of representation can feed back to revisions made at more abstract levels of representation.

Coding has several advantages. One is that the code is typically much more portable and replicable than the construction. Thus we lug around our genetic codes in compact multiple copies for progenerative purposes. Another related advantage is that generating variations in or modifications of the code is typically far more economical of energy than their construction (§2.1). Another is that the code usually sits in a fitness landscape of its own – for example, through memes evolving in a technological community (Vincenti, ch.13; Constant, ch.16) – reflecting enough information about the fitness landscape of the constructions for some adaptive work to be done without the huge investment of actually producing a construction. Thus inventors tinker with their codes and exchange information about them with other inventors (Fairtlough, ch.19) without needing to produce a prototype, and so gene repair processes function short of producing an organism.

In summary, the notion of Klondike fitness landscapes and the three general search strategies of adaptation by revision, selection, and coding offer an account of the evolution of adaptive form that cuts across the biological world and the many worlds of human discovery. The account makes room for human consciousness, intentionality and imagination in managing such searches. However, the basic search strategies, and even some rather complex mixes of them, do not necessarily require consciousness, intentionality or imagination. The approach also applies to much of contemporary research into artificial intelligence programs that are in some sense 'creative',[21] and as we shall see (Miller, ch.15) research into genetic algorithms and artificial life.[22]

It should be added that the three strategies often occur not in their pure forms but in subtle mixes. Imagine, for instance, an artist painting. The artist may be working from some initial conception, perhaps a set of sketches. Yet the development of the final work on the canvas is far from a straightforward process of rendering. Rather, the artist enters into a dialogue with the emerging work, noticing problems and opportunities that were not anticipated, making adjustments, and perhaps even undertaking a radical change of direction. Such a process – not at all uncommon in human creativity – plainly involves a complex weave of search by revision, selection and coding at multiple levels.

12.4   How biological evolution is Klondike smart

Darwinian evolution in the biological world allows a straightforward analysis in terms of Klondike spaces and the three strategies of search. Biological evolution (Jablonka & Ziman, ch.2; Jablonka, ch.3) mainly occurs through adaptation by coding mediated by adaptation by selection. The phenomenon of punctuated equilibrium (§2.6) suggests that biological evolution often operates as a process of search through fitness landscapes with strong Klondike characteristics. To be effective as a process, biological evolution has to be 'smart' about coping with those Klondike characteristics. In what ways is this so?

The constrained variation of adaptation by selection provides the core strategy of evolution. As noted earlier, constrained variation offers a way to escape canyon traps by finding paths past the boundaries of canyons and a way to escape oasis traps by moving away from oases. Variability in biological evolution occurs principally by mutation and recombination – the mixing of genes from parent organisms (§2.2). Under conditions of environmental stress, some bacteria show increased variability (§3.2). Moreover, selection for variability can yield bacteria where greater variability itself is carried in the genetic make-up of the organism.[23] All this suggests that biological organisms have evolved in ways that favour more efficient evolution.

What is more, organisms carry in their genomes considerable silent code, not dominant in the current phenotype but available when activated by selection processes. Thus, certain bacteria adapt much more quickly to various adverse conditions than it seems reasonable to expect. This may be because they are already carriers of relevant latent adaptations, expressed in a few instances of the organism that under the right selection conditions quickly expand to dominate the population.[24]

Building on core mechanisms of variability, biological evolution adds a straightforward strategy of massive parallelism over long periods of time (§2.2). Many genetic experiments are conducted during the same generation, and generation follows generation for thousands or millions of years. This parallel long-term strategy addresses effectively the wilderness and plateau gaps. The wilderness gap, recall, refers to the sparseness of viable forms in a large fitness landscape of possibilities. The massively parallel search mechanism of biological evolution simply explores a lot of those possibilities. The plateau gap refers to the lack of directional signs toward greater fitness in large regions of the fitness landscape. The massively parallel search of biological evolution meets this challenge head-on by radiating in all directions across a plateau.

As to canyon traps – search bounded by regions of low viability and other factors – the radiating wave of variation from an initial genotype-phenotype can fill canyons and spill over into neighbouring canyons in the fitness landscape, provided that all the trial organisms along the way are viable. Likewise for oasis traps, natural selection has no particular bias for more fit rather than less fit organisms, so long as the less fit organisms are sufficiently fit to survive and breed. Variants, and their variants, and theirs in turn, will radiate out away from an oasis of relatively high fitness, perhaps eventually to reach greener pastures.

With all this recognized, biological evolution in some ways is not very Klondike smart. Perhaps most fundamentally, evolution can only search through regions of some viability. To lead anywhere, each speculative venture of biological evolution must survive and breed. Thus evolution gets trapped in canyons and oases surrounded by regions of non-viability in the fitness landscape.

Relatedly, evolution invests heavily in each trial. While processes like gene repair and spontaneous abortion do some editing short of birth, the principal search of evolution gets carried out through actual organisms striving to survive in the real world. Compare this with the exploratory thinking of a human inventor (Carlson, ch.11), who may examine dozens, even hundreds, of 'virtual variants' at the code level before constructing a physical prototype. A smarter process of evolution would involve a better model of the fitness landscape and a more elaborate search process within the genome and genetic manipulation, so that selection could proceed to a greater extent at the code level.

However, the biological coding strategy makes this difficult. As noted earlier, the genome codes for the structure of an organism indirectly by shaping the processes that produce it, more like a recipe than a blueprint. The 'genic' style of coding (Jablonka, ch.3) does not lend itself to modelling well the ultimate construction landscapes. Imagine, for example, trying to choose the best recipe for an angel food cake simply by inspecting the competing recipes, in contrast with choosing the best design for a house by inspecting competing floor plans.

These observations notwithstanding, in many ways biological evolution *is* Klondike smart. It is a process itself well-adapted to the discovery of adaptive form in the shifting coevolutionary fitness landscapes of the biological world.

12.5.   How human invention is Klondike smarter

Although biological evolution is Klondike smart, human invention is Klondike smarter. Or at least, it can be. Depending on the inventors and the context, human invention can proceed in clumsy and obtuse ways, but also in ways remarkably alert to Klondike hazards.

The most obvious contrast between biological evolution and human invention is human invention's use of adaptation by revision – for example, 'recursive practice' (Constant, ch.16). This helps meet the challenges of a Klondike topography in two ways. First of all, adaptation by revision implies efficient hill-climbing towards a fitness peak when the local gradient of the fitness landscape points in a clear direction. However, by definition Klondike landscapes contain large apparently clueless regions – plateaus. Here, the human capacity for analysis comes into play. Often, diligent inquirers discover clues where at first there seemed to be none, in effect transforming the Klondike landscape or parts of it into a more tractable Homing landscape. For example, the scene of a crime may appear to contain little information at first, but the skilled criminologist, knowing what to look for, finds abundant clues in traces such as fingerprints, hairs, and skin scrapings.

In general, opportunities for discovery may be better ordered and more accessible than at first they seem. Weber[25] writes of a number of heuristics that point up potential inventions, for instance the heuristic of combining mechanisms with their inverse mechanisms, as in the claw hammer. Langley *et al.*[26] identify a number of heuristics useful in various kinds of scientific discovery. The eye sensitized by such resources sees not clueless plateaus but contours of opportunity.

Even so, adaptation by revision is not the principal strategy for dealing with the Klondike challenges to human invention. This is because adaptation by revision is fundamentally limited in the Klondike context because of its problems with plateaus, canyons, and oases. Adaptation by selection, the mainstay of biological evolution, also is a powerful resource for human invention. Indeed, inventors can do many of the things that biological evolution does, only better. Just as some organisms increase their rate of variation under environmental stress, human inquirers can conduct deliberate far-ranging brainstorms to churn up possibilities (§11.8). Just as evolution benefits from the haphazard chance of organisms migrating to new ecosystems, human inquirers can cultivate chance strategically, exposing themselves to likely sources of stimulation through reading, contacts with colleagues in related fields, and the like.

Also, human inquirers can deal with Klondike topography in ways not accessible to biological evolution. They can recognize oases and canyons as such, noticing that they seem to be clinging to an idea that will not quite work or travelling in circles within the boundaries of unrecognized presuppositions. Armed with such process insights, human inquirers can generate deliberately divergent variations of current approaches to break this mind set.

Besides the superior handling of adaptation by selection, human invention benefits from smarter use of adaptation by coding. Inventors do a great deal of search by manipulating sketches and images – adaptation by revision and by selection in the coding space rather than through actual productions (Carlson, ch.11; Solomon, ch.14). Nowadays they are usually members of technological communities, where memes such as 'design concepts' undergo independent evolution (Constant, ch.16; Stankiewicz, ch.17).

As discussed earlier, adaptation through search solely in the code space is difficult for biological evolution, because the genome specifies organisms indirectly, through procedures for constructing them. Inventors have the best of both worlds, using codes that specify the structure of an artefact or the procedure for fabricating it, as convenient. Moreover, as emphasized earlier, human invention takes advantage of multiple tiers of code-construction relationships. A concept suggests an exploration of alternative sketches, two of which lead to competing prototypes, the problems of which recommend not only revisions in the sketches but an adjustment in the initial concept. Articles by world-class inventors[27] offer anecdotal evidence of the complex dialogue between concept and prototype.

Finally, human invention benefits from its ability to use non-viable forms as stepping stones at any level in the code-production sequence. A bad concept may lead to a good one, a fundamentally flawed sketch to a more promising one, an unworkable prototype to one that does the job. Boundaries of non-viability – impenetrable walls for biological evolution – are simply regions of low promise for human discovery that adventurous inquirers may pass through or leap over.

The actual patterns of search reported by successful inventors range over a wide spectrum.[28] While biological evolution is stuck with a blind process, human inquirers can navigate with different styles for different occasions. Nevertheless, human invention largely lacks two major advantages of biological evolution – massive parallelism and geological time. Although inventors not uncommonly resort to a more or less parallel examination of multiple possibilities, as in Edison's draghunts, no such efforts approach the profligacy of nature.

12.6   Is invention Lamarckian or Darwinian?

This chapter began by pondering the possible analogy between biological evolution and human invention. Did it make sense to speak in both cases of the evolution of adaptive form, or was the comparison superficial? The analysis offered here suggests a deep analogy rather than a superficial parallel. Both biological evolution and human invention can be described within a common framework. The framework asserts that the fundamental challenge of the discovery of adaptive forms lies in the Klondike character of the fitness landscapes that must be searched – the wilderness and plateau gaps, the canyon and oasis traps that make search difficult. The framework identifies three broad strategies of search – adaptation by revision, selection and coding – and notes how some versions of these strategies deal better than others with the Klondike challenges, both in biological evolution and in human invention.

A related question concerned the contribution of Lamarckian versus Darwinian processes to human invention. It might seem that the analysis so far resolves the issue in favour of a Lamarckian view. Human invention turns out to be decisively smarter about handling search in Klondike fitness landscapes than biological evolution, albeit lacking the brute force resources of the latter. However, such a conclusion would be hasty. The issue is not whether human invention handles Klondike fitness landscapes better, but whether human invention does so in a more Lamarckian or more Darwinian way.

The case against a Darwinian view of human invention might run as follows. The primary thrust of human invention, some might say, appears to be Lamarckian. Human inquirers rely overwhelmingly on single-level and multiple-level Lamarckian procedures – that is, 'recursive practice' (Constant, ch.16) and many other varieties of adaptation by revision. Certainly history teaches us that chance enters at almost every stage. But this can be seen as a secondary perturbation of the principal flow of human invention, which is essentially a Lamarckian progression.

Artificial intelligence simulations of scientific discovery[29] point in this direction, although not organized along the Darwinian–Lamarckian axis. Thus a number of past scientific discoveries could be reproduced through relatively methodical search techniques that foregrounded hill-climbing and minimized the role of chance. Langley *et al.* take the view that accounts that emphasize leaps of intuition or inspiration should be regarded as non-explanations that mask the real phenomena.

I share this scepticism about intuition and inspiration as well as the paradigm of search through possibility spaces. However, my argument here is that at least some scientific discoveries, technological innovations and inventions of other sorts present a different sort of challenge. The Klondike reality of fitness landscapes will not allow a Lamarckian strategy to win the day. As argued earlier, Klondike landscapes present troublesome features that straight Lamarckian mechanisms do not deal with well. They require vigorous constrained random variation because they offer only sparse and often deceptive clues to direct search, as well as presenting traps that tend to capture adaptation by revision. In Campbell's words,[30] there must be a process of continual breakout from boundaries. In Boden's terms,[31] creative discovery requires changing the rules of the game – and it's not generally clear in advance what rules to change to. The need for a somewhat Darwinian response derives from the Klondike characteristics of the possibility spaces that human inventors must search.

Consequently, human invention displays a thrust toward smart Darwinian processes as well as smart Lamarckian processes. The comparison between human invention and biological evolution gratifyingly counts humans as Klondike smarter – but not at Darwin's expense. Humans are Klondike smarter about biological evolution's specialty: adaptation by coding coupled with adaptation by selection. To the question 'Is human invention Lamarckian or Darwinian?' a clear answer emerges: 'It's irreducibly both.' Efficiency invites a smart Lamarckian process but the cantankerous character of Klondike fitness landscapes demands a smart Darwinian process. The balancing act between the two is a fundamental characteristic of the human quest for adaptive form.

13

Real-world variation-selection in the evolution of technological form: historical examples

WALTER G. VINCENTI

13.1   The constraints of the real world

My message here is one that should be obvious: *any complete model of technological evolution must include the physical real world.* Artefacts, by definition, are made to 'work' (in some sense) in the world around them. It follows that they must conform to real-world constraints, such as laws of nature, that are not open to alteration by human agency. As a philosopher acquaintance puts it, 'One thing that determines the fate of an idea or technology is how its structure [for present concerns, the physical form of an artefact] relates to the objective structure of the rest of the world.' Most scholars outside technology subscribe to this idea in the abstract. When the analytical chips are down, however, it seems to me that many of them (my acquaintance and some others excepted) tend to neglect or underplay it. This chapter may provide warning, if such be needed, against that temptation.

I shall illustrate my point by three historical examples of technological variation-selection, two from bridge engineering and one from my professional field of aeronautics. As we shall see from even this limited sample, the real world enters in diverse and complex ways; theorizing about the mechanisms of variation and selection, the locus of their application, and the implications for non-technical issues will not be trivial. In the present examples, the bridge stories show the action of the real world in the failure of specific artefacts, unintended in direct use in the first case, intentional in the course of design in the second. The aeronautical example deals with a component common to a class (or genus) of varyingly successful artefacts and ultimately with a generic design concept. All have historical significance outside the topic of this book.

13.2   Variation-selection in direct use

A spectacular instance here came in the wind-driven failure in 1940 of the suspension bridge across the Tacoma Narrows in the northwestern United States.[1] Many readers will no doubt have seen the famous film of this 'selection by the real world'. The century of variation leading to the failure, however, also needs attention.

The importance of wind forces for suspension bridges had been apparent from at least the early 1800s. Even Thomas Telford's graceful and much admired structure across the Menai Strait in northwestern Wales, whose unprecedented tower-to-tower span of 580 feet set the pattern for modern suspension bridges, experienced wind-driven troubles. Its slender and flexible deck moved up and down severely in high winds following the bridge's opening in 1826; in a storm ten years later it was seen to oscillate vertically through a distance of 16 feet. Finally, in 1839, part of the deck came down in a hurricane and had to be rebuilt and stiffened.

By John Roebling's testimony, this and similar failures were in his mind when, in the mid-1800s, he designed his eminently successful, though longer-spanned, Niagara Gorge and Brooklyn bridges (respectively, 820 ft, opened 1855 and 1,595 ft, opened 1883). To borrow from one of Henry Petroski's fine treatments of these matters, 'by identifying and confronting failure modes [including wind-driven] as the principal part of his design charge', Roebling devised stiffer and heavier decks and diagonal anchoring stays that avoided the earlier problems.[2]

In the 40 years after the Brooklyn Bridge, variants on Roebling's ideas evolved as a successful series of long-span suspension bridges. As shown by Petroski, however, the necessity for designing against dynamic wind failure was to become dismissed or forgotten when the ever longer, more slender, and hence more flexible, bridges of the 1920s and 1930s were built.

The design climate of the period shows itself in chief engineer Othmar Ammann's account of his design of the 3,500-foot-span George Washington Bridge, opened across the Hudson River at New York City in 1931. Though aware of failures and their design implications, Ammann avoided discussing them in his report. Instead, he stressed past successes and dwelt on the cost and (especially) aesthetic advantages of a shallow and hence more graceful deck. To the extent that flexibility receives mention, he conceived that the wide deck dictated by the abundant automobile traffic would afford sufficient stiffness and the accompanying dead weight enough inertia to avoid wind oscillations. As it turned out, he was right.

Unfortunately, the success and grace of Ammann's masterpiece and his

Figure 13.1 Tacoma Narrows Bridge, south of Seattle. From Ammann, O.H., Karman,
T. von & Woodruff, G.B. 1941. *The Failure of the Tacoma Narrows Bridge*
(Washington DC: Federal Works Agency), frontispiece.

account of its design led fellow engineers to take aesthetics increasingly as the
criterion for design selection and neglect questions of stiffness and dynamic
motion. Warnings surfaced in the late 1930s in wind undulations in several new
bridges, including Joseph Strauss's Golden Gate Bridge at San Francisco and
Ammann's Bronx-Whitestone Bridge in New York City. Designers, however,
failed to heed the message – Petroski speaks of 'hubris'.[3]

Real-world selection came in 1940 at the Tacoma Narrows. The Tacoma
Narrows Bridge in the state of Washington south of Seattle (fig.13.1) was
designed by Leon Moisseiff, who had acted as a consultant for Ammann on the
George Washington Bridge and whose 'deflection theory' had enabled the
analysis of slender flexible bridges in the first place. At its opening in 1940, its
span of 2,800 ft was the third longest in the world, after the Golden Gate
(4,200 ft) and the George Washington (3,500 ft). More important, its untypically
narrow two-lane deck gave it a width-to-span ratio of 1:72, unprecedentedly
slender compared with the Golden Gate (1:47) and the George Washington
(1:33). In addition, vertical stiffening at each side of the deck was afforded by a
shallow solid-plate girder in place of the stiffer open-work trusses of the earlier
bridges. In the event, these variants, motivated by aesthetic and to some extent

economic concerns, made the bridge more susceptible to the dynamic wind-driven motions that had caused trouble at the Menai Strait. Moisseiff's static theory gave no attention to such movements, however, and no analysis was made of them on their own.

Undulations of the bridge appeared early during construction and, more dramatically, after the bridge was put into use. People who came for the roller-coaster thrill of driving on it spoke of it fondly as Galloping Gertie. Their enjoyment was short lived, however. On 7 November 1940, only months after the bridge's opening, one of the checking cables that had been added to limit the undulations slipped at centre span, and the deck began to twist and oscillate alarmingly in a 40-mile-per-hour wind (fig.13.2).

The structure was quickly closed to traffic, and Professor Frederick Farquharson of the University of Washington, who was already studying the bridge's behaviour, came to observe and record the events. His film of the movement and destruction (fig.13.3) of the structure – Petroski describes it as 'the most famous film footage in structural-engineering history'[4] – provides Darwinian selection for all to see (figs.13.2 and 13.3 are frames from it).

Fortunately, no lives were lost. Following the disaster, the bridge was re-designed and rebuilt taking wind-driven oscillations into account. Suspension bridges since have been analysed and models dynamically tested in wind tunnels to avoid such problems. Traffic-limiting movements have appeared in a few cases in especially strong winds, but no failures have occurred.

Thus, at Tacoma Narrows, in the course of everyday use and the absence of adequate selection within the design process, real-world phenomena selected a variant for literal elimination. As a consequence, the criteria for selection and scope of analysis were fundamentally revised. The result was a less flexible replacement bridge, together with design practices that take account of the dynamic interaction between the external forces of wind and gravity and the internal forces of inertia and elasticity. Real-world selection had taught engineers a lesson that, one hopes, will not be needed again.

13.3   Variation-selection in design

Our example here is the design of the Britannia Bridge, built across the Menai Strait near Telford's span in the years 1845–50 (fig. 13.4).[5] Traversing the Strait on the way to the port of Holyhead for the London-to-Dublin train connection presented the railway's engineer-in-chief Robert Stephenson with a formidable challenge. As with Telford's bridge, not only was a large span required, but the use of arches was ruled out by the Admiralty's need to navigate the Strait with tall-masted sailing vessels. In addition, in Britain, contrary to

Figure 13.2 Wind-driven oscillations of the Tacoma Narrows Bridge. From Ammann *et al.* 1941, p. 6.

Figure 13.3 Failure of the Tacoma Narrows Bridge. From Ammann *et al.* 1941, p. 15.

Figure 13.4 Britannia Bridge, Menai Strait. From Knight, E.H. 1877. *American Mechanical Dictionary* (New York NY: Hurd & Houghton), vol. 3, plate 62.

what Roebling was to show a few years later at Niagara Gorge, a suspension bridge was thought too flexible for railway traffic. Known types of long-span bridges were thus eliminated.

It was under these circumstances that Stephenson conceived of an unprecedented variant: a tubular beam of 450-foot maximum span, constructed of thin riveted plates of the increasingly cheap wrought iron and large enough for a train to pass through it. Nothing of this form or size of railway bridge had been attempted before.

Realizing he would need help, Stephenson enlisted the versatile William Fairbairn, an experienced shipbuilder and iron fabricator as well as an engineer practised in design and testing. Together the men projected a series of experiments to explore the load-carrying properties of thin-walled tubular beams before attempting to design the bridge.

Though the theory of beams under vertical loads had been established, little experience was available with it in practice, and what there was for metal beams was mostly for cast iron and for simple, thick cross-sections. Little reason existed for confidence in theory. To help carry out the tests at his shipbuilding plant near London, Fairbairn brought in his friend Eaton Hodgkinson, a mathematically minded engineer who had worked for him previously on tests of cast-iron beams and columns. Over four months in 1845, Fairbairn, with Hodgkinson assisting in analysing the data, tested 34 thin-walled wrought iron tubes of circular, elliptical and rectangular cross-section and varied dimensions. The tubes (up to 31 feet long) were supported at the ends and loaded by a weight at midspan; the weight was then increased until the beams failed, a variation-selection process typical in design research.

The deliberate failures in the experiments put engineers on notice of a type of problem not encountered before. As they had expected from tests of previous solid beams, the stretched bottom of a tube gave way by tearing of the material, much as happens to a bar pulled apart by pure tension. The compressed top, however, failed, not by crushing of the material as had occurred with thick cross-sections, but by structural crumpling (i.e. *buckling*). Failure by buckling was by Fairbairn's time reasonably understood in the case of the total collapse of a long strut or rod with its ends pushed toward one another. Localized buckling of a tube or other thin-walled structure, however, was completely unforeseen – in Fairbairn's words, 'anomalous to our preconceived notions of the strength of materials, and totally different to anything yet exhibited in any previous research'.[6] The real world had manifested itself in a previously unrecognized way.

The buckling problem, though at first intimidating, was quickly solved in the exploratory experiments. Reasoning that what was needed was an increase in flexural rigidity of the top of the tubes, Fairbairn tested three tubes with cellular tops. The most successful (fig. 13.5) had a rectangular section with a top of corrugated plates assembled to form tubular cells extending longitudinally along the beam. This beam failed by tearing of the top and bottom of the tube away from the sides (i.e. with local buckling suppressed) at a load almost twice the buckling load of a comparable beam with a flat top. Real-world selection had pointed the way to an intentional, rational variant that could get the job done.

For design development of the bridge itself, Fairbairn, in mid-1846 and guided by the foregoing experience, began tests on an unprecedentedly large model (one-sixth-scale, 75-foot span) having a rectangular section and a top flange of six square cells. The plate of the bottom flange he deliberately made weak, so that its tension failure would take place first. He then proceeded by successive trial and error: after failing in a test, the flange was repaired to strengthen against the observed failure, and the resulting variant tested again. Finally, in April 1847 in the sixth test, failure occurred by buckling of the top under conditions judged close to the ideal of simultaneous failure of top and bottom.

The large-scale tests also put the engineers on notice of unforeseen diagonal-buckling problems with the thin sides of the tube, fortunately easily solved by addition of vertical stiffening bars. Design of the bridge, which went on simultaneously with the tests because of time pressures, utilized the test findings as they became available. Details of the complex engineering trade-offs,

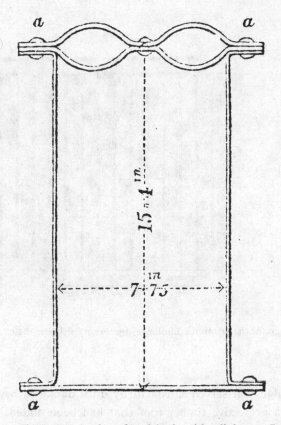

Figure 13.5 Section of model tube with cellular top. From Fairbairn, W. 1849.
An Account of the Construction of the Britannia and Conway Tubular Bridges
(London: J. Weale), p. 19.

while interesting and instructive, are not essential for our purpose; the full story can be found in the references. In the final design, the bottom, like the top, was made of cells, mainly because the 3-inch plate that would have been needed could not be rolled or otherwise fabricated at that time. The tube as finally built and put to successful use – the final variant (fig. 13.6) – provided the most impressive metal structure to that time, the crucial part of what British economist William Jevons in 1865 called 'our truest national monument'.[7]

In design of the Britannia Bridge, two successive variation-selection processes thus helped to bring into being an unprecedented structure. Selection took place experimentally within constraints set by the real-world interaction of gravitational and elastic forces as revealed through vicarious testing of laboratory models – 'vicarious' (§4.1; §5.3; §12.3; §16.4; §20.4) that is, in contrast to direct use of the artefact as at the Tacoma Narrows. Vicarious 'testing' can also occur, of course, by analytical methods on paper or computer.[8]

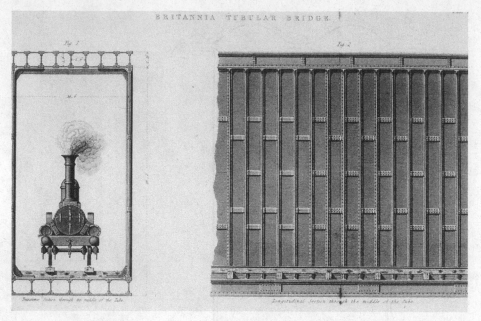

Figure 13.6 Final design for the Britannia tubular bridge. From Fairbairn 1849, plate 4.

The outcome was a bridge that served satisfactorily until damaged beyond repair in 1970 by fire in a protective timber roof that had been added. (The replacement, incidentally, is an arch bridge – no sailing ships now!) Stephenson's tubular bridge, however, turned out to be heavy and expensive, more so than subsequent types of metal bridge; with a few exceptions designed by Stephenson, it was not imitated. The enduring retention was knowledge of the existence of a previously unrecognized mode of failure of thin-walled structures. Systematic study of buckling, under the label 'theory of elastic stability', has since become a recognized subfield of engineering science.

13.4   Variation-selection in a design community

The third case has nothing to do with bridges or failure, either directly in use or vicariously in design. We deal here with the real-world-mediated selection, through design substantiated by successful operation, of the landing gear for specific aeroplanes. The result in the end is a communal, generic solution – a 'techno-meme' (Stankiewicz, ch.17) – for high-speed aeroplanes generally.

The process in question took place mostly in the United States in the first half of the 1930s when, through a number of innovations, especially increased

Figure 13.7 Northrop Alpha, original model. Courtesy of John B. Kimball.

engine power, everyday aeroplane speeds of over 200 miles per hour did not appear out of reach. As a move in that direction, designers faced the challenge to reduce the aerodynamic drag of the various types of unstreamlined, fixed landing gear, like that on the Northrop Alpha of 1930 (fig. 13.7). This end could be pursued by two options: (1) retracting a folding gear into the interior of the aircraft, or (2) streamlining a fixed gear. Designers pursued both options through a range of variants.

Early tries at retraction (see next two pages) were the Boeing Monomail and the Lockheed Orion, of 1930 and 1931, respectively. For the Monomail (fig. 13.8), rearward retraction left the wheels partially exposed to cushion the impact in the event a malfunction forced a landing with gear retracted; in the Orion (fig. 13.9), inward retraction hid the entire gear within the aeroplane structure. Streamlining of fixed gear followed two paths. One, exemplified by the Lockheed Sirius built for Charles Lindbergh in 1931 (fig. 13.10), refined the support structure and enclosed the bluff, high-drag wheels inside streamlined, low-drag 'pants'. The second, called 'trousers', covered the structure and wheel on each side with a vertical, streamlined enclosure, as on John Northrop's modification of an original Alpha (fig. 13.11).

Whatever the designer's choice, more than aerodynamic drag was at issue. Streamlining, and especially retraction, contributed additional weight, requiring additional lift and consequent drag, thus reducing the original gain. They also entailed increased cost and, for retraction, especially at first, decreased reliability and increased maintenance. The trade-offs were not at all clear, even after a number of years of experience.

Influenced by these factors, and given the continued increase in aeroplane

Figure 13.8 Boeing Monomail. Courtesy of the Boeing Company and Paul G. Spitzer.

Figure 13.9 Lockheed Orion. Courtesy of Richard S. Allen.

Figure 13.10 Lockheed Sirius. Courtesy of Richard S. Allen.

Figure 13.11 Northrop Alpha, modified. Courtesy of Richard S. Allen.

speeds from various sources, designers tried every variant they could think of. Photographs in the annual volumes of *Jane's All the World's Aircraft* of the early 1930s show aeroplanes with trouser gear, various kinds of retractable gear, and a variety of gear with wheel pants, some as part of an external wing-support structure. Selection in individual cases was made on the basis of vicarious testing (Northrop experimented with models in the wind tunnel at Caltech), testing in flight (Boeing tried trousers as well as retraction on the Monomail), and seeming intuition. The noted Giuseppe Bellanca's aircraft of the period exhibited unstreamlined fixed gear, wheel pants, trousers and retraction, in no discernible relation or pattern.

The various approaches to streamlining and drag reduction all had their temporary successes. Northrop's outstanding trousered design for civil use, the much admired Gamma, set transcontinental speed records across the United States, while his similarly equipped XFT-1 was the fastest Navy fighter of its day. A variation-selection process was obviously and widely at work. With similar processes going on for propellers, wing flaps, and metal structures, plus continued increases in engine power, the outcome was not at all clear. The full story, with focus on Northrop, appears in my paper of some years ago in *Technology and Culture*.[9]

In the years following 1935, however, retraction prevailed. As speeds went up sharply above 200 mph from engine and airframe developments and engineers learned through experience to deal with reliability and maintenance, aerodynamic drag became overriding, and retraction won out. In this ultimate result, I believe we can see – contrary to current fashion – a kind of technological determinism.

If such were the sole goal, fixed gear could, of course, be forced through the air at any speed, given unlimited power. Since drag grows as the square of the flight speed, however, the weight of the increasingly powerful engine and its fuel will rapidly overwhelm the lifting capacity of an actual vehicle. It is not an exaggeration, I believe, to see a real-world imperative at work here. In any event, since the early 1940s, high-speed aeroplanes invariably have retractable gear; no aeronautical engineer today would think of designing a jet transport or military combat machine with anything else.

Variation-selection based on aerodynamic reality thus appears unmistakably in the evolution of high-speed aircraft. Designers, struggling with specific design problems and using both vicarious and direct testing, selected trouser gear, wheel pants, or retractable gear for their individual aircraft. Though varying degrees of success were achieved, no failures like those of the earlier cases entered the process. In the end, as speeds went up and aerodynamic drag became overriding, the design community was driven to retractable gear as the long-term solution.

Solution of special problems for individual aeroplanes thus contributed cumulatively, through a complex communal activity, to solution of a generic problem for all high-speed aircraft. Variation-selection may be said to have produced over time a concept – a fixed techno-meme – that has since been passed on, so to speak, from aeroplane to aeroplane. Landing gear vary widely in their details from case to case, but they all disappear inside the vehicle in flight.

13.5   Observations

The foregoing examples contain a wealth of implications about the variation-selection process as it applies to technological form. Space limits me here to some brief observations. The examples, besides illustrating the impact of the real world, reflect a typical diversity regarding the loci of variation and selection and the nature of the outcome.

In the case of suspension bridges as a class, variants evolved from one bridge to the next, with successive refinement in slenderness. An analytical variation-selection process, of course, went on within the design of each bridge, but that does not concern us here except insofar as it overlooked wind-driven motion. Finally, at the Tacoma Narrows, real-world selection took place through unintended failure in everyday use. The result, aside from a redesigned bridge, was renewed attention thereafter to an essential aspect of selection within design.

With the Britannia Bridge, our concern was for the process *within* the specific design. In the experiments leading to the final variant, failure was sought deliberately as a means of design selection; the variant for each cycle in the process was devised by modifying the failed result of the previous cycle. The new type of bridge that resulted, though successful for its purpose, turned out to be little imitated; the lasting outcome was discovery of an unanticipated mode of failure for the thin-walled structures then beginning to appear.

In the case of retractable landing gear, variants in the design process for specific aircraft were devised initially in the designer's imagination and selected either through testing directly in flight or vicariously in wind tunnels, or through some unrecorded, perhaps less rational means. Deliberate failure was not involved; unintended failure is difficult to document, but it does not appear to have played a significant role. Over time, the individual selections cumulated as variants in a kind of wider, communal process that is not as well understood as it needs to be. In the process here, unlike the previous cases, the artefact of interest (the landing gear) was a component of a larger artefact (the aeroplane);

its real aerodynamic world – the flight speed – thus changed as this technical environment evolved. In the end, the design community tacitly selected retraction as the long-term generic solution. The outcome was thus a techno-meme for high-speed aircraft – the knowledge, today taken for granted, that such aircraft should have retractable landing gear.

Our examples can also be looked at in terms of the recursive process described by Edward Constant (ch.16). The suspension-bridge case is especially interesting for being somewhat unusual. After Roebling, the primary focus of designers in their pursuit of longer-span bridges shifted gradually from structural behaviour to aesthetic refinement. The recursive criteria for successive variants shifted accordingly. Designers, seeing the success of a previous slender bridge, were encouraged to try a still slenderer one and neglect adequate analysis of possible modes of failure. At the Tacoma Narrows, the real world finally said 'Enough!' Designers had neglected a critical aspect of the real world in their recursive practice and had to revise their practice accordingly.

The other two cases were more typical. Both Fairbairn and the aircraft designers kept their focus firmly on their essential goal – to design a structurally safe bridge of an innovative type in the first case and to increase aeroplane speed in the second, both with due regard for the physical constraints of the real world. Fairbairn methodically followed recursive practice in moving from one variant to the next improved one. Aircraft designers observed the performance of their own and others' landing-gear variants as the level of aeroplane speed increased and recursively altered or continued their choices accordingly. Even Northrop after a few years gave up his temporarily successful trouser gear and went over to retraction.

Sorting out technological variation-selection, as it involves the real world, can thus not avoid diversity and complexity. One can plead, of course, that the distinctions here are matters of the internal microevolution of technology that have little to do with technological macroevolution in its social and cultural sense. The constraints the real world puts on such microevolution, however, can have consequences for what and how technology can contribute to the macroevolution. To theorize about that larger evolution, we may have no choice but to face at least some of the technical complexities.

In contrast to the above diversity, however, a further commonality does exist as a corollary to that of real-world selection: as implied above and in my title, such selection relates in all cases to the *form* of the associated artefact. For the Tacoma Narrows Bridge, the failure (i.e. negative selection) was influenced by the bridge's slender form, and that failure in turn influenced the form of the replacement. Similar statements could be made for the other cases. Such relationship between selection and form is of the nature of things in the real world – action of natural forces on or within an artefact is strongly inter-dependent with the shape and arrangement of the artefact. I do not mean that the physical world is the sole arbiter of form; the form of a consumer product, for example, is closely related to its popular appeal (Nelson, ch.6). As a general tendency, however, the more rigid the real-world constraints, the more that decisions about form must be based on technical rather than social concerns.

The *direction* of decisions, however, is another matter. The ultimate selection between retractable and streamlined fixed gear for aeroplanes, for example, was made on the basis of higher speed – once faster aeroplanes had become credible technically, a desire for speed for military and commercial purposes had taken over. Again, similar statements could be made about the other cases. Once technical credibility is established, I would expect the direction of variation and selection generally to be more socially than technically shaped. A distinction between form and direction (or something similar) may have a place in the analysis of technological evolution.[10]

The concepts and tools for such analysis must obviously be selected with care. As I remarked earlier, to the extent that the real-world environment prevails, technological selection can be seen as 'Darwinian'. For the retractable gear, however, the production and subsequent inheritance of an acquired solution to a generic problem can also be thought of as 'Lamarckian'. To insist on analysing technological evolution in terms of the natural-selection metaphor, however, may only complicate matters. Seeing evolution of technology on its own terms might hasten the day (Ziman, ch.1), when the study of technological change could contribute to evolutionary biology instead of the other way round.

Details and complexities aside, I return to my original thesis: technological artefacts generally must function at least adequately and at best efficiently in the physical real world around them, and this places inevitable constraints on their design and evolution. In contrast to social, economic and cultural con-straints, which are characteristically negotiable and contingent, such real-world constraints are (or at least must be taken by engineers to be) non-negotiable and universal. Such constraints and the concepts needed to incorporate them must have a fundamental role in any theory of technological evolution. With them, of course, must be combined a host of non-technical constraints and forces – I do not mean to suggest in any way that real-world concerns are sufficient. Along with non-technical matters, however, they are necessary.

14

Learning to be inventive: design, evaluation and selection in primary school technology

JOAN SOLOMON

14.1   Education

Education comes into this study under two headings. On the one hand, a better understanding of technological innovation ought to show us how to teach children to be personally more inventive, and thus to make a greater contribution, as citizens, to industrial and economic change. On the other hand, by watching in the classroom how children actually go about the task of 'inventing something', we can observe technological creativity in action, albeit in an artificially simplified form. My finding is that research of this latter kind is producing results that are directly applicable under the first heading. In particular, this analysis and its applications are facilitated by the identification of many of the evolutionary characteristics of technological change even amongst primary school pupils.

This research was undertaken in the course of teaching primary pupils who were working on technology projects stimulated by a series of stories written to embed technological need in the historical periods they were studying. What it has to offer scholarship in other fields is not any simple kind of recapitulation theory. I do *not* believe that our children in their cognitive growth proceed stage-by-stage through the phases of epistemological development recorded within the histories of philosophy, as the human embryo was once thought to recapitulate the evolution of our species.

Educational studies are interesting for quite different reasons. In the first place school children meet subjects like science and technology almost *de novo* in their school setting. Of course they may well come with naive personally or socially constructed ideas on the operations of, say, light, electricity and levers, and most make things, often with construction sets, at home. However the 'subject', as it is called in school, with its ways of working, aims and accolades, is often quite a new entity. It has a novel but wholly implicit epistemology.

For example, despite the frequency with which children's drawings of weird and balding scientists are presented as research data about their views on the nature of science,[1] there is much more compelling evidence[2] that it is the encounter with school and the teacher which is most powerful in building up pupils' understanding of the nature of the science. So, in the primary school right at the beginning of the educational process, we can observe the conflict between lightly held lifeworld ideas about science and technology, with those more authoritative ones imbibed at school.

But there is a much more important way in which educational studies can be valuable. They record the impact of a new discipline on a fresh young mind, often at a time when it is at its most receptive, creative, and communicative. This encounter, drawn out over a regular series of lessons, will include impediments to progress that can be seen not only as emanating from undeveloped cognitive skills, but also as challenges to personal creativity which may 'ring bells' (to use a school metaphor!) with all of us. The children are very ready to talk about their intentions and frustrations, so informal research carried out in the working classroom can cast light on processes which, in the mature technologist, are in danger of becoming so internalized and taken-for-granted as to be almost completely opaque, even to the technologist.

14.2   The nature of technology

If you ask pupils 'What is technology?' they give different answers depending, to a large extent, on their school experience. Those from small schools with a cross-curricular topic approach usually answer 'making things'. Those who learn in a workshop set up with tools for working wood, and are taught by a craftsman with carpentry skills, say 'making things out of wood'.

> [P₁]   It's to do with wood.
> [P₂]   Yea. Wood and equipment.

Pupils from a middle school where technology was taught within the Art & Design department answered 'designing and painting things'.

> [Teacher]   What is technology, M?
> [M]   Well, it is a type of art, it's like you make things. Several things. Like building. And when you are a builder you have to know about technology. When you are a builder.

> [*Teacher*]   *You said technology was using art?*
> [*M*]   *It is* not *using* art, *it is a type of art.*

And there are huge gender influences. A teacher in a First school[3] found recently that even in her reception class the girls and boys used the word slightly differently. The girls applied 'technology' to everything that was made – curtains, chairs and tables, as well as computers, and washing machines. The boys tended to choose only computers, washing machines and toys with moving parts. Some of the older pupils and most of the adult general public[4] associate the word with appliances which are '*high-tech*' like computers, microwave cookers, and electronics generally.

14.3   Technology and history

We began our research by asking pupils if they thought '*there was technology in the old days*'. Usually they answered '*Yes, there was*'. The discussion that followed served two useful functions:

(a)   it steered the pupils away from the gendered 'hi-tech' image of technology

(b)   it began a discussion of *need* in everyday life at other times.

The second of these is important for at least two reasons.

Firstly it addresses Arnold Pacey's 'unrestricted' view of technology,[5] rightly advocated by Janet Burns (ch.21) as a part of all modern education. Of course it is important to introduce our pupils to the place of technology in the cultural landscape: only a few of our pupils will become practising technologists but all will need to grapple with technology-based dilemmas during their lifetimes.

Secondly, social need fits the role of evolutionary selection. Any model for learning technology in the classroom depends upon *almost* blind variation of design as these young pupils come up with a diversity and creativity which is wild and even wacky. The humorous detail can be hilarious, and may be so intended. Perhaps real engineers also do something similar in an early brainstorming session. (We shall return to this in a later section on drawing.) Before the 'real-world constraints' (Vincenti, ch.13) close down upon these young pupils as they struggle with resistive materials, we need them to begin to operate their own more sober selection process by considering the needs of people in the story.

The idea of linking school technology and history is not new. There have always been teachers who have got their pupils to make model Saxon villages or Viking ships when teaching history. For them history was the primary objective and there was little if any problem-solving of a technological nature. In Design & Technology lessons there are enormous practical advantages in beginning technology lessons with a historical story. Firstly stories are much enjoyed by children and very well remembered; secondly the hindsight that history offers shows pupils the power of technology to change the ways in which we live; and thirdly it prolongs their attention to the people and the context of their needs.

Under the heading *Progression in designing skills*, the National Curriculum says '*relate the ways things work to their intended purpose . . . people's needs . . .*' Usually that is rather a tall order for young children. However, while using historical resources – a water wheel for grinding corn in the Anglo-Saxon times, and one on printing from Tudor times – there was a significant shift in the way pupils spoke about technology. They began to focus less on making just for its own sake, and more on thinking about people's needs. Nearly every answer about the nature of technology included the word 'need'.

> '*it's like (saying), "if only I had something like this I could do the job so much better" . . . like* needing something *which makes things a bit easier'.*[6]

14.4   Starting the design process

Even the clash between the different social processes which are encouraged in the classroom is reminiscent of the commercial world of technology (Nelson, ch.6). Teachers often believe that the best way to promote reflection is to begin with a class discussion (brainstorming). Talking in public, like written publication, takes ideas from the quick pupils, shares them with those who are slower to start and, in the hands of the organizing teacher, begins the process of selection and modification to suit perceived need. This fits in well with the ethos of primary teaching which sees itself as much as a social and moral undertaking as an intellectual one. The term '*sharing*' has very high prestige with teachers, but in technological practice it presents the children with two serious drawbacks which may also mirror the adult world:

(a) breaking secrecy over cherished personal ideas, and
(b) the shattering effect of public criticism from those who 'know best', which crushes the creativity of the less confident.

In recent years many employers have been urging schools to teach their students to work in groups in imitation of the industrial world. Although little actual teaching of group skills takes place, at least it is now commonplace for teachers to allow it and for educational researchers to study it. In science lessons group work is the norm, largely because of the perennial shortage of equipment. In technology, however, it is more difficult to put into effect, especially amongst boys who cooperate less well, at least at primary school age.

Why is this? Although the reasons may have most to do with the psychology of children and their sense of ownership, the problems seem also to be sharpened by the very nature of the technological process. It will be argued later that creativity is central to learning technology, in a way that it is not in many other school subjects. To be creative requires listening to inner speech, as Vygotsky called it,[7] and builds on seeing internal images. Interior personal quietude, even in the midst of the tumult of a classroom, may be essential for both processes. In addition to those reasons there is the secrecy of the inventor which is endemic among adults and belongs in the world of pride and patents. In the words of the fifteenth-century inventor Francesco Martini:

> I am reluctant to show them forth to all, for once an invention is made known not much of a secret is left. But even this would be a lesser evil if a greater did not follow. The worst is that ignoramuses adorn themselves with the labour of others and usurp the glory of an invention that is not theirs.[8]

14.5   Problems with drawing for selection

Designing distinguishes 'technology' from the older craft lessons. Without it, the whole thing could turn into a messy, unplanned, trial and error with egg-boxes, or anything else which comes to hand. Since young children may find drawing out their ideas on paper very difficult, some middle, and most secondary, schools still teach technical drawing to their pupils before they are allowed to begin making any artefacts at all. This seems the very antithesis of creativity, and can make the pupils understandably restive and impatient. *'Technology is supposed to be for making things'*, as they say.

Young pupils' drawings are understandably clumsy and messy. There are three reasons for this. Firstly our youngest pupils find 'de-centering', as Piaget called it, very difficult; secondly three-dimensional drawing is difficult even for those who are ready to use it; and thirdly the design process is still going on in the child's head while the drawings are being sketched and also when the artefact is being made. Even the drawings of mature technologists (fig. 14.1) are often half-way between a doodle and a technical drawing. So we should expect, welcome, and even preserve, the evidence of change and selection in the design drawing, including a lot of 'rubbing out'. Unhappily this sometimes runs counter to the ethos and practice of school teachers.

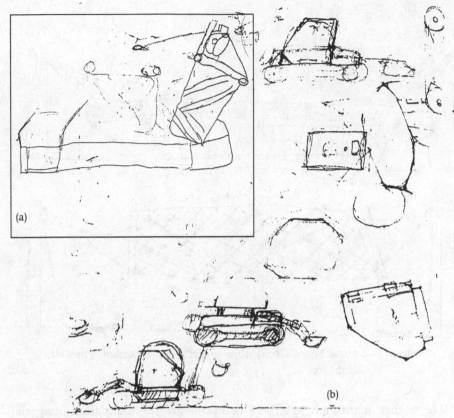

Figure 14.1 Drawings by (a) child; (b) technologist.[9]

14.6   Spatial ability and mental modelling

In our research we found that children's spatial ability, a fundamental development skill, increased very rapidly while they were learning technology, in a way which focused on moving mechanisms in three dimensions.

Children in reception classes, as well as those much younger, draw *all they know* (e.g. objects hidden from their point of view) and what they think important (e.g. the number of fingers on each hand) *rather than what they can actually see*. The drawings in fig. 14.2 were made by a pupil in Year 5, Tessa, when asked to draw the hammock she had just made from the top and from the side. She has been unable to draw the hammock other than how she *knows* it to be. Not only can she not decentre; she has not yet developed the power to control and rotate the image of her hammock in her mind.

Tessa argued with the teacher saying that she could not see anything wrong with her drawings. She did not seem to be envisioning the artefact and so was

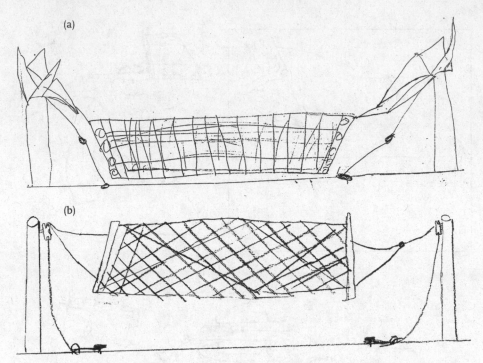

Figure 14.2 A Year 5 pupil's (Tessa's) drawings of a hammock from (a) the side,
(b) the top.

unable to turn it through 90º. These two skills are the foundation of spatial ability which should be developing between 6 and 10 years, during primary school.[10] Some four months later in this active technology course Tessa could not only draw model cars from the side and the top, but also bend up paper 'nets' in her mind to see if they would make a pyramid or not. And here is Tessa talking about the nets, along with several of her friends.

[Tessa]   *The first one will make a short one (pyramid). I think all the edges will match.*
[Teacher]   *And number 2?*
[Tessa]   *No. It won't*
[Teacher]   *Why?*
[K]   *Because they won't reach. They are too small.*
[Teacher]   *What did you say for number 3?*
[Tessa]   *Yes it will.*
[Teacher]   *Why did you write that?*
[Tessa]   *Because they will meet.*
[Teacher]   *Can you see it in your mind's eye?*
[Tessa]   *Yes, I can. Now I can.*

Do other primary age children have the capacity to 'see' images in their mind's eye and can they practise both design and selection inside their heads before drawing? This was the basis of discussion held with some of our older pupils, aged 9/10.

> [Ste]   *You don't always need to plan. You might have it in your mind.*
>
> [Teacher]   *Maybe you can plan it in your mind?*
>
> [H]   *Yea. That's what I do sometimes.*
>
> [Teacher]   *Where do you think people get ideas for making things in technology?*
>
> [H]   *From other things.*
>
> [Sam]   *Miss, if you see a tricycle – you might think of a bike.*
>
> [Teacher]   *But where do people get ideas from?*
>
> [E]   *From their heads.*
>
> [M]   *And they put some things together and they make . . .*
>
> [Ste]   *They think about things. They see things and they think 'Oh yea, well I probably won't copy them or anything, but I've got a similar idea.'*
>
> [Teacher (to Ste)]   *Do you have pictures of ideas in your mind?*
>
> [Ste]   *Yea I have that all the time. When I go to bed as well. I dream about pictures and things like that.*

So it seems that at least Ste, Sam and E (all girls) and H (a boy) do rely on images seen in their minds, during their designing. They regularly practise not only design and selection in their minds, but also a simple kind of recombination of ideas or perhaps 'memes' from other sources. This is very reminiscent of Edison's work as explored by Bernard Carlson (ch.11).

At a young age, imagining or modelling in the mind is taxing.[11] A First School teacher did some controlled action research on her children's learning. She taught technology with an emphasis on drawing the moving mechanisms that her children made using construction kits. At the end of the year she tested them, and also a parallel class which had not done any technology. The children had to draw a plastic cup which had a small ball half immersed in it. While 82% of her pupils could draw the cup with just the protruding part of the ball visible above its rim, even though they knew the whole ball would be round, in the other class only 37% of the pupils could do this.[12] They still drew the ball as they 'knew' it to be. I take this as some indication that the endeavour to sketch three-dimensional mechanisms on paper draws so strongly on mental activity and imagery (§11.4) that it is instrumental in developing spatial ability, decentering and modelling. Drawings in technology, however unskilled, can thus provide evidence not only of design, but also of the process of mental modelling which it encourages.

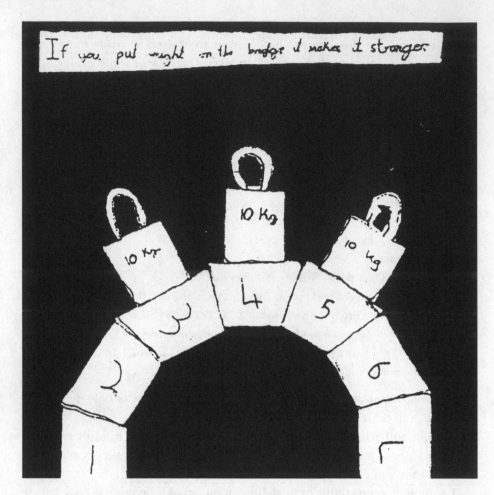

Figure 14.3 Year 4 pupils' drawing of a bridge.

Visualizing forces within a stationary structure is another kind of mental modelling. The national inspectorate (HMI) say that good structures are very rarely made or properly understood in primary schools. We also tried to extend the pupils' developing spatial ability into this area. The drawing in fig. 14.3 was produced by Year 4 pupils (age 7/8) who were working on a story about the Romans building a bridge in Celtic Britain. It is clear that their teacher has got them to think carefully about the force of weights on top pushing down and forcing the stones of the bridge closer together. The bridges that the pupils made afterwards, to their own design, usually incorporated features from the Roman design, but also extended them to include modern bridges across the motorways. This may be another example of the recombination of learnt ideas, or 'techno-memes' (§1.2; §13.5; §17.3). The development of drawings of hammocks also shows how the imaging of tension forces develops.

14.7   Evaluation

The Design and Technology Order speaks about both '*quality*' and '*evaluation*'. 'To evaluate' is explained as:

> to distinguish between how well a product has been made and how well it has been designed;
> to consider the effectiveness of a product, including the extent to which it meets a clear need, is fit for purpose, and uses resources appropriately.

The first definition asks pupils to look back on the design plans that they drew in the first place. Now that the artefact, with all its warts, is standing fully realized in front of them, they should be able to see how the real-world constraints of making (Vincenti, ch.13) have impaired, or improved upon, their original intentions. It is quite difficult for these young technologists to admit that their prized object has not turned out quite as well as they intended, or imagined. As we have seen, the artefact may be a deeply personal creation, and pupils easily become angry and defensive when it is criticized.

Nevertheless we tried to get them to face up to seeing where the differences between the idea and its realization lie, and to explain why this apparent discrepancy occurred. Research has shown[13] that it is the practitioner's reflection on action, especially where it may have previously been tacit, which makes the knowledge durable and accessible. In particular the pupils need to be able to see, in principle, how they would plan and make the same artefact if they started all over again. Sometimes the material chosen was not quite right, or did not handle the way the pupils thought it would (e.g. wood and clay are often more difficult to control than pupils imagine). Sometimes they have already modified the design continuously and almost imperceptibly, as they went along. There are very useful lessons to be learnt from changes to the design whether they were the result of clear decisions or '*just happened*'. If the development of technology is the story of a diversity of forms followed by selection by need (Mokyr, ch.5) and use (Vincenti, ch.13; Constant, ch.16) it may be incumbent upon teachers to make this process open and explicit to their pupils.

The second part of the 'quality' appraisal which aims to find out how well it fulfils a need or purpose, and hence the 'evaluation' of it, calls for a test. Sometimes this test is easy to put in place. A bridge must support weight, a printing press must print, and a pottery drinking mug must be able to hold

EVALUATION SHEET (D&T)

Name . . . .

I made a . . . .

Does your finished model look like your first drawing? . . . .

In what way is it different? . . . .

Which part are you most proud of? . . . .

How did you test the model? . . . .

What happened? . . . .

. . . .

If you made it again what might you do differently?

. . . .

. . . .

. . . .

Figure 14.4 Evaluation sheet.

water and be drunk from without spilling. Other products are less easy to test. A sailing boat will not sail in water unless the keel is deep and straight. If the base is a flat piece of wood, then it must be tested without floating. We tested the sails by placing 'the boats' on rollers made from pencils and using a hair drier to see if they would fill with wind and push the boat along. A water-wheel made from cork and plastic is not likely to grind grain. It was a great moment when some of our pupils suggested, for themselves, that they were going to make their water-wheel do work by winding up a bobbin of thread!

Readers will notice that the questions in the evaluation sheet (fig. 14.4) move **from the past** (the original design drawing) **to the present** (what you are proud of and what happened in the test), **to the future** (how you might do it again). Designing is a long continuous process and if the various selection processes are not explored and reflected upon then the chances are that little or no durable learning will take place.

14.8   Creativity and conclusions

In this final section I want to speak about creativity in a sense that is important to education. This is certainly not confined to the study of geniuses, as so much literature on creativity has been. Indeed there is a broad strand in European educational thinking, often called naturalism, stemming back to Rousseau and Steiner, that sees the cultivation of children's 'creativity' as the chief objective of all education, as opposed, for example, to the rationalist tradition where the transmission of externally established knowledge is the chief educational objective. For these purposes creativity includes two rather similar processes which have in common a gestalt shift within personal thinking:

- moments of creativity in learning when a theory or concept *'makes sense'* or *'clicks'*;
- moments of creativity in design when a *'new solution'* to a problem strikes us.

It is important to notice that being a genius, or even 'original' in an external socially recognized sense, is not necessary to either of these processes. Creativity can be, indeed should be, experienced by any pupil. It is characteristic of professional teachers that they prize the moment when a slow and passive child suddenly perceives what was hidden before, as much as the insightful comment of a 'brilliant' pupil. The glow of personal satisfaction, as well as the discontinuous leap in comprehension, may be essentially the same for both.

Learning theory[14] often speaks of *surface* (rote) and *deep* (meaningful) learning. This gross dichotomy is probably as misleading as such sharp divisions so often are, but the varieties of deep learning, and ways of identifying how they take place, do repay study. One such piece of work[15] reports the language of students' realization of their own understanding *'Everything fits into place . . . It clicks.'* The interesting point to note is how often this realization of coherence and understanding occurs rather suddenly. Polanyi[16] reports how, as a young medical student, the visual input from the X-ray photograph and the verbal explanation by a doctor, both previously incomprehensible to him, quite suddenly made sense together.

Both Polanyi and Johnson-Laird[17] can only explain the suddenness of new creative thought as the coming together of articulated and non-articulated knowing. The latter type of knowing, both agree, is likely to be process (or procedural) knowledge. Johnson-Laird's example is that of a musician improvising, and he comments that *'The fascination of improvisation is that the musicians may surprise themselves.'* This suggests that the ability to recognize and critique is present, as well as the wordless serendipitous improvisation. Koestler[18] also believed that creativity involved crossing from one domain of knowing to another. In evolutionary terms, this may be a special case of recombination (§2.2; §4.4; §5.2; §12.4), where well-founded knowledge meshes with different ways of knowing, to form a new phenotype.

In learning technology, young pupils acquire useful knowledge of many kinds. There is also at least one learnt skill, a kind of inner process knowledge, which enables them to put together different images, perhaps techno-memes, and to manipulate them. This is the aspect of spatial ability, already discussed, which leads to mental modelling. That wordless procedural skill, along with at least some of the critical evaluating techniques we have taught them, can bring about the delight of personal creativity. Nine times out of ten there will be no applause from friends or teachers and indeed little may come out of the occasion. This provides an example, in microcosm, of the importance of the social system for the recognition of creativity, for it to grow and flourish.[19]

In the case of children there is special difficulty in recognition because the child's discovery is not original in the absolute sense. Thus few of their fellow pupils ever acclaim the creative moment because, as they coolly comment, '*I thought of that ages ago!*' But it certainly is original to the pupil and may be powerful. This is where the teacher's recognition and praise may make all the difference between a habit of exploration and discovery which can enrich a lifetime of thinking and doing, or the reimposition of the sad habit of con-forming. This, I suppose, is why great achievers so often point to the encouraging or inspiring words of a parent or teacher as a turning point in their lives.

Some of the findings from this study of young children learning technology, such as their slow development of complex imaging and spatial skills, are instructive in showing what necessarily precedes adult technological capability. Other findings could be claimed to be surprisingly general to the whole techno-logical process.

One of these has been the effect of strong new structural knowledge on diversity. This was not an effect expected in the study, but became obvious once the work was in progress. It has been in the gift of this multi-disciplinary programme that the meaning of events in technology have been clarified or exemplified in studies as far removed from it as biology and economics. Thus, the biochemical instructions for organ development can strongly suppress diversity in embryonic development (Jablonka, ch.3), whilst a knowledge-based market can suppress creative diversity in supercomputer technology (Nelson, ch.6). It is a curiosity and a difficulty in the school teaching of technology that the provision of strong instructional knowledge can have much the same damping effect on creativity. Where pupils had to be taught directly how to use pulleys or other components which none of them had met before, little diversity in their use was observed. Other parts of the artefact may have been original, but not the environment of the new item. One may surmise that the formalism and context of the knowledge had not yet entered into the imaging process and hence there were constraints on the children's creativity.

15

Technological evolution as self-fulfilling prophecy

GEOFFREY MILLER

15.1   From genetic algorithms to Darwinian engineering

Engineering and computer science are undergoing a Darwinian revolution. In the last ten years, computer scientists have hijacked the idea of 'technological evolution', transforming it from a metaphorical model of historical change into literal methods for doing evolutionary engineering using explicit processes of random variation and selective replication inside computers (Ziman, ch.4). These methods, including genetic algorithms, genetic programming and evolution strategies, have attracted exponentially increasing interest as powerful ways of finding good engineering solutions to hard, complex, real-world problems.

Surprisingly, these developments remain almost unknown to scholars interested in evolutionary models for technological progress and to evolutionary epistemologists interested in more general applications of Darwinian theory to human culture and knowledge. Conversely, computer scientists working on genetic algorithms or genetic programming know very little about studies of technological innovation that use explicitly evolutionary models. This mutual ignorance is unfortunate, because there is so much each field can learn from the other. Genetic algorithm research, for example, has developed powerful insights into the way that evolution works as a stochastic search method for exploring design spaces and finding good solutions, and these insights may hold even for technological evolution outside the computer. On the other hand, studies of evolutionary processes in invention, market competition and historical change reveal a rich, diverse, multi-level interplay between design and selection that may hold valuable lessons for attempts to automate this process. This chapter is a first attempt at match-making between these two fertile fields.

15.2   How computer science turned Darwinian

When computers were slow, expensive and unreliable, they were no good for simulating evolution. But in the early 1960s, they became fast enough for evolutionary biologists[1] to try solving some of the harder problems in theoretical population genetics using simulation rather than paper-and-pencil proof. Typically, this meant simulating how the allele frequencies of just one or two genes might change over evolutionary time in response to various selection pressures.

Biologists made no attempt to use such evolutionary simulation to actually design something useful in the computer. However, some researchers in the field of artificial intelligence realized that the same kinds of computer programs could be used as an engineering method rather than a tool for scientific simulation. In 1966, Fogel, Owens and Walsh[2] suggested that if real biological intelligence evolved through real evolution, perhaps simulated intelligence could evolve through simulated evolution – that is, by *evolutionary programming*.

In the late 1960s and early 1970s, John Holland and his students at University of Michigan developed a new type of computer program called a *genetic algorithm* (§4.3). Populations of *bit-strings* – strings of zeros and ones – could be randomly generated to form an initial generation and then each bit-string could be interpreted as a particular design according to some *development scheme*. Designs could be tested according to a *fitness function* that determined how good they were for solving a particular problem. The bit-strings that made good designs get to have many copies – 'offspring' – in the next generation, while those that make bad designs are eliminated. And so on.

Holland[3] proposed a general formalism for representing evolutionary processes, showing how they could be implemented inside a computer and proving various theorems about how those processes will operate as search procedures for finding good solutions. The most important 'schema theorem' showed that if bit-strings can recombine their sub-strings ('schemata') by a method analogous to biological 'crossover' (§2.4), then selection on individual bit-strings is sufficient to increase the frequency of good schemata and drive out the bad. This theorem suggested that genetic recombination rather than mutation generated the most important variations that selection acts on during evolution. Accordingly, Holland's students made the study of recombination a major focus in genetic algorithms research.

Meanwhile, in Germany, engineers such as Rechenberg and Schwefel[4] were independently developing a type of Darwinian engineering called 'evolutionsstrategie'. For example, they constructed an aerofoil composed of many moveable sub-segments that could be adjusted and immediately tested in a wind tunnel.

To optimize the aerodynamics of the evolving wing, they used a carefully controlled mutation and selection process to generate and test better wing shapes. They then generalized this method into a stochastic optimization technique centred around continuous mutation and testing of a single design.

None of these methods made much progress until computers became much faster and cheaper in the 1980s. The problem in all evolutionary simulation using ordinary sequential computers is that the time it takes to produce a single 'run' of evolution equals the time required to test each individual design, multiplied by the number of designs in the population, multiplied by the number of generations that the population evolves. With population sizes around 100 to 1000 and around 100 to 1000 generations per run, a typical genetic algorithm run requires somewhere between 10,000 and one million evaluations to produce a decent result. For a complex problem, testing each solution using 1970s computer technology might take several minutes, so that an entire evolutionary run would require about a year of computer time.

Genetic algorithms became viable as research and engineering tools only when computers became fast enough for each fitness evaluation to take only a few seconds. Also, the development of massively parallel computers (such as the revolutionary CM-2 Connection Machine, with over 64,000 processors) cuts out one of the loops, because an entire population can be evaluated at once, with each processor testing a different individual. Once it became possible in the 1980s for risk-averse computer science graduate students to complete hundreds of simulation runs during their PhD programme, the field of evolutionary computation flourished.

1985 saw the first of a successful biannual series of International Conferences on Genetic Algorithms.[5] The first genetic algorithm textbook appeared in 1989[6] just in time for this author to take one of the first graduate courses on the subject at Stanford University. This course was given by one of Holland's students, John Koza, who shortly afterwards published a magnum opus,[7] detailing how a modification of the genetic algorithm concept could be used to evolve Lisp programs to solve various computational tasks. Instead of representing designs using bit-strings, genetic programming applied selection and recombination directly to the Lisp programs, which can be represented as tree-like structures.

Since 1992, when the key journal, *Evolutionary Computation*, was founded, the whole field has taken off academically. The half dozen main journals complement the proceedings of nearly 30 periodic conferences, grouped in a number of regular series. The field has also produced three main textbooks,[8] which are used in the more than 30 graduate courses on evolutionary computation currently taught around the world.

In addition to research on genetic algorithm theory, research has flourished on real-world applications in a number of fields, including: industrial design such as aerospace, automobiles, robotics, civil engineering and factory layout; control systems engineering such as job shop scheduling; neural network design and system identification; pharmaceutical engineering such as molecular design and protein folding; and financial optimization in spreadsheet programs. Several corporations have recently begun marketing genetic algorithm software packages for corporate and personal use, or consulting about genetic algorithm applications.

15.3   How genetic algorithms work

Classic genetic algorithms[9] have five key components: a *genotype format* that specifies how genetic information is represented in a data structure; a *development scheme* that maps that information into a phenotypic design; a *fitness function* that assigns a fitness value to each phenotype; a set of *genetic operators* that modify and replicate the genotypes from one generation to the next; and a set of *evolutionary parameters* such as population size and mutation rate that govern how evolution runs.

The *genotype format* specifies the type of data structure that will represent the genetic information. In place of the 4-letter nucleotide alphabet of DNA (§2.2; §3.2), genetic algorithms mostly use binary 'bit-strings'. Typically, these have a fixed length (e.g. 1,000 bits), interpreted as a fixed number of 'genes' (e.g. 100 genes) each composed of a fixed, equal number of bits (e.g. 10 bits per gene). In the last few years, however, researchers have explored a wider variety of genotype formats, including strings of real numbers, branching tree structures, matrices, directed graphs and so forth. The initial generation of genotypes is usually produced randomly, for example by assigning a zero or one with equal probability at each point in a bit-string.

The *development scheme* maps a genotype into a phenotype according to some algorithm or recipe. Thus each gene might be interpreted as a binary number specifying some parameter of a possible engineering solution. For example, 20 genes might be sufficient to specify (or 'parameterize') a design for a jet engine turbine blade, which could then be selected for its aerodynamic efficiency in a simulation. Alternatively, successive segments of a bit-string might be interpreted as successive rows in a matrix specifying possible connections between the processing units in a neural network.[10] In more recent work on evolving dynamic neural networks capable of controlling robots that pursue or evade one another, we used a much more complex development scheme where some genes specify the spatial locations of neurons in a two-dimensional 'brain', whilst other genes specify their interconnections.[11]

The trick in genetic algorithms is to find schemes that do this mapping from a binary bit-string to an engineering design efficiently and elegantly, rather than by brute force. Good development schemes map from small genotypes into complex, promising phenotypes that already obey fundamental design constraints. Bad schemes require large genotypes and usually produce phenotypic monstrosities. Of course, the smaller the genotype a development scheme can use to specify a set of phenotypes to be searched, the faster evolution can proceed.

The *fitness function* maps from phenotypes into real numbers that specify their 'fitness' and hence the probable number of copies the underlying genotype will be awarded in the next generation. The fitness function is the heart of the genetic algorithm: it is at once the environment to which all designs must adapt and the grim reaper (or 'selective pressure') that eliminates poorly adapted designs. As with development schemes, fitness functions can range from the trivial to the astoundingly complex. Early exploratory research on genetic algorithms often used literal mathematical functions, such as $y = x^2 - \cos x$, to map from a real-number phenotype (x) into a fitness score (y). For real applications, however, fitness functions are usually computer simulations of how a phenotype design would perform at some task. In our research on evolving pursuit and evasion strategies,[12] for example, each neural network pursuer was tested in about a dozen simulated chases around a virtual arena and awarded fitness points for catching different randomly selected opponents as fast as possible. In a civil engineering application, a fitness function might assign points to bridge designs based on their structural integrity, estimated cost, traffic capacity and resistance to wind-induced oscillations (cf. Vincenti, ch.13).

But if the fitness function does not realistically reflect the real-world constraints and demands that the phenotypic designs will face, the genetic algorithm may deliver a good solution to the wrong problem. Again, if each fitness evaluation takes too long, a genetic algorithm that relies on millions of evaluations to make evolutionary progress will not be practical. Most difficult in practice is the 'multi-objective optimization' problem: giving just the right weight to each design criterion in the fitness function so that the evolved designs reflect intelligent trade-offs rather than degenerate maximization of one criterion over all others. For example, giving too much weight to the traffic capacity criterion in a bridge-evaluation program might result in 1,000-lane bridges with no structural integrity and exorbitant cost.

In effect, the fitness function must embody not only the engineer's conscious goals, but also her common sense. This common sense is largely intuitive and unconscious, so is hard to formalize into an explicit fitness function. Since genetic algorithm solutions are only as good as the fitness functions used to evolve them, careful development of appropriate fitness functions embodying all relevant design constraints, trade-offs and criteria is a key step in evolutionary engineering.

The *genetic operators* copy and modify the genotypes from one generation to the next. Classic genetic algorithms used just three operators: *fitness-proportionate reproduction* – genotypes are copied in proportion to the fitness scores that their phenotypes received; *point mutation* – each bit in a bit-string is flipped from a 1 to a 0 or vice versa, with some very low probability per generation; and *crossover* – 'offspring' are formed by swapping random genotype segments between two randomly matched 'parents'. Mutation and crossover thus generate 'blind variation' and fitness-proportionate reproduction provides 'selective retention'.[13]

Much genetic algorithm research has focused on making these basic genetic operators work well together and trying new, quasi-biological genetic operators such as 'gene inversion', 'duplication', 'deletion' and 'translocation'. Getting the right balance between mutation and selection is especially important. If selection pressures are too strong relative to mutation, genetic algorithms suffer from 'premature convergence' on to a genotype that was better than any other in the initial, random generation, but which is far from optimal.

The typical evolutionary problem of getting stuck on a 'local fitness peak' (Perkins, ch.12) can be especially acute with genetic algorithms, where crossover between nearly identical parents does not introduce significant genetic variation and the vast majority of mutations tend to make even sub-optimal designs worse, so they get 'selected out' almost immediately. Significant genetic diversity can be preserved by spreading the population across a simulated geographic area, allowing subpopulations to evolve 'allopatrically' (§2.6) to different solutions and then exchanging innovations via migration and crossover.[14] Alternatively, if 'assortative mating' is favoured, so that crossover is programmed to occur more frequently between similar 'parents', then the population tends to split apart into divergent 'sub-species' with different adaptations.[15]

Finally, the *evolutionary parameters* determine the general context for evolution and the quantitative details of how the genetic operators work. Classic genetic algorithm parameters include the population size (usually between 30 and 1,000 individuals), the number of generations for the evolution to run (usually 100 to 10,000 generations), the mutation rate (usually set to yield around one mutation per genome per generation), the crossover rate (usually set around 0.6, so three-fifths of genotypes are recombined and two-fifths are replicated intact) and the method of 'fitness scaling' (e.g. how differences in fitness scores map onto differences in offspring number).

Deciding the best values for these parameters in a given application remains a black art, driven more by blind intuition and communal tradition than by sound engineering principles. For example, there is a trade-off between population size and generation number: the larger your population, the fewer generations you can run for a given amount of computer time. The genetic algorithm community has no consensus yet about how best to allocate these computer cycles.

15.4   Some strengths and weaknesses of genetic algorithms

Conjointly, the five components outlined above determine a 'design space' (Stankiewicz, ch.17). Genetic algorithms search these spaces using a massively parallel, stochastic, incremental strategy called 'evolution'. They are not an engineering panacea. Their performance is only as good as their ability to search a particular design space efficiently and inventively. This in turn depends critically on a host of subtle interactions between genotype formats, development schemes, genetic operators, fitness functions and evolutionary parameters. Genetic mutations should tend to produce slight but detectable alterations in phenotypic structure that open the way for cumulative improvement. Genetic crossover should tend to swap functionally integrated parts of phenotypes to yield new emergent properties and behaviours. And so on.

For very simple problems, one can be a bit sloppy about bringing all five components into alignment, because genetic algorithms are rather robust search methods for small design spaces. But for hard problems and very large design spaces, designing a good genetic algorithm is very, very difficult. All the expertise that human engineers would use in confronting a design problem – their knowledge base, engineering principles, analysis tools, invention heuristics and common sense – must be built into the genetic algorithm. Just as there is no general-purpose engineer, there is no general-purpose genetic algorithm.

Most obviously, there is no general-purpose development scheme because different applications require completely incommensurate types of designs. The design spaces of possible bridges, neural networks, proteins, factory layouts, jet turbines, computer circuits and corporation financial strategies cannot be translated into a common language (Stankiewicz, ch.17) and, even if they could be, searching that generic design space would be vastly less efficient that searching a more focused subset.

Genetic algorithms tend to work best when the design space they are searching has already been rather well-characterized – as in the fine tuning of a rough working system[16] – or, ideally, fully formalized into a kind of design grammar. For example, genetic programming[17]seems to work well because the design space of computer programs in a particular programming language is clearly structured by that language's formal grammar. Genetic programmers favour languages like Lisp because the 'S-expressions' that constitute Lisp programs are branching tree structures that remain interpretable when their end-nodes or sub-trees are mutated or crossed over.

By contrast, there is no design grammar yet for fully re-useable ground-to-orbit spacecraft – indeed, there remain wildly disparate strategies for solving this difficult problem, each of which require some components that go beyond current technology. In such a case, using a genetic algorithm to generate promising new design solutions would be vastly more difficult than in genetic programming. Still, it might be useful, because by forcing engineers to think about characterizing the design space as a whole rather than perfecting one particular solution, the discipline of setting up the genetic algorithm may yield new insights and ideas.

One of the most disturbing features of genetic algorithms is that they often produce solutions that work, but one cannot quite understand how or why they work. Whereas traditional engineers are constrained to working on designs that they more or less understand, genetic algorithms select only for performance, not for clarity, modularity or comprehensibility. Like insect nervous systems that have been under intense selection for millions of generations to get the most adaptive behaviour out of the smallest, fastest circuits, artificial neural networks evolved for particular tasks[18] almost never do so in a way that makes any sense to human minds that expect modular decomposition of function.

Is this a problem? It depends more on the social, cultural and legal context of engineering in particular domains. The patent office may be sceptical of a design delivered by a genetic algorithm if you cannot explain why it works. A client corporation may reject an optimal marketing strategy designed by a management consultancy using a genetic algorithm if you cannot explain why this strategy makes sense. The automaticity of the genetic algorithm and the opacity of its solutions may create problems of accountability, liability, safety and public confidence. Also, well-understood solutions can be easily modified and generalized to other problems and other contexts, whereas specifically evolved solutions may not.

On the other hand, animal and plant breeders have been content for thousands of years to use artificially selected products without knowing exactly how they work. Likewise, many traditional technologies such as Japanese sword-making (Martin, ch.8) evolve by trial-and-error experimentation and cultural imitation, without any theoretical understanding of why the production techniques work. Computer scientist Danny Hillis has commented that he would rather fly on an airplane with an autopilot evolved through a genetic algorithm than one with a human-engineered autopilot. The reason: the evolved autopilot's very existence was at stake every time it confronted the simulated emergencies used to test it and breed it, whereas the human designer's existence was never at stake. For better or worse, genetic algorithms break the link between innovation and analysis that has been considered a fundamental principle of modern engineering.

15.5   Fitness evaluation in Darwinian engineering

As I noted earlier, the crucial practical issue for real 'Darwinian engineering' is whether the process of evaluating candidate designs can be automated. *Fitness evaluation* is the computational bottleneck in simulated evolution. For example, in our project on pursuit and evasion, it took 95–99% of the computer time.[19] Applying the genetic operators to breed one generation from the previous generation is usually computationally trivial by comparison. This is because physical reality has a lot of detail that needs simulating, and most serious applications require evaluation in many tests under many different conditions. Consider the computational requirements for evaluating the aerodynamic efficiency and stability of a single jet fighter in simulation under a reasonable sample of different altitudes, speeds, weather conditions and combat scenarios. Now multiply by the million or so evaluations needed to evolve a decent jet fighter design. Serious Darwinian engineering seems to require prohibitive amounts of computer power.

Is there any alternative to doing all fitness evaluation in simulation? Engineers didn't always test things in computers. Technological progress used to depend on visualizing or sketching design solutions, mentally imaging how they would fare in various tests and hand-building prototypes for testing in the real world (Vincenti, ch.13; Constant, ch.16). Before the twentieth century, much of engineering and architecture also depended on building scale models and testing them in various experiments (§13.3). Prototyping and model-making are very efficient ways to let the physics of the real world do much of the evaluation work for you. But how could this sort of real-world testing be incorporated into an automated fitness evaluation method for Darwinian engineering?

A possible future strategy for Darwinian engineering is suggested by a project at the University of Sussex[20] to use genetic algorithms to evolve computer vision systems for guiding mobile robots. To evaluate how well each vision system would work, they originally had the robots moving around in a virtual environment, using very time-consuming computer graphic ray-tracing methods to determine what visual input the robot would get at each position in its little world. They decided to let the real world take care of the ray-tracing for them, and put a video camera on to a gantry that could move around in a tabletop model of the test environment. Each robot vision system to be tested was downloaded to a small computer that could translate the robot's simulated movements into gantry movements, with the digital video input then being used directly as the input to the simulated robot vision system. Apart from graduate students needing to untangle the video input cable once in a while, this hybrid between simulated evolution and real-world testing could automatically evaluate a few dozen robot vision systems per day and led within a few weeks to the evolution of a system capable of navigating towards triangles in preference to circles.

Carried to its logical extreme, this strategy interfaces with *combinatorial chemistry*, where innumerable variant molecular entities are synthesized automatically by the random combination of segments drawn from different populations and then screened automatically for the part they play in a particular chemical or biological process.[21] Efficacious new catalysts, enzymes, therapeutic drugs, electronic materials, etc. can thus be discovered and systematically improved by an evolutionary process that is closely akin to a genetic algorithm.

New methods of computer-controlled manufacturing, robotic assembly, rapid prototyping and automated lab testing may allow candidate designs to be incarnated and evaluated without human intervention. Automating certain aspects of research and development in this way will be much more challenging than automating the manufacture of standardized products, because candidate designs are worth evaluating only if they are unique. Modifying factories to do automated prototyping, testing and fitness evaluation will be a major challenge for Darwinian engineering.

Another alternative is to put human judgement into the evaluation loop. Interactive artificial selection can be used to guide evolutionary search through a design space.[22] Thus, computer graphics artists[23] have applied human aesthetic judgement to evolve fantastic images based on compact 'genotypes' combining mathematical formulae and computer-graphic primitives, with the advantage that favoured 'phenotypes' can be easily copied and recreated without having to store an entire multi-megabyte image.

Darwinian engineering could extend this sort of artificial selection in two main ways. First, human engineers could use their common sense and expertise to rate candidate designs for their overall plausibility as they are generated by a genetic algorithm on the computer screen in front of them, with computer simulation to test design details. Moreover, the computer could keep track of the human responses to candidate designs, learning how to make its own ratings – for example by training a neural network to emulate human judgement. The outcome might then be an automated evaluation function that combined common sense assessments with sound engineering principles.

A more revolutionary way to put humans in the evaluation loop is to use consumer judgements directly to evolve customized products through interactive, online Darwinian engineering. Modern businesses usually try to make money by second-guessing average consumer taste, manufacturing a limited range of products to span that taste and trying to attract mass sales (cf. §21.2). Genetic algorithms with interactive evaluation may permit a radically different strategy that integrates design, manufacturing, marketing and sales in a single system.

Suppose individual consumers could log on to a company's 'interactive catalogue' directly. Each product line would be a genetic algorithm for evolving a customized product design. It would start with an initial population of candidate designs that could be rated by the consumer and then mutated and recombined to yield successive generations of new, improved designs. Allow each consumer to evolve their preferred designs, subject only to certain safety, functional and legal requirements and to being capable of being manufactured profitably by the automated production system. A few minutes of interactive evolution should lead to a most-preferred design, needing only to be priced and debited to a credit card number before being manufactured and shipped to the consumer. Up-to-date consumer-preference data could be fed into the interactive catalogue to bias the interactive evolution for each consumer towards areas of design space that have recently proven popular with other consumers with the same demographics and tastes.

This sort of interactive evolutionary consumerism would be most appropriate for fairly low-tech, easily modularized products such as wallpaper, furniture, holiday packages and standard financial services. Yet even for relatively high-tech, complex products that must function safely and reliably, like automobiles, cardiac pacemakers, nuclear-powered aircraft carriers and automated stock-trading systems, where companies normally market only a few carefully optimized, thoroughly tested designs, interactive evolution might allow consumers to explore the design space around these designs and observe the fitness trade-offs and constraints for themselves.

In any case, by bringing consumers directly into the design loop as agents of interactive evolutionary selection, the diversity, originality and richness of human material culture might be substantially increased.

One benefit is that it would introduce an analogue of sexual selection. Consumers select products for more than their functional fitness; they also impose various aesthetic and symbolic criteria. These two modes of selection produce quite different evolutionary dynamics and can powerfully complement each other as search and optimization processes.[24] For example, selection for apparently maladaptive 'sexual' traits is a very efficient way for populations to escape from local optima in which purely functional selection would otherwise leave them trapped. By allowing consumers to bring their apparently frivolous aesthetic judgements to bear on product development, they may stumble upon promising new areas of design space that more utility-minded engineers may have overlooked.

15.6   The future of technological evolution

Most historians, psychologists and engineers who have studied techno-logical evolution in the past agree that the evolutionary process has always been 'automated' to some extent, both in the unconscious mental processes of inventors searching design spaces for innovative solutions (Carlson, ch.11; Perkins, ch. 12) and in the competitive market processes that sift good product designs from bad (Nelson, ch.6; Vincenti, ch.13; Stankiewicz, ch.17; Fairtlough, ch.19). The rise of genetic algorithms as engineering methods adds a third, more explicit, type of automation: the evolution of designs inside computers. Clearly, the utility of genetic algorithms will depend heavily on their ability to comple-ment these other two types of highly efficient, massively parallel search processes – human creativity and economic markets.

There are good reasons for expecting this complementarity to prosper. Minds and markets are excellent at combining vast amounts of diverse, distributed information under multiple constraints into workable solutions (e.g. ideas that solve the problem or prices that clear the market). Genetic algorithms do some-thing similar, combining vast amounts of information about the fitnesses of different design components (i.e. genes and their phenotypic effects in the simulation) into good designs that are at least locally optimal.

But genetic algorithms work by principles rather different from human creativity and market competition. Human minds are not as good as computer simulations at detailed, quantitatively accurate fitness evaluation. And markets are not nearly as fast, or as free from social, cultural, political and legal biases. There is a niche then, between minds and markets, for genetic algorithms to contribute to technological evolution. This will probably lead to a division of labour in technological evolution, where problem-solving is often automated using genetic algorithms, but problem-framing still depends on individual and group creativity and solution-verification still depends on market processes and social history. Engineers will have to think more like selective breeders (§11.10) who design fitness functions, set population parameters and oversee evolution inside computers and automated design-testing facilities.

Such a development sounds strange, but would simply mark a return to the earliest, most important technological revolution in human history: the domestication and selective breeding of animals by Neolithic pastoralists, followed by the domestication and selective breeding of plants by farmers around five thousand years ago.[25] The early pastoralists and farmers did not know how pre-existing biosystems – wolves, wild cattle, seed-headed grasses, etc. – worked or where they came from; they simply substituted their own selection pressures for those of nature and reaped the benefits. As the complexity of manufactured technologies begins to approach that of natural biosystems, we may be forced to revert to this humbler form of engineering qua selective breeding. The second millennium may have been exceptional as a period in which humans were able to comprehend their own technologies sufficiently to design them through engineering principles.[26] In future, technological evolution may rejoin the main stream of biological evolution, with humans breeding designs whose operational details lie far beyond their comprehension.

IV
INSTITUTIONALIZED INNOVATION

16

Recursive practice and the evolution of technological knowledge

EDWARD CONSTANT

16.1   Introduction

Virtually all the fundamental principles of biological evolution have proved troublesome when applied to technology. It is not at all clear what evolves: devices, artefacts, techniques, systems, sociotechnical systems, knowledge or memes? It is not clear whether, or on what grounds, 'selection' might be said to occur, or at what level: devices, designs, production techniques, firms, regions, even nation-states? It is not even clear what counts as a 'useful adaptation' in technology in the first place, or upon what grounds we believe that such 'useful adaptations' in fact arise, persist, or are conserved over time.

Several notable but related challenges to the notion of efficacious technological evolution centre around these issues (Mokyr, ch.5; Nelson, ch. 6; Fleck, ch.18). To one degree or another, these all attack the ideas of selection, fitness or adaptation. Thus, social constructivism in technology (SCOT) argues that all technology is socially constructed, and therefore reflects purely the social interests of relevant social groups rather than any 'selection' on rational technical or economic criteria.[1] In contrast, actor network theorists erase altogether the distinction between the social and the material. For them, technologies are only (precariously) stabilized networks of human and non-human actants. The fitness or adaptedness of technology has no meaning outside the persistence of the 'alliances' between actants, and has no objective construal whatsoever.[2]

The usual contrarian move in response to these sorts of critiques is to persist in separating technology from its context, and to stipulate that while the goals, intentions, purposes or interests that technologies are meant to serve are purely negotiated social constructs, such a context so defined nevertheless does

219

serve as a selection environment in which various technologies or devices are more or less fit, according to traditional 'rational' technological criteria, such as efficiency or cost-effectiveness. But as Donald MacKenzie and Thomas P. Hughes have each demonstrated, in very different ways, this move will not wash either.[3]

Broadly conceived notions of technological evolution by efficacious selection or adaptation thus run into serious conceptual difficulties. Nevertheless, despite these difficulties, I believe the original quest of evolutionary epistemology – the journey that Donald Campbell set out on nearly forty years ago – remains viable.[4] That quest, as I conceive it, is to understand, in Campbell's memorable phrase, the progressively better 'fit of phenomena to noumena', or the fit between our perceptions of, or here our beliefs about, the world (phenomena) and the otherwise unknowable world-as-it-is (noumena, or the 'thing-in-itself,' Kant's 'Ding an sich').[5] I want to defend that programme for technology and, by implication, for science (although not here), by focusing very narrowly on the variation and selection processes in technological practice that enhance the reliability, and, I will argue, we have reason to believe, the validity of techno-logical (and scientific) knowledge.

In his two seminal papers on the methodology of the social sciences,[6] Donald Campbell conceived of reliability as 'the agreement between two efforts to measure the same trait through maximally similar methods', while validity is 'represented in the agreement between two attempts to measure the same trait through maximally different methods'. Thus reliability entails the consistency and internal consilience of results from doing the same thing over and over again, while validity entails that results be both convergent (different measures of the same variable converge to the same result) and discriminant (measures can discriminate between variables). The broad claim advanced in this chapter is that recursion in technological practice, as characterized below, leads to increas-ingly reliable and valid knowledge, as Campbell conceives it.

To substantiate that claim involves four subsidiary arguments: 1. I will argue (cf. Mokyr ch.5; Stankiewicz, ch.17) that what evolves in technology (as in living systems) is *information*, not artefacts, devices or social practices (cf. Fleck, ch.18). What matters for evolution is ancestor-descendant lineages. Technological and scientific change – continuous and discontinuous – is thus interpreted in terms of combinations, recombinations or radical saltations of 'memes' (§1.2). 2. The efficacious *selection* – of information – does occur in technology through the social process embodied in *recursive practice*. 3. This recursive practice in tech-nology yields *reliable knowledge* in the Bayesian sense, which is equivalent to reliable and valid knowledge in Campbell's sense. 4. This reliable knowledge, and the processes giving rise to it, likely can produce, in the limit, veridical knowledge about the world as it is.

In what follows, however, these arguments are not presented in the above order. Rather, I begin by characterizing the key concept of recursive practice, first empirically, then in evolutionary terms. Only then do I try to locate these processes epistemologically, and to assess their importance for the 'rationality' and evolution of technological knowledge.

16.2   Recursion in engineering science and practice

The essential claim in this chapter is that the social practices of engineers and scientists are typically *recursive*, in that they comprise a highly nuanced hierarchical structure of alternate phases of selection and of corroboration by use. In effect, this is how 'Popperian' testing actually occurs.[7] The result is strongly corroborated foundational knowledge: knowledge that is implicated in an immense number and variety of designs embodied in an even larger population of devices, artefacts and practices, that is used recursively to produce new knowledge. That new knowledge, in turn, if it forms the foundation for successful practice, itself can be used recursively in still further innovation. The two exemplars related here represent nearly the opposite ends of a continuum of technological practice, ranging from local, specific and routine at one end, to general, creative, and extraordinary at the other, and are intended to show the breadth of problem-solving power, as well as the creative, novelty-producing potential, of recursive practice.

The first example is taken from Bucciarelli,[8] who describes an engineer, pseudonymed 'Beth', writing a computer program to analyse a solar-array-powered salt-water desalination system, a prototype of which was operating in Saudi Arabia. Beth deploys a standard repertoire of scientific and engineering concepts: conservation of energy, Ohm's and Kirchhoff's laws for circuits, the generalized metaphor of flow and control volume theory.[9] Initially, Beth's simulation doesn't match the experimental data from the prototype system. Significantly, however, Beth does NOT behave like a stereotypical Popperian[10] by shouting 'Eureka' and running through the corridors of her company proudly announcing her refutation of the First Law of Thermodynamics.

Rather Beth adopts a recursive strategy. She keeps plugging quietly away, rerunning her simulation, double-checking known, independently established parameters of photo-voltaic cells, batteries, evaporators, always presuming that foundational theories are inviolate. Eventually, she discovers that a 'faulty' ammeter had been replaced about a month into the test-run: with appropriate ad hoc corrections for the ammeter, the computer simulation perfectly matches the system's performance before the change; without those corrections, the program perfectly matches performance after the change. Happily for all, the phenomena, and the foundations of engineering science, are saved.

Although the Duhem-Quine thesis[11] has long held that no specific theory or conjecture can be tested independently of the whole network of related theories, assumptions and background knowledge supporting the test situation, and that there is no logical way to tell which specific claims are wrong when such theory-and-practice complexes are 'falsified', that is not how Beth, or any other sensible engineer or scientist, goes about her business. Contra what the relativist critics of science and technology would have them do, such practitioners believe the universal and abstract and doubt the local and specific. However counter-intuitive, this behaviour is no different in principle from the problem-solving or analytical strategies used daily by motor mechanics, computerized diagnostic systems, expert systems, or high-energy physicists.[12]

But the same recursive processes not only solve normal everyday problems; they also can lead to radically new technological inventions – indeed to new technological types, species or genera. Before the mid-1920s, for example,[13] *all* axial-flow turbine systems – water turbines, steam turbines, internal combustion gas turbines, and axial-flow air compressors – were designed in accordance with hydrodynamic theory for flow through passages, on the assumption that energy was neither withdrawn from nor added to the flow medium. By the eve of the Second World War, virtually all turbine systems were designed according to aerofoil theory, with concepts of lift and drag used explicitly to account for energy additions to or withdrawals from the medium.

Mature aerodynamic theory for subsonic aerofoils derived largely from the work of Ludwig Prandtl and the students he trained at Göttingen University. Over a period of some 25 years – from 1904 to 1930 – classical hydrodynamic theory was extended to provide aeronautical engineers throughout the world with the theoretical means to rigorously analyse and precisely design subsonic aerofoils, including propellor blades. But these impressive advances in aerofoil theory were completely separate from work on normal turbine systems. They occupied a different community of practitioners, with traditions and interests quite distinct from those of the heavy machinery engineers who designed water turbines and marine and industrial steam turbines.

For example, Charles Parsons himself, the original inventor of the axial-flow steam turbine (1884), had quite deliberately adopted very small pressure drops across each of the many stages of his steam turbine design exactly to mimic nineteenth-century water turbine practice, and thus to avoid issues of com-pressibility and expansibility in each stage (another example of recursion in technological invention). This fidelity to traditional analysis of hydrodynamic flow through passages, when applied early in the twentieth century to axial-flow air compressors for industrial internal-combustion gas turbines, doomed all attempts to build such machines, including Parsons's own, to abject failure. That the devices failed was unequivocal and undisputed: their efficiencies never broke 3%. Why they failed is only apparent retrospectively, from the vantage point provided by later aerofoil theory.

Given the weight of tradition in the heavy engineering community and, with the notable but largely ancillary exceptions of axial-flow air compressors and industrial gas turbines, its extraordinarily successful practice, it is hardly surprising that application of aerofoil theory to turbine elements and systems was pursued not within that community but rather within the younger aeronautical community. In the mid-1920s, trained aerodynamicists began work independently on axial-flow compressors and turbines in Britain, Germany and Switzerland. In each case, they came to two critically interrelated realizations: that the primary purpose of turbomachinery was in fact the transfer of energy between blades and working medium, and that aerofoil theory provided a way of conceptualizing and designing turbomachinery components such that this energy transfer (analogous to lift) could be maximized, while energy loss (analogous to drag) could be minimized.

The resulting investigations were recursive in the sense defined above. Investigators uniformly held foundational theories – conservation of energy and mass, aerofoil theory and all its simplifying assumptions and mathematical conventions – inviolate. In each case, theoretical calculation was followed by construction of small-scale, non-rotational test rigs ('cascades' of aerofoils), then development of experimental multi-stage rotary compressor and turbine units. This work permitted definition of limiting parameters, determination of interference effects among closely spaced aerofoils, vortex effects, and so on: results of experiment fed back into emerging theoretical description, and sustained further cycles of conceptual and practical progress. By the early 1930s, these new results were just beginning to be disseminated among those attuned to leading-edge aerodynamic research.

Indeed, all four of the successful inventors of the turbojet (as well as some others) based their design assumptions, recursively, on these insights, and the new aerofoil theory for axial components undergirded development of all successful turbojet and turboprop engines. From turbojet practice, the new design approaches then diffused backward to steam turbine practice. Although casual observation of morphology would not likely reveal to the uninitiated the difference between an archaic and a successor modern turbine, in fact their memetic genotypes are radically, indeed revolutionarily, different. A good cladist would mark the divergence of all 'modern' turbine systems from their primitive ancestors exactly at the point where this new strand of 'memetic DNA' evolved.

What these two examples of technology in action share is recursive practice. In each example, from relatively routine (but hardly trivial) problem-solving analysis and design to the invention of truly revolutionary technologies, the underlying process – recursive use of foundational knowledge – is the same. And the result is the same: practice, and knowledge, get 'better'.

16.3   What evolves?

Recursive practice is clearly a highly effective process in technology. But sticky issues still remain. What exactly does get 'better', in what sense, and why? What is it that evolves? What is the unit of heredity – what is it that is conserved and reproduced over time (Mokyr, ch.5; Fleck, ch.18)? And what is the level of selection, or how is, or even is, that unit selected? As we show throughout this book, these are contentious questions even within evolutionary biology, and are even less settled for technology.

Moreover, as evidenced by other chapters, there are many possible ways to interpret these issues. But because I think it is faithful to the original intent of evolutionary epistemology, and because I think it makes it possible to articulate a sense in which 'technology' not only evolves but indeed progresses, I have chosen to focus on the evolution of technological knowledge or information. This harmonizes with the neo-Darwinian principle (Jablonka & Ziman, ch.2; Jablonka, ch.3) that what counts in the evolution of living systems is ancestor-descendant lineages of information – genes metaphorical if not genes literal.

What in technology evolves then – in the very specific and demanding sense of 'becomes better fit' – is information, the fit between phenomena and noumena, that is, reliable knowledge.[14] Although scientific and technological development do not simply mimic biological evolution in all its particulars, each is a particular instantiation of spatiotemporally universal selection processes. In general in such processes, what matters is inheritance, not superficial morphological similarities. Thus, on this view, the evolution of technology becomes the descent and modification, or recombination or mutation or saltation, of fabrication techniques, designs, design techniques, and, more recently, engineering science and science itself – information.

16.4   A quasi-Bayesian solution

To sustain this argument requires a Bayesian interpretation of reliable knowledge and rational belief, and an account of the peculiar population structure of technological knowledge. The Bayesian approach to knowledge is essentially probabilistic. The quest for certain knowledge – 'truth' – is replaced by the pursuit of maximally credible hypotheses.[15] Given some non-zero prior subjective probability, or belief, that some proposition is indeed 'true', a Bayesian rational agent will simply adjust his degree of belief, up or down, as evidence comes in for or against the proposition. In principle, if this evidence is strong enough (whatever that means in practice), and uniformly supports the proposition, then this probability estimate tends towards unity. For Bayesians, the 'truth' of a hypothesis can never be determined for sure, but its *reliability* is clearly indicated by this concentration of the probability distribution more and more tightly around one as evidence increases without bound.[16]

This epistemological strategy obviously raises a number of objections, which are rehearsed at length in the philosophical literature.[17] The main point is that, although most people – including most scientists – are not adept at *formal* Bayesian reasoning,[18] it does model qualitatively the way in which they adjust their beliefs in light of evidence.[19] What is more, this mode of reasoning applies equally in everyday life, in sophisticated technological practice, and in high science.

But what, in the recursive practice depicted above, counts as a Bayesian hypothesis? What in technology might evolve by such a process? As we see throughout this book, devices or artefacts are concatenations of an immense diversity of heterogeneous elements whose behaviour and fate is in turn contingent on a whole host of environmental factors, including those ordinarily considered to be purely 'social' or 'cultural'. For this reason, technological evolution is perhaps not best seen as direct selection of devices or artefacts, any more than the evolution of species can be seen as the fate of phenotypes. What matters in each case is the differential survival and reproduction of heritable variation – that is, in each case, the information encapsulated and passed on in genes or 'memes'.

In technology (and similarly in science) this selection process has an unusual structure. The more general or foundational knowledge or theory is, the more severely it is winnowed. Foundational knowledge, almost by definition, is implicated in a much larger population of artefacts, designs, and practices than more specific knowledge. As a result, not only is it subjected to more numerous tests (or attempted refutations), but it must also survive in an even greater diversity of testing environments. Duhem-Quine reservations are thus largely vitiated, and Campbell's standard for reliable and valid knowledge through convergent and discriminant measures is fulfilled.

This claim can be justified by formal Bayesian reasoning. The gist of the argument is as follows. Note first that modern technological cultures comprise very large populations of very diverse artefacts whose individual designs each embody a large but finite set of 'memes'. These 'memes' are drawn from a much larger set of theories, production processes, materials, tacit knowledges, local conventions, and so on. The fact that any particular artefact in this culture actually 'works' can be interpreted as the successful outcome of a Popperian experiment – i.e. the failure of an attempted refutation of its constitutive 'memes'. By the Duhem-Quine principle, this only means that all the members of this particular set of 'memes' remain unrefuted.

In the normal course of everyday life, however, this test is applied (more or less independently) to all the innumerable artefacts in the population. A great many different artefacts – doorbells, automobile starters, soup kettles, coffee pots, turbojet engines, and so on – are found to 'work'. Thus, the overall outcome is corroboration (i.e. non-refutation) of the 'memes' that are common to all these different designs embodied in this huge population of artefacts. This common set is quite small, constituting the knowledge basic to essentially a whole technological culture. Thus recursive practice, to the degree that it embodies basic theories and concepts in huge sets of highly diversified artefacts, yields reliable knowledge in the Bayesian sense.

There is, of course, no guarantee that knowledge that is incontrovertibly reliable technologically is true ontologically. Indeed, some of the most basic theories used in the making of artefacts that 'work' are known to be false in some more or less fundamental sense. For example, the 'law' of 'the conservation of energy' that underlies almost all engineering practice has been replaced in the scientific canon by the 'conservation of mass-energy'. At 'ordinary' temperatures, velocities, and scales, this so-called 'reduction' may be a very good mathematical approximation, entirely reliable in practice, but it remains ontologically and metaphysically doubtful. Similarly, the Prandtl-von Kármán boundary layer theory that underpins the calculation of sub-sonic lift and drag also may turn out to be false in precise detail.[20] The reliable Bayesian knowledge confirmed so lavishly in recursive practice may be in fact not true.

What is happening, of course, is that scientific research, although not directed toward the production of useful artefacts, contrives experimental or hypothetical situations that extend beyond the bounds of normal technological practice. The behaviour of such contrived natures or artefacts may, on occasion, challenge the Bayesian credibility of even the most basic theories, however reliable these have proved to have been so far in technological practice. For example, Albert Einstein was forced to amend the law of conservation of energy because 'thought experiments' involving fast-moving objects did not 'work' as previously expected.

These circumstances lead to a bizarre epistemological irony: true theoretical revolutions – those in which older basic theories are overturned and replaced – actually serve to increase Bayesian confidence and reliability in practice, by both extending and bounding practical domains. After the relativistic conservation of mass-energy 'revolution', both the inner and outer bounds for considering conservation of mass and conservation of energy as separate are clearly defined and empirically unproblematic. Similarly, although computer programmers do not believe the universe is 'really' Ptolemaic, they can, safely assume that the heavens move around a fixed Earth to write programs for nuclear missile guidance.[21] In addition, new theories open entirely new domains for technological exploration and exploitation in the light of this new knowledge.

The appearance of genuine novelty seems to produce some difficulties for orthodox Bayesian reliability. In principle, a truly novel conjecture, scientific or technological, born bereft of any empirical evidence, should have an initial Bayesian prior probability of zero, from which it can never escape.[22] But this is a purely formal objection, since no meaning or attention can be given to an apparently unmotivated hypothesis, that is, one that does not give at least some preliminary indication of why it might just possibly come eventually to be believed.[23]

In technological practice, moreover, novel conjectures (true inventions) often result from anomalies: unexpected failures (Vincenti, ch.13). Once perceived, such a failure lowers the expected value assigned to a theretofore highly reliable design (and its theoretical rationale), and may thus provide the inspiration for assigning a non-zero subjective prior probability to an unproven, apparently very unlikely, radical alternative. In other cases there may be a 'presumptive anomaly',[24] where recursive use of other information from outside a traditional domain – new scientific theory for example – may both reduce the expected value imputed to traditional designs or processes and indicate a non-zero prior probability for some radical alternative. The role of aerofoil theory in turbine design shows that 'rational' but corrigible extensions or conjectures of radically novel technological systems, to the extent that they are recursive, are consistent with Bayesian confirmation theory.

Despite the great and well-placed confidence inspired by these Bayesian confirmation processes, they remain, in Donald Campbell's phrase, fallible, hypothetical and corrigible. Indeed, it is not uncommon for the beliefs of a Bayesian rational agent to converge in favour of a hypothesis that is later found to be quite false. This can happen (David, ch.10) where the boundary conditions are very restrictive. First, some beliefs are maintained no matter what the evidence: if the true hypothesis is the negation of some such maintained belief, it is by definition assigned an initial prior probability of zero, and no evidence can ever rescue it. Second, any one agent experiences a strictly limited number of trials, and has recourse to very limited, and imperfect, information about other agents' experiences.

In contrast, for most of the technologies of interest here, the boundary conditions are much less burdensome. Historically, industrialized countries have enjoyed two great advantages over less developed countries. They have had the marginal resources to experiment, which simply means they have enough surplus above subsistence to absorb the costs of the inevitable failures that experimentation entails (Martin, ch. 8). They have also enjoyed culturally positive evaluations of technological change (Macfarlane & Harrison, ch.7), stemming perhaps from the contingencies of their intertwined military, political and economic histories.[25]

What may be even more important (Stankiewicz, ch.17), since about the beginning of the nineteenth century, technologists in these countries have constituted a large, diverse, heterogeneous, internally differentiated, but still interconnected, community of learners whose very number and variability provides the evolutionary foundation for collective learning. Historically institutionalized open communication systems (patents, patent journals, scientific and technological associations) have promoted what Donald Campbell called 'collective omniscience', that is to say 'all the objectivity there is to be had'.[26] Unlike the earnest but epistemologically deprived propagators described by Paul David (ch.10), technological communities thus meet the minimal necessary conditions for efficacious Bayesian confirmation processes, even though, by their very nature, those processes can never *certainly* produce positive outcomes.

16.5   Recursion and rationality

This quasi-Bayesian analysis directs attention back to the question of whether recursive practice can be considered to be 'rational'. In the context of technological change, 'rational' is conventionally interpreted to cover some or all of the following meanings, all of which are problematic:

(a) *Technological design and practice is a logical or deductive enterprise.* This claim has four principal variants.

The first is that technology follows, depends upon, is deduced from, or exploits, scientific discovery. In the motto of the 1933 Chicago World's Fair,[27] 'Science Finds, Industry Applies'. Although still current in some circles, this view is almost completely rejected both among practising engineers and among historians of technology.[28] Technology is simply not deduced in its entirety from science – or, for that matter, from anything else.

The second variant espouses the belief that technology has its own internal 'autonomous' logic or teleology, unknown or perhaps only dimly grasped by us, and certainly not controlled by us. While the 'real-world constraints' (Vincenti, ch.13) do play a powerful selective role in shaping the course of technological development, even vicariously through our phenomenological beliefs about that world, there is simply no evidence that technology acts autonomously or seeks goals not of our making: that much, surely, social constructivists have established.

The third variant, rarely explicitly invoked, but often implicit in notions of technology's optimality, objectivity, or even efficiency, holds that search for solution to technological problems is logically complete: all relevant prior experience is taken into account and appropriately discounted or weighted. In Dr Pangloss's rosy view, this is the best of all possible worlds. This claim is patently false: there is no reason to believe that any satisficing search process (§4.1) in real time is logically complete or globally optimal.[29]

The fourth, somewhat weaker variant holds that technological problem-solving, like scientific discovery, is logically amenable to algorithmic capture. It is true that artificial intelligence (AI) algorithms such as BACON, SOAR, or PASCAL have been developed that claim to be capable of rational problem-solving and discovery by comprehensive search. These programs not only depend on initial input of problem specification and solution criteria, as well as massively preprocessed data, but also search relatively compact and logically well-ordered 'spaces' for potential solutions

(b) *Technological practice is intentional, purposeful and goal directed.* This construal of rationality suggests a means–ends calculus absent in classical Darwinian evolution. But any careful reading of the empirical evidence shows that technological practice, as well as its goals, purposes and intentions, is time and path dependent, contingent and emergent, uniformly subject to what Andy Pickering so nicely describes as the 'mangle of practice'.[30] Neither is the course of technological evolution, whether of the relatively mundane sort discussed by Bucciarelli, or fundamentally innovative, as in application of aerofoil theory to turbines and axial compressors, in any serious sense foresighted.

Nevertheless, even though technological practice does not conform to such conventional notions of 'rationality', the search for solutions to technological problems is not just random groping in the dark either. Although there are seldom sufficient logical premises for the creative leaps that so often lead to clever solutions or radical discoveries, problem-solving still is undertaken with open eyes and alert intelligence, in the full light of (Bayesian reliable) well-winnowed knowledge, which, of course, is itself a function of past recursive practice. Satisficing problem-solving processes first search familiar, well-understood domains for candidate solutions (Carlson, ch.11; Perkins, ch.12). A great many features of technological practice, ranging from mundane learning by trial and error to elaborate programmes of 'RDD&D' (research, development, design and demonstration) all involve strenuous ratiocination in planning, performance and interpretation (Stankiewicz, ch.17; Fleck, ch.18; Fairtlough, ch.19). The various strategies and processes covered by 'vicarious selection' (§4.1), when applied to technology – that is, to the purposeful variation and selection of virtual artefacts and the evolution of technological memes in academic communities – are surely as 'rational' as any human action ever could be.

In general, this type of vicarious exploration is a consciously structured search process, or problem-solving activity, in which prior learning recursively directs attention away from negatively evaluated or avoided search spaces (for example, perpetual motion is normally precluded), and at the same time directs attention toward probabilistically positive – trophic – search spaces. But because what is being explored is always to some degree unknown, there is necessarily uncertainty both about what constitutes relevant knowledge and about its applicability and reliability in a new domain. Indeed, part of the rationality of recursive practice is to continually take advantage of new information as it emerges in the search process.

For these reasons, then, the process of technological change is not fully 'rational' in the everyday sense of the term. But, by the same token, it is certainly not 'irrational' in the sense of ignorant, uninformed, mystical or purportedly inspired by some supernatural revelation. Technological evolution includes irreducibly 'arational' insights – no doubt themselves products of blind variation and selective retention at some deeper psychological level – such as, 'If I treat a turbine or axial compressor as a set of aerofoils, . . .,' together with highly rational, exactly in the sense of deductive, steps, 'then I can analyze them according to the theory of . . .'. The 'If . . .' is highly creative – blind or random in Campbell's sense (§1.3; §4.1). The 'then . . .' is highly analytical, deductive and rational – recursive in the sense used here.

But here we are running into a barren controversy over terminology. My view is that Bayesian agents are rational in their overall procedures for producing reliable knowledge, and that technological change through recursive practice is clearly rational in this broader sense, even though some of its essential elements are necessarily 'arational' according to narrower criteria.

16.6   Recursive practice and the evolution of technological knowledge

Well-winnowed knowledge in science and engineering science provides a rich repertoire from which to craft solutions to novel analytical and design challenges. More importantly, such well-winnowed knowledge permits efficient vicarious exploration of alternative solution or design spaces.

Artefacts, of course, and by implication their designs, face direct environmental selection: aeroplanes crash, engines explode, computers freeze. But extinction is a hopelessly blunt instrument. Learning, in the sense of reliable and valid knowledge, comes not from the brute fact but – with limitations[31] – from the autopsy. The use of established knowledge to probe the apparently mystifying or anomalous experience is the same whether the goal is to create something new or to understand something old. Beth isn't trying to do something novel; she is simply trying to sort out an apparent 'failure' (anomaly) in a prototype system.

Similarly, just before the Second World War, when test pilots first encountered terrifying 'control reversals' as aircraft neared 'the sound barrier', the presumption was not that theories of lift and drag had been refuted. Rather, aerodynamicists and aeronautical engineers understood that they had encountered empirically a phenomenon that wind-tunnel experiments and idealized supersonic theories had recognized for some time. The consequence – control reversal – was unexpected; the phenomenon was not. As Walter Vincenti demonstrates,[32] creating reliable and valid knowledge of supersonic and transsonic regimes was an extraordinarily challenging, indeed daunting, task. But the recursive process invoked in response to this challenge was no different in principle from those invoked in the examples examined earlier.

Vincenti's Britannia Bridge example (ch.13) illustrates the 'directed' expansion of technological knowledge to meet specific design challenges. Although not quite analogous to the (highly disputed[33]) biological concept of a 'directed mutation', this is clearly far from a classical mutation in which 'there is no known correlation between a particular set of environmental conditions and the particular allele among many potentially possible ones to which a gene will mutate'.[34] It seems to me that the key element is William Fairbairn's 'Reasoning that what was needed was an increase in flexural rigidity at the top of the tubes . . . ,' – that is, his recursive use of his prior learning and experience, which resulted in 'real-world selection' pointing 'the way to an intentional, rational variant'. Significantly, however, what Fairbairn had discovered, localized buckling, was actually both novel and unforeseen. Moreover, Fairbairn's 'discovery' was in turn dependent on Stephenson's entirely novel, and wholly inexplicable – except as the result of blind variation at some level – conception of a tubular bridge.[35]

In general, *all* learning, however focused or directed, still depends, at some level on unforesighted trial and error. As Donald Campbell argued in his definitive statement of his version of evolutionary epistemology,[36] all increases in knowledge, or of fit between phenomena and noumena, depend upon a nested hierarchy of variation-selective retention processes that incorporate, *at some level*, random variation. Indeed, in a naturalistic evolutionary account which excludes supernatural revelation, novelty must arise blindly, since it would not be genuinely novel if it were the foresighted result of prior, accumulated wisdom or information. Thus, both classical genetic mutations and the 'unconscious' random associations of ideas in the creative thought of scientists' or engineers, are blind, or unforesighted, in Campbell's sense.[37]

But this random-variation element cannot be separated from the nested hierarchy of processes within which it is embedded. What matters is that these processes are embedded in increasingly remote and abstract processes of vicarious exploration and selection. Even our friends the amoebae 'know' (through genetically encoded learning) how to recognize and avoid the toxic gradients they are most likely to encounter, and, through their chemical receptors, are fully capable of non-lethal vicarious exploration. Computational aerodynamicists are in principle no different: they just command vastly greater stores of accumulated wisdom, and vastly more sophisticated, and, of course, more expensive, instruments.

To establish the relationship between this evolutionary epistemology, with its primary emphasis on error elimination, or selection, and Bayesian confirmation, the latter must be thought of as nested within the already nested hierarchy of random variation-selective retention processes characterized by Campbell. Any particular cycle of variation and selection is undertaken against an enclosing background of 'confirmed', but always fallible and hypothetical, beliefs. In practice, beliefs that have attained high levels of Bayesian confirmation are implicitly or explicitly assigned what Campbell terms a very low 'doubt-trust ratio'.

For this reason, even the best results of scientific research are (in principle, at least) corrigible, and our trust in them is always provisional: 'Away from the research frontier, these doubts [what in any novel experimental situation can in fact be 'taken for granted'] soon vanish: all well-established theory is taken to be valid, and all that can be deduced from it is treated as an empirical "fact." But even though there is no incentive to observe such a fact directly so as to test the theory any further, its epistemological status does not stay constant. On the one hand, every theory that contributes to any prediction is corroborated by the success of that prediction.'[38] Bayesian reliability is thus a trustworthy indicator of which knowledge to trust, and which to hold suspect, or to suspect when something goes awry. In well-developed scientific and technological systems, this well-corroborated, or strongly confirmed, knowledge provides a powerful guide to parsimonious and efficient vicarious exploration and selection.

It is precisely this feature of recursive practice that gives variation and selective retention in science and in technology its rational, reasoned, intentional character. These characteristics suggest that the evolution of technology is neither Lamarckian nor Darwinian, but rather comprises a unique, or at least a different, instantiation of more general variation-selective retention processes. This view is consistent with the general argument of this book, that biological evolution is an exemplar of selection processes, not the essential model in all its details.

Despite its substantial epistemological purchase, recursive practice still cannot be presumed to enhance the reliability of technological knowledge, or the fit between phenomena and noumena, except with respect to the physical (including the fabricated) world. It does not speak directly to social or cultural 'fitness', whatever those might be (Nelson, ch.6), nor does it necessarily entail social reproduction – adoption, diffusion, or 'reproductive' success.

Nevertheless, however contingent, local, or indeed 'relative', those social processes might be, the simple fact remains that recursive practice in technology, over several thousands of years, has led to progressively better 'fit of phenomena to noumena', to an ever larger population of scientific and technological 'memes' that command progressively greater Bayesian reliability. This progressively better 'fit' is all that can be asked of any evolutionary process, and more than many promise or can deliver. That, after all, is the way we learn:

> Only when the system is expected to work, that is, to achieve something in relation to the external world in which the real and species-preserving meaning of its whole existence does indeed consist, then the thing begins to groan and creak: when the shovels of the dredging-machine dig into the soil, the teeth of the band saw dig into the wood, or the assumptions of the theory dig into the materiel of empirical facts . . ., then develop the undesirable side-noises that come from the inevitable imperfection of every naturally developed system . . . But these noises are just what does indeed represent the coping of the system with the real external world. In this sense they are the door through which the thing-in-itself peeps into our world of phenomena, the door through which the road to further knowledge continues to lead.[39]

17

The concept of 'design space'

RIKARD STANKIEWICZ

17.1 A conceptual framework for technological evolution

An evolutionary research programme should focus on the description and analysis of specific 'evolutionary regimes', and pay close attention to their 'heredity' aspects. Existing evolutionary interpretations of technology (Mokyr, ch.5; Nelson, ch.6; Constant, ch.16) tend to show much more interest in the broad patterns of technological change than in the regimes involved. Discussion of the mechanisms typically[1] focuses on variation and selection processes and only to a limited extent on information accumulation and transmission – that is, cultural heredity.

We therefore need better understanding of (1) the processes of accumulation and transmission of technical knowledge; and (2) the relationship between the knowledge base of technology and the character and dynamic of technology development processes (R&D in the broadest sense). In this chapter I will sketch out a tentative conceptual framework where technological change is seen as the evolution of 'design spaces' and associated 'design languages'.

17.2 Cognitive dimensions of technology

The character of technological change is a function of many factors, of which the most important is the body of accumulated technical knowledge. Every engineer is embedded in a particular technological tradition characterizing his profession, the company he works with, or the team he is a part of. These 'technological communities' (Martin, ch.8; Vincenti, ch.13; Constant, ch.16; Fairtlough ch.19) are the main source of the engineer's 'object world'.[2] Mixed with this core of shared knowledge are the idiosyn-

234

cratic individual experiences accumulated by the engineers in the course of their careers.

Technology-as-knowledge has suffered from the excessive identification of technology with artefacts. It tends to be viewed as 'cook-booky', and hence not worth much reflection, or else as essentially derivative from science and therefore reducible to it (cf. §16.5). In recent decades, however, some historians of technology,[3] cognitive scientists[4] and students of technological innovation[5] have begun to study technology as a cognitive system in its own right. But we are still far from having a model of technology-as-knowledge, which could be used directly in the construction of a theory of evolutionary regimes of technology.

Clearly the closest in spirit to such a model has been the concept of a *technological paradigm*.[6] This may be defined as 'an "outlook", a set of procedures, a definition of relevant problems and of the specific knowledge related to their solution'.[7] In this model, each technological paradigm defines its own concept of 'progress', based on its specific technological and economic trade-offs along a *technological trajectory* (§4.1; §18.10).

Unfortunately, despite its great success at the rhetorical level, the concept of paradigm has serious weaknesses when used in the context of technological change. It fails to recognize the eclectic character of technology and overemphasizes technological discontinuities. In fact, real life technologies appear to be so 'multi-paradigmatic' that the concept virtually loses its meaning. Whilst the rate of technical change undoubtedly varies greatly, the overall trend is cumulative. Technology grows less through radical substitution of one paradigm for another than through absorption and integration of both new and old bodies of insight and know-how.

In other respects, however, the notion of a paradigm does come close to the concept of an evolutionary learning regime (Constant, ch.16). I believe that some of its drawbacks can be eliminated by introducing a similar but more flexible concept of a 'design space'.

17.3   Design spaces

Technical problem-solving is predominantly (though not exclusively) a constructive/synthetic activity and this informs the way in which technological knowledge is accumulated and structured. This is reflected in the concept of *design space*. As an elementary illustration, consider a simple 'toy', such as *Meccano* (*Erectors* in the US) or *Lego*.

Meccano consists of a set of simple elements which can be assembled to form a variety of structures: the set generates a certain universe of technical possibilities. A person playing with Meccano gradually acquires a certain type of knowledge about that universe: its vocabulary and grammar. This includes knowledge of the properties of the various elements as well as the relationships among them. Simultaneously, one develops skills required for the manipulation of the components. At a somewhat higher level one discovers *assemblies* of components which tend to recur in many structures. These assemblies become a part of a *repertoire* which is used repeatedly in many design situations. Finally one gradually discovers the various functions which can be performed by Meccano structures, that is, the *application domain* of the Meccano design space.

It is not hard to see close parallels between the Meccano design space and an engineering discipline. An engineer is a person who has mastered a particular design space. Normally that space is not his personal invention. He acquired it, both in its software and hardware aspects, from his predecessors. He rarely ventures far from its confines, but is likely to give it a certain personal twist. Some of these innovations will be picked up by the members of his community and become a part of a technological tradition.

Formally, a design space is the combinatorial space generated by a set of *operants* – for example, components, unit operations or routines. Operants in their turn are defined as the structure-function (or process-function) relationships which are used in the designing and assembling of artefacts. Any technical object (artefact, system) is either an operant in its own right or a configuration of operants. Decomposing an artefact into its constituent operants we will eventually arrive at a point where we cannot proceed any farther. The operants at this boundary are our 'primitives'.

In evolutionary terminology, the operants are the 'techno-memes' (§1.2) at the core of technological inheritance. They are heterogeneous information packages, in that they have both declarative and procedural dimensions. This information is encoded and transferred in a variety of ways and contains both coded (symbolic) and tacit elements. The degree of articulation and codification influences the ease with which operants can be transferred and symbolically manipulated. Equally importantly, operants can be embodied in artefacts and transferred in that form. The elements of a Meccano set are just such embodied operants. Indeed, Meccano is a designed design space whose 'primitives' are artefacts generated within the more inclusive design space of its original inventor.

Design spaces shape problem-solving processes. They do so mainly by generating the domain of possibilities within which the search for technical solutions is undertaken. The richer and more finely grained the design space, the more precisely it can be used to map the corresponding fitness landscape (Perkins, ch.12), and thus optimize the actors' ability to identify and articulate goals.

17.4   Dynamics of design spaces

Design spaces undergo change over time. There is a large difference between the stone-and-bone space of the cave men and the vast and highly differentiated space of, say, the modern electrical engineer. Analytically one can distinguish two dimensions in that change:

- the expansion of spaces though the addition of new operants, and
- their progressive structuring and articulation.

The Meccano space, for example, can be expanded through the addition of new components, or by modifications of the existing ones. This space is also structured/articulated internally, in the sense that its complexity is reduced to a functional minimum through the development of higher-level (composite) operants and problem-solving heuristics reflecting its fundamental properties. Although these two processes tend to be closely interwoven, their relative importance may differ considerably from one field of technology to another.

The evolution of a design space involves sifting from the pool of individual and collective experience those elements which make good operants, that is, which *jointly* facilitate adaptive problem-solving. In general, such operants might be expected to have certain characteristic properties, such as *stability*, *reliability* and *transparency*, and to be easily *decomposed* and *maintained*. What is required is a core set of *primitives* capable of generating a *fine grained*, *hierarchically structured*, *generic* design space with a multifunctional domain of applications. These basic operants should be capable of being described, represented and manipulated *symbolically* for ease of *communication* and sharing in a large community of practitioners – and so on. But there can be no *a priori* formula for what eventually emerges from such a process.

17.5   The evolutionary regimes of technology

Although the evolution of design spaces is essentially unpredictable, it seems to follow four distinct patterns, corresponding to four more or less distinct technological regimes: (1) the *craft regime*, (2) the *engineering regime*, (3) the *architectural regime*, and (4) the *research regime*. These may be characterized as follows.

The craft regime

Crafts are generally viewed as 'traditional' technologies. They have evolved slowly through the piecemeal accumulation of experience and thus fit easily into a simple evolutionary model. Their products tend to be relatively uncomplicated artefacts, poorly standardized and of high unit cost. Craft techniques and craft products tend to be conservative (Macfarlane & Harrison, ch.7), often to the point of ritualization, as with Japanese swords (Martin, ch.8). They are context-dependent and difficult to scale up.

In a craft regime, technology development is typically very gradual. There is relatively little distinction between design and production activities, and the small amount of deliberate off-line experimentation that does occur is largely of the 'cut-and-try' variety. Chance discovery and serendipity thus play a large role. Selection processes at the artefact level are relatively straightforward and take place primarily in the final use.

Technology accumulation and transmission are centred around the learning and passing on of the procedural knowledge and skills. That kind of knowledge is predominantly tacit and requires face to face interaction to be effectively transferred (Turnbull, ch.9). Consequently an important mechanism of selection is the choice of 'masters' to be imitated. This in turn is reflected in the institutionalization of technological activities through guilds, apprenticeship, etc.

The design space of a craft regime is poorly developed symbolically, and the basic operants are not well articulated (Turnbull, ch.9). Knowledge accumulation is strongly artefact bound and local, and the 'object worlds' of individual practitioners are highly idiosyncratic. This means that the design space is socially fragmented and incoherent. All these features limit the potentialities of craft technologies in terms of the complexity they can handle, the flexibility with which they can be deployed, and the cost/effectiveness which can be achieved.

The engineering regime

The emergence of engineering regimes signifies a radical expansion of technological capabilities. It usually coincides with the development of indus-trial modes of production, and is marked by a sharp increase in the sophistica-tion and performance of artefacts, their greater complexity and standardization. In particular specialized *design* activity is separated from production. This presupposes the existence of a certain repertory of relatively standard operants which are well understood and capable of being represented symbolically. Artefacts are now planned to a large extent at the symbolic level. The design space can, at least in part, be explored hypothetically using various graphic and computational devices. It is characteristic of engineering artefacts that their performance characteristic and critical constraints are relatively few and can be stated with some precision (cf. Vincenti, ch.13). The design activity can then be conceived as a process of quasi-optimization within given constraints.

Knowledge accumulation and transmission also change in character. Systematic symbolic representation of technology becomes an important tool of technology storage, documentation and transfer. Systematic professional training emerges and eventually becomes macadamized. Significant technical knowledge is no longer tied to specific artefacts. Instead it becomes increasingly compartmentalized, analytical and 'scientified'. As the design space becomes more highly articulated and codified, attempts are made to reduce it to a limited set of well understood and defined operants – for example, by using theoretical scientific concepts as basic building blocks of an 'engineering science' (Constant, ch.16).

The architectural regime

Architectural regimes emerge in the design of complex multifunctional systems. The architect–designer takes the existence of basic technical capabilities for granted and concerns himself with the exploitation of the design space created by the craftsman and engineer. Thus the scope of an architectural regime depends directly on the level of development of the two regimes discussed above.

Architectural development is very nearly a pure design activity. It differs, however, in important respects from classical engineering design. While the latter is concerned predominantly with the internal coherence and reliability of the artefact, taking functional requirements as exogenous givens, the architect–designer is predominantly concerned with the very definition of function and the user interface. In most cases, complex multifunctional systems cannot be optimized along engineering lines (Miller, ch.15). Architect–designers specialize in reading/anticipating user needs and designing complex systems which mobilize the existing technological resources to meet these needs.

Complex architectural artefacts are few in number and often unique. They usually serve large and highly heterogeneous groups of users. The difficulty of specifying their performance characteristics combines with the extreme complexity of the selection process. Conventional trial-and-error is generally out of the question, and is replaced by 'vicarious selection' amongst 'virtual' representations of the proposed architectures by users (§15.5; §20.5).

Architectural knowledge is rather like craft knowledge, in that it is harder to define, accumulate and communicate than engineering knowledge. There will generally be a large tacit dimension and a high degree of 'subjectivity'. In place of 'engineering science' we have 'design philosophies' and 'schools of thought'. But even though the design spaces of architectural regimes are typically vast and ill defined, there are strong pressures to articulate and standardize them. Since much (probably most) technological development today occurs under architectural regimes it is important to understand their nature.

The research regime

Design spaces can be expanded through the discovery of new or modified basic operants. In traditional technologies such discoveries were accidental by-products of regular technical activities such as construction and manufacturing (§13.3; §16.5). Since the middle of the nineteenth century a technology development regime has emerged in which new operants are discovered/developed through deliberate search activities.

Research regimes depend for their effectiveness on the existence of a science base. It is science – that is, 'natural science' as distinct from 'engineering science' – that conceptually defines and structures their search spaces as well as supplies methodologies and instrumentation required for effective search. Science is also the main source of the symbolic apparatus required for encoding and manipulating the information generated. Hence the processes of accumulation and transmission of knowledge (including education) associated with research regimes are often indistinguishable from those found in science. Indeed the design space of a research regime can appear virtually coextensive with a conceptual space generated by science. In reality, however, the operants supplied by science are not necessarily optimal from the standpoint of design.

Relatively few technologies wholly rely on research regimes. The main exceptions are biotechnologies, including pharmaceuticals, which are driven by 'discovery' rather than by 'design'. In most cases, however, research regimes are closely linked with the engineering regimes whose design spaces they function to expand.

17.6   Structuring design spaces

The technological regimes described above are not paradigms. The boundaries between them are blurred, since they are not mutually exclusive and often complement each other. Historically the craft regime is primary, whilst research regimes are generally late to appear, since they depend on the maturation of the underlying sciences.

What is more, they evolve into each other. In what follows I will briefly discuss certain important transitions between technology development regimes. The first two transitions involve structuring the design space, typically through the emergence and development of 'design languages'. The third transition shows how a regime changes when its design space is radically expanded.

17.7   Design languages

Even a relatively modest set of operants is capable of generating a vast design space. For example, it takes quite some time to master Meccano. In fact the process never ends: different uses and application domains of the space are constantly discovered, particular artefact lineages emerge and evolve, etc. Along with such specific application knowledge, the user of a design space is likely to acquire a generic knowledge of the space – its *grammar*.

The accumulation and structuring of technological knowledge – in evolutionary terms, its 'heredity' – is similar in important ways to the growth and development of a *language*. This process can be found in both primitive and advanced design spaces. But it first becomes clearly visible in the transition from a 'craft' to an 'engineering' and/or 'architectural' regime. This is well illustrated by a particular episode in the history of mechanics.

The traditional mode of training craftsmen by apprenticeship 'on the job' (Martin, ch.8; Turnbull, ch.9) prevailed in most technologies until the Renaissance. It then began to change, largely through certain important innovations in 'information technology'. The introduction of the printing press and the invention of new techniques for presenting three-dimensional objects made it possible to reproduce text and pictures on a scale that enormously increased the diffusion of technical knowledge.[8] Starting with the sixteenth century, large numbers of books cataloguing all manner of mechanical devices and production processes began to appear. Another popular method of representing technology was the setting up and public display of various collections of machines.

This 'encyclopaedic' approach to technical knowledge eventually gave way to a more analytical one. In the eighteenth century, the celebrated Swedish engineer Christopher Polhem (1661–1751)[9] invented a 'mechanical alphabet', which had a strong and direct impact on the training of Swedish engineers until the mid-nineteenth century, and is still indirectly influential. This alphabet consisted of a large collection of mechanical devices. However, these were not complete machines, but rather machine components – or, as we would call them, operants. Polhem believed that with just five 'vowels' – the lever, the wedge, the screw, the pulley and the winch – and more than 70 'consonants' he could construct every conceivable machine. In other words, he claimed to have identified and fully described the entire mechanical design space of his day.

Polhem's use of the language metaphor is significant. The same metaphor occurs frequently in much talk about technology – as in 'libraries of peptides', 'the language of architecture', 'computer languages', or Bucciarelli's notion of design as 'story telling'.[10] It suggests that the 'memetic systems' of technology are also *generative* cognitive systems, capable of applying 'transformation rules' to combine a finite set of 'building blocks' into an indefinite variety of messages/forms. Such systems are familiar in music, mathematics, chemistry, molecular genetics (§2.2; §3.1; §21.3), and so on. Indeed they are essential to all rational discourse about truly complex systems.

Technology is no exception – witness the early development of design languages in architecture ('the canons of classical architecture') and alchemy. The strategy of the 'engineering sciences' was to define the design space in terms of concepts taken from science – predominantly physics. But this effort to transform the art of technology into a science has never been fully successful. The science of the day rarely produces operants that are optimal, or even adequate, for technological application, so that the 'scientification' of a technology does not necessarily produce a satisfactory generative language.

There has always been a considerable tension between engineering science and practical engineering, which has always included a strong element of craft. But that is not the whole story. The creation of an effective design language requires a different type of conceptual development – one that is capable of reducing the vast complexity of technological design spaces to manageable forms. 'First principles' modelling of complex systems tends to be quite impractical. Effective design work requires 'high-level' languages which cannot be logically deduced from lower ones, and must be allowed to evolve in their own right.

17.8   Hierarchies of design languages

A well-developed design space is likely to be spanned by a *hierarchy* of design languages. This is reflected in the 'top-down' strategy typical of design processes.[11] The broad features of the artefact are first outlined and then used as constraints/specification for more detailed designs at a lower level, and so forth. For a relatively simple artefact produced by a single person, different languages are not required to describe the various stages in this process. But with increasing complexity the responsibilities of overall design and detail are separated, and require different skills and different languages. Thus, software engineers have had to develop a whole range of lower and higher design languages, from the simplest primitives of the binary digits up to a very general 'metalanguage' that can be used to describe – and hence structure – almost *any* design space.[12]

The articulation of architectural design space and the creation of architectural design languages differs significantly from what occurs in the engineering regime. The representation and storage of architectural knowledge typically runs into problems of complexity and individuality. The more complex the system the more idiosyncratic and context-dependent it is. How does one extract the re-usable elements from past experience with complex, highly individual, structures?

In architecture itself, the *styles* and *orders* were constituents of a high-level design language based on well-developed craft technologies, whose slow evolution contributed to its traditional stability and continuity. In the nineteenth century, however, the engineering infrastructure of architecture began to change very rapidly, as did a host of social and economic forces governing construction activities. This led to a more pronounced differentiation of the craft, engineering and architectural dimensions of the building design space. The virtual collapse of the traditional architectural languages – primarily classicism – paved the way for Modernism. The heroic efforts of several major architects failed to turn this into a new, generally accepted design language – for example, by integrating it, as 'functionalism', into the engineering regime. In spite of the interesting work of, for example, Christopher Alexander,[13] the present situation in architecture is a 'postmodern' Tower of Babel – as it is also, if the truth be known, in software engineering.

But the efforts will continue. Alexander has suggested that, if we had a properly articulated design language for buildings, the job of designing them could be left to the users/inhabitants themselves (cf. §15.6). This somewhat romantic idea has not worked very well in practice, but it is not as odd as it may at first appear. Many products that we buy are not self-contained unit artefacts with a narrowly defined function. Like our Meccano set, they are *embodied design spaces*. The best examples are computer systems of various kinds. A PC with a software package in it defines a design space. Development of a successful product becomes tantamount to the creation of a design language which is shared by the user and producer – an aspect of technological evolution which is worth pondering.

17.9   The expansion of design spaces

In almost all technologies, design spaces are expanding and deepening as well as becoming better structured and articulated. They acquire new operants, piecemeal by accident or serendipitous discovery, by deliberate search, or by fusion with other pre-existing design spaces.

The design space of a traditional craft technology grew chiefly by accident or empirical discovery – for example, through gradual mastery of naturally occurring materials such as wood, organic fibres, stone, clay and metals. The learning process was exceedingly slow. The achievements of such technologies were occasionally extremely impressive – witness Japanese swords (Martin, ch.8) and medieval cathedrals (Turnbull, ch.9) – but they tended to be isolated. Because the operants generated in that manner were usually quite complex and heterogeneous they did not form an effective design process.

The deliberate search for new operants implies a research regime, which presupposes a certain minimum knowledge base – usually imported from science. More specifically, it requires vicarious testing techniques, a bounded and structured search space and effective search heuristics. Nevertheless, even under a research regime, technical solutions are often 'discovered' rather than 'designed' – typically in the form of a 'final' artefact or formula solving a specific problem. The field of pharmaceuticals has been, and largely continues to be, a good example of this process. Indeed, biotechnologies in general have tended to be driven by discovery rather than by design, although this is now changing, as we shall see.

At the other pole of the spectrum of technological research regimes are those which aim at the expansion of the design space to enable a shift from discovery to design. This involves several different processes.

Reverse engineering of nature: the science-based analysis of a natural system which we wish to control, as in much biomedical research.

Analytical deconstruction: the science-based analysis of an existing technology to elucidate and explain scientifically the principles on which it depends for its effectiveness (Constant, ch.16). This mode of 'scientification' amounts to the creation of a design space ex-post. As we all know, technology often leads science. Trying to discover why a given device, such as a steam engine, actually worked led to the knowledge of conceptual 'building stones' which enabled the design of quite new devices, such as steam turbines.

Fusion of two hitherto separate design spaces[14] is a more radical event. It often occurs when the evolution of each reaches the point where their functional commonalities are becoming visible (Carlson, ch.11). This fact may be heralded by a spectacular breakthrough in some highly specific area. However, the long-term impact can be much wider and more dramatic, as the resulting space explodes combinatorially. 'Mechatronics' and 'optoelectronics' are contemporary examples.

An important feature in the development of modern technology is the fusion of an existing technology with a particular field of science. The instrumental infrastructure of the latter may then play an especially prominent role. Thus, chemistry brought its sophisticated instrumental paradigms into the fusion with mechanical engineering that produced chemical engineering. Today's biotechnology, to take another example, is a product of two major fusions: (1) chemical engineering with microbiology; and (2) the 'old style' biotechnology with biochemistry and molecular biology.

Technological development in these newly emerging design spaces is typically very research intensive. But 'discovery' may at first be so dominant over 'design' that it takes a long time to establish an effective engineering regime. Strangely enough, research regimes are in some ways closer to craft modes of technological development than their claims to 'scientific' status would suggest. Indeed, the 'new biotechnology', with its emphasis on 'genetic engineering' and 'protein engineering', is a conscious effort to move into a regime where new medicines or nutrients can be rationally designed (Fairtlough, ch.19). I forebear to comment on how far it can or ought to succeed in its attempts to 'reverse engineer' biosystems and design complex molecules, *de novo*, to function as desired.

17.10   The convergence of design spaces

Two general evolutionary trends are of particular importance. On the one hand, design spaces tend to become *more finely grained*. Thus, traditional technologies employ highly complex and heterogeneous operants, arising as specific adaptations to particular circumstances. Gradually, however, there occurs a shift towards simpler operants corresponding to the lower levels of organization of matter. During the last 150 years there has been a steady development in that direction. The design spaces of virtually all high technologies today have molecular or submolecular dimensions – and the downward trend continues.

On the other hand, this trend facilitates the appearance of *commonalities among previously unrelated design spaces*. This explains the increasing frequency and pervasiveness of technological fusions. One does have the impression that we are heading for a 'universal design space' spanned by a variety of design languages corresponding to different application domains. This is not as utopian as it may seem. The shape of things to come can be glimpsed in fields such as the 'new' biotechnology and nanotechnology, as well as in the ubiquity of information technology and in the widespread use of genetic algorithms envisaged (Miller, ch.15).

Nanotechnology illustrates very clearly the remarkable change that is occurring in the process of technological development. The atom-by-atom design space being opened up by Drexler and his colleagues[15] is obviously enormous, and like software design would be unmanageable without a hierarchy of structural languages – including, ironically, Polhem's codification of the traditional mechanical grammar of beams, gears, etc. The stability and reliability of these tiny operants can well be questioned. Nevertheless, this development indicates a move away from the situation where design spaces have been generated mainly as by-products of attempts to solve *particular* problems in *particular* circumstances towards one where we make deliberate efforts to expand and structure *new* design spaces, in the expectation that multiple specific applications will follow.

17.11   Technological change as conceptual evolution

This chapter began with the proposition that evolutionary accounts of technological change required greater attention to the mechanisms of 'technological heredity', that is, the processes through which technological knowledge/capabilities are accumulated, organized, transmitted and used in adaptive problem-solving and learning. What I have tried to show is that these mechanisms are seen most clearly in the evolution of 'design spaces' and their associated 'design languages'.

In other words, technological change, like all other forms of cultural change, involves *conceptual* evolution (Mokyr, ch.5; Constant, ch.16). Here we enter into largely uncharted territory, which is only now being explored by a few brave cognitive scientists and evolutionary epistemologists.[16] But this emphasis on the conceptual features of technology should not be taken to mean that it is 'just knowledge'. The opposition of 'artefactual' and 'cognitive' aspects is misleading (Fleck, ch.18). As we have seen, Meccano is a prime example of a complex design space 'embodied' in apparently simple hardware. Artefacts play a role in technology transmission which the biological phenotypes do not (§2.1; §3.2; §5.4). Yet the 'fossil record' of artefacts gives an incomplete picture because so much of the variation and selection that drives technological evolution leaves no material traces.

17.12   Organizational and institutional implications

I have also said little about the *social* dimension of technological evolution. Cultural evolution (Nelson, ch.6) cannot be properly understood except in social terms. Cultural selection is largely driven by the coevolution of entities in different cultural domains. The accumulation and transmission of knowledge occurs in, and through the formation of, technical communities, and is strongly affected by their structure and dynamics.

Technological evolution is closely correlated with social structures (Fleck, ch.18; Fairtlough, ch.19). The increasing complexity and sophistication of technological learning regimes demands corresponding development in the social infrastructures supporting them. Each of the technological regimes discussed in this chapter requires a different set of institutions. The emergence of a new regime is usually accompanied by the appearance of new roles. This process of differentiation and specialization calls for compensating institutional innovations to maintain a minimum of integration of the system. Cultural and institutional inertia (Macfarlane & Harrison, ch.7) will retard technological change; technological change undermines established institutional patterns.

The whole notion of a socially 'disembodied' design space is too abstract. In real life, design languages create technological communities. Social segmentation encourages emergence of new languages and 'dialects'. The design space accessible to an individual is determined by his/her location in a social structure. The same is true at corporate level. The design space accessible to a company will be partly a function of its accumulated technological resources and partly a function of its location in a particular technological system.[17] The general character of the design space (its degree of codification, its unification and generic character) will in turn influence the mechanisms of appropriation of technological benefits, the R&D strategies and even the structures of entire industries.

A well-developed model of the 'heredity mechanisms' in technological evolution would have to incorporate these social dimensions. There is clearly a lot more to learn from the recent literature on the social construction of technology, organizational learning, the economics of knowledge, technological systems and systems of innovation, etc.

We have noted, moreover, that the emergence of a new technological regime in the sixteenth century could be attributed to a major innovation in the handling of information. Information technologies are themselves subject to evolutionary learning and are thus caught up in a boot-strapping process of accelerating evolutionary change. The question is: are we now witnessing the emergence of a fifth major technological regime (Miller, ch.15) which, for want of a better term, might be called *computational*? Is not such a development implied by the radical expansion of the design space towards molecular and submolecular level, in combination with the very powerful design languages made possible by modern computing and communication technologies? We seem to be living in interesting times!

18

Artefact↔activity: the coevolution of artefacts, knowledge and organization in technological innovation

JAMES FLECK

18.1 Introduction

There are several crucial elements in the process of technological evolution, namely, artefacts, knowledge and organization. All of these are clearly necessary but none is sufficient on its own. This chapter explores the proposition that *the* crucial unit for technological evolution is the *artefact–activity couple*. This couple, rather than either the artefact by itself or the knowledge which 'produces' it, is the essential building block of technology development and the basic element of technology practice.

In essence, this couple is the mutually supporting combination of artefactual elements with the immediate human activities in which those elements are used and produced. The focus is therefore on the dynamic characteristics of technology in use. Moreover, certain characteristics of this unit, such as its inherent stability, are analogous to those of a neo-Darwinian gene (§2.2; §3.1). It thereby raises the prospects of setting up an evolutionary theory of technological innovation which is more like its biological counterpart than is usually suggested. Even if, in the end, this argument cannot be sustained, it sharpens our understanding of precisely how technology evolves – and indeed whether it is useful to consider it as 'evolving' in the sense being explored in this book.

18.2 Technology development

Technological development has been examined in a range of settings and from a wide variety of disciplinary perspectives. 'Innovation/Technology Studies' is an eclectic, inter-disciplinary and multi-disciplinary field. At the margins it overlaps with more conventional single disciplines such as the history of technology or the economics of technological change. There are also a number of well-developed but independent and sometimes contradictory literatures around topics of direct relevance to technological change – for example, the labour process, business strategy or creativity. Substantial achievements have been made in the understanding of technological change, but insights are distributed across different specialist communities and tend to be couched in localized jargons and terminologies. This disciplinary fragmentation around the topic has not helped the emergence of a coherent or comprehensive conceptual framework for understanding innovation and technology development.

Consequently, there are many approaches to understanding the development of technology. Moreover, given the huge variety of technologies it is unlikely that we will find any one mechanism or pattern of development capable of accounting for every case. Yet the opposite contention, that every technology is unique, is just as debatable. Empirically, we find a remarkable commonality across disparate technologies. Although the specific *forms* taken by different technologies vary considerably (i.e. the artefacts are quite distinct) there are common patterns to their development. We find broadly similar patterns of diffusion into use, and strong similarities in the way that improvements are realized in successive 'generations' of the artefacts (often identified as various forms of 'learning by doing'[1]).

Many of these similarities suggest that technology development is 'evolutionary' in a general sense. 'Selection processes', in particular, seem easy to observe. Thus, evolutionary theory ought to be a useful guide for the identification of a core set of underlying common processes, or mechanisms. Yet, relatively few studies take the evolutionary approach seriously. The evolutionary metaphor is widely noted (Nelson, ch.6), but, with rare exceptions,[2] is not developed as a systematic theory that explicitly takes advantage of what is known about biological evolution.

18.3   The units for technological evolution

Direct comparisons between biological and technological selection processes can be misleading. It is easy to get lost in a maze of verbal parallels and to naturalize social problems which in fact arise from deliberate human agency: the illegitimate equation of the 'fittest' with the 'right' or the 'best'. But as we have seen elsewhere in this book (Jablonka, ch.3; Mokyr, ch.5; Constant, ch.16), the nature of the evolving units is central to any understanding of a Darwinian process. The question is: 'what are the entities that replicate, vary, are selected and differentially transmitted?'[3] Are they artefacts,[4] are they ideas of some sort ('technological memes')[5] or are they forms of organization?[6] This question of unit is related to another important issue: the scale at which the evolutionary framework becomes most revealing. This scale ranges from the individual artefact, through ideas, sets of ideas such as the knowledge base of the individual firm,[7] to even wider industry segments.

To be more specific, we follow the formulation of David Hull. 'In order for selection processes to operate, the entities must be organized into populations integrated through time by descent'.[8] Moreover, entities are needed with two distinct functions – though these functions may coincide in the one entity (§4.1). The function of a *replicator* is to 'pass on its structure largely intact in successive replications'; an *interactor* is an entity that 'interacts as a cohesive whole with its environment in such a way that this interaction causes replication to be differentiated'. This then leads to: 'selection – a process in which the differential extinction and proliferation of interactors *cause* the differential perpetuation of the relevant replicators'. This process in turn can give rise to a *lineage* – an entity that 'persists indefinitely through time either in the same or altered state as a result of replication'.

In the biological case (Jablonka & Ziman, ch.2; Jablonka, ch.3), DNA provides a powerful base for replication, common to all living forms. There appear to be a myriad of forms of interactors. Furthermore, the genetic code is remarkably stable and provides the ultimate material base which progressively differentiates under the pressures of selection. It also strongly integrates lineages over time. Paradoxically, the biological mode of evolution – at least in its modern neo-Darwinian version – requires the basically *unchanging* substrate provided by these very stable replicators.[9]

As we have seen (Ziman, ch.1; Nelson, ch.6), this is a fundamental problem for all evolutionary models of cultural change, which is so manifestly fluid and lacking in long-term stability. To develop a strong evolutionary approach to technological change, we must, therefore, carefully examine the different candidates for the role of technological replicators. Can we find any entities equivalent functionally to DNA and the genetic code? If not, that would not rule out an evolutionary theory of technology development, but it would suggest that this would have to be quite different from the biological version. After outlining his own approach, Basalla commented:

> . . . it is important to remember that because novelty is a fact of material culture, the evolutionary theory developed here remains intact despite our inability to account fully for the emergence of novel artifacts. Modern theorists of technological evolution indeed face the same dilemma that confronted the Darwinists in 1859. The latter could

point to reproductive variability as a fact of nature but were unable
to explain precisely how and why variants arose because they did not
possess a knowledge of modern genetics. We who postulate theories
of technological evolution likewise have our Darwins but not our
Mendels.[10]

It is this putative Mendelian component that I would like to explore in this
chapter.

18.4   The role of artefacts

It is abundantly clear from many examples of technology development
that artefacts by themselves do not evolve in the same way that organisms
evolve. Artefacts by themselves are static and lack the dynamic vitality necessary
for evolutionary development. Even computer viruses, modelled explicitly on
biological analogues, are ultimately static: they survive and persist only in larger
contexts of human activity with information technology. Take away this human
element and they would perish.

The genetic code for an organism is written *within* the organism. In contrast,
the equivalent for an artefact is written *outside* the artefact proper, but within
assemblages of people involved in producing the artefact, that is, within
organizations (Constant, ch.16; Stankiewicz, ch.17; Fairtlough, ch.19. The artefact itself may be an important source of information as a model or exemplar,
but it is far from sufficient. Many different forms of knowledge are required in
technology, including design techniques and production methods.[11] And many
of these forms of knowledge are embodied tacitly as skills within the human
agents involved in the use and/or development of the technology.[12] The complete
units required for the effective replication and use of technology thus extend
beyond the artefact to embrace a whole complex of factors (see fig. 18.1).

This 'technology complex' includes various elements ranging from the basic
purpose for which the technology is used, to the broad set of values which
underpin and sustain its existence in its cultural context. *All* of these factors
have a bearing on the creation, propagation and use of technology. Indeed, one
of the liveliest debates within technology development is about where to draw
the line between the technical (or that which is seen as specifically technological)
and everything else. Arnold Pacey, for instance, argues for a broad notion of
'technology practice':

> We would understand much of this more clearly, I suggest, if the
> concept of practice were to be used in all branches of technology as it
> has traditionally been used in medicine. We might then be better able

PURPOSE
MATERIALS
ENERGY SOURCE
ARTEFACTS / HARDWARE
LAYOUT
PROCEDURES (programs, software)
KNOWLEDGE / SKILLS / QUALIFIED PEOPLE
WORK ORGANISATION
MANAGEMENT TECHNIQUES
ORGANISATIONAL STRUCTURE
COST / CAPITAL
INDUSTRY STRUCTURE (suppliers, users, promoters)
LOCATION
SOCIAL RELATIONS
CULTURE

Figure 18.1 The technology complex.

to see which aspects of technology are tied up with cultural values, and which aspects are, in some respects, value-free. We would be better able to appreciate technology as a human activity and as a part of life. We might then see it not only as comprising machines, techniques and crisply precise knowledge, but also as involving characteristic patterns of organization and imprecise values.[13]

But even his 'restricted' category (fig. 18.2) includes knowledge, skills and resources extending far beyond the artefact proper.

A number of other conceptual frameworks, such as *strategic frameworks*,[14] *technology audits*,[15] and *systems analysis* approaches,[16] have emerged from work on the practical management and actual use of technology. In all of these frameworks, the artefact is only one part of a far wider structure. It is in this wider structure that one must look for the primary unit of analysis, the crucial building block of technology development. Here we will find entities that pass on their structure largely intact, that is, replicate. It is also in the wider structure, but probably at more inclusive and varying levels, that we will find the cohesive wholes which interact with their environment to cause that replication to be differential (see fig. 18.3).

Nevertheless the artefact remains the focal point for the dynamic expression of the technology and provides a ready and visible index of development. Furthermore it is the crucial differentiating feature which distinguishes technology from other forms of culture. Basalla, for example, focuses uncompromisingly on artefacts:

> The artifact – not scientific knowledge, nor the technical community, nor social and economic factors – is central to technology and

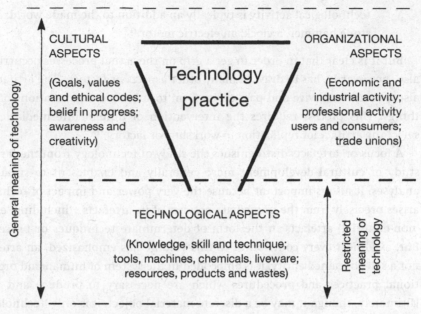

Figure 18.2 Pacey's model of technology.

```
                       PURPOSE              )
                      MATERIALS             ) ) ) ) ) ) ) ) )
                   ENERGY SOURCE            ) ) ) ) ) ) ) ) )
(              ARTEFACTS / HARDWARE         ) ) ) ) ) ) ) ) )
(                     LAYOUT                ) ) ) ) ) ) ) ) )
(           PROCEDURES (programs, software) ) ) ) ) ) ) ) ) )
(        KNOWLEDGE / SKILLS / QUALIFIED PEOPLE ) ) ) ) ) ) ) ) )
(                WORK ORGANISATION          ) ) ) ) ) ) ) )
            MANAGEMENT TECHNIQUES           ) ) ) ) ) ) )
           ORGANISATIONAL STRUCTURE         ) ) ) ) ) )
                  COST / CAPITAL            ) ) ) ) )
   INDUSTRY STRUCTURE (suppliers, users, promoters) ) ) ) )
                     LOCATION               ) ) )
                SOCIAL RELATIONS) )              ) )
                     CULTURE)                   )
Replicator                                   Interactors
```

Figure 18.3 Replicators and interactors in the technology complex.

technological change. Although science and technology both involve cognitive processes, their end results are not the same. The final product of innovative scientific activity is most likely a written statement, the scientific paper, announcing an experimental finding or a new theoretical position. By contrast, the final product of innovative technological activity is typically an addition to the made world: a stone hammer, a clock, an electric motor.'[17]

But it is clear that in order to get a grip on the actual processes occurring, he also, necessarily, has to draw in many other aspects or factors. 'The artifact may also be said to survive and pass on its form to subsequent generations of made things. This process requires the intervention of human intermediaries who select the artifact for replication in workshop or factory.'[18]

A focus on artefacts distinguishes the study of technology from the far wider study of cultural development more generally and enables us to situate our analyses. It also is important because the very power and impact of technology arises precisely from the immediate potency of its artefacts – including 'empty', 'non-material' artefacts in the form of determinate techniques or procedures. But, as almost every contributor to this volume has emphasized, an artefact is not a self-contained entity. It is embedded in the system of human and organizational practices and procedures which are necessary to produce and use it. Ultimately, though it may be a discrete material object, it is only 'technological' by virtue of the web of meanings and cultural practices in which it is enmeshed.

Moreover, it is the richness of this web which provides creative material for further innovation and for the development of related products. This has led some authors to talk metaphorically of 'crystallization', 'congealing' and 'embodiment', to describe how new artefacts emerge from their broader, more fluid contexts. Industrial software systems for example, are often explicitly modelled on existing organizational procedures, translated into sets of rules, and then coded as computer programs, before being finally realized and sold as physical products. Consequently, an evolutionary theory casting artefacts into the role of 'genes' will not work. It is like focusing on the bones and ignoring the soft tissues of a living creature.

18.5　The role of knowledge

Instead of concentrating on artefacts, other studies (Mokyr, ch.5; Constant, ch.16; Stankiewicz, ch.17) focus on entities in the domain of concepts and ideas – for instance the *knowledge base* of the firm[19] or its *core competencies*.[20] This is often argued explicitly – for example:

> In interpreting technical change as related sequences of innovation it is useful to shift the focus of attention from technology as an artefact to technology as knowledge. More specifically, technologies can be regarded as knowledge which can be used to generate a number of designs 'in the pursuit of conceivable ends' irrespective of whether the

design principles have other origins in prior scientific discovery or in prior technological practice.[21]

But this is the implicit focus in laying stress on the continuity between different forms of culture such as poetry and technology. Indeed, from an evolutionary epistemology perspective, both technological artefacts and living creatures are viewed simply as particular forms of knowledge.[22]

As elsewhere in this book (e.g. §5.1; §10.6; §11.4; §12.3; §13.4; §14.6; §16.4; §17.3) we follow Dawkins[23] in designating as a 'meme' any conceptual entity that behaves like a 'gene' in conceptual evolution. However, memes do not capture the full sense of technology. Here the distinction between ideas and practices/activity is crucial. If technology is about anything, it is about effective action in the real world, and not just about ideas as ideas. Technological practice is part and parcel of the material and social interaction out of which come artefacts, while memes are purely cognitive. 'Memes' or ideas are therefore only one part of the technology complex – and in practice a much smaller part than is usually assumed in the scholarly imagination. Consequently, a focus on the evolution of ideas *per se* simply misses the mark as far as technology is concerned.

Technology is not simply a subset of the realm of ideas. Its real-world interactions (Vincenti, ch.13) and their organizational necessities (Nelson, ch.6; Fairtlough, ch.19) generate quite distinctive characteristics. These strongly constrain the plasticity inherent in ideas *per se*, while at the same time introducing further features that cannot be captured conceptually – except after the event, as an explanation of what occurred. The ability to manipulate technology effectively is often described as a 'black art' – that is, a skill that is opaque to formal analysis.

A focus purely on knowledge also makes the evolutionary problem very tough. It is very difficult to put boundaries around an idea. It is very difficult to identify changes or indeed to ascertain whether ideas are the same or not, especially between different groups using different expressions and languages, or over time. Moreover, because ideas can change so easily and rapidly, it is often unclear, in practice, what processes are occurring. The artefactual elements render this problem more tractable, especially in a material sense, as well as ensuring that the issues addressed are authentically technological.

Moreover, the most telling objection to ideas as evolutionary units is that they, like artefacts, are inherently static. Although it might seem that they are more dynamic than artefacts, they still depend on human agency for expression and vitality, and so are still subject to the same kinds of limitations. Consequently, a focus on the knowledge underlying the development of technology is also necessary but not sufficient.

18.6   The role of organization

Organizations provide both the context for the use of technology and the matrix for the generation and preservation of the knowledge that underpins it. In exploring the logic of focusing on the knowledge underlying technology, Georghiou *et al.* find themselves emphasizing the role of the firm, that is, the organization, rather than technology *per se*, or even technology as knowledge. They identify as a crucial unit the 'design configuration': 'the basic design-parameters which underpin products and processes and constitute the knowledge base of the firm'.[24] Moreover, they claim: 'each design configuration will be identifiable in terms of a set of product and process characteristics within a technological regime which, in principle, are measurable in a precise way'.[25]

The focus on the firm and other organizational forms is common amongst evolutionary economists in general (Nelson, ch.6). In their seminal work, Nelson and Winter emphasize the crucial diversity of firms' characteristics and experience, and liken 'organizational routines' to genes, in that 'firms behave over time with continuity in routines'.[26] They also use the term 'technological regime' (cf. Stankiewicz, ch.17) much as others use the term 'meta-production function', as 'a frontier of achievable capabilities, defined by the relevant economic dimensions, limited by physical, biological and other constraints, given a broadly defined way of doing things'.[27]

This broad view of technology as necessarily having an organizational aspect is also advocated by other authors such as MacDonald, who suggests that 'technology may be regarded as simply the way things are done'.[28] Such a view is a strong challenge to a narrow and exclusive focus on artefacts. Indeed, there may even be utility in seeing technology as a by-product of organizational forms. By concentrating on the artefact itself, technologists, philosophers and others tend to see the organizational level as accidental rather than crucial – if, that is, they notice it at all!

Study of the organizational correlates of technology has a very respectable pedigree. It includes Adam Smith's analysis of the division of labour and its relationship to technology, Marx's analysis of the development of machinery for manufacture, and Babbage's analysis of the division of intellectual work. More recently, Woodward showed a strong relation between organization structure and the workflow technology employed.[29] Historical analyses also suggest that modern organizational forms have evolved along with the development of modern technologies in four long waves of development since the industrial revolution.[30] Indeed, some argue (cf. Miller, ch.15) that we are now embarked on a fifth such wave.[31] Currently there appear to be moves towards flatter organizational structures, associated with the further development of information and communication technologies such as document image processing.

This is very much chicken-and-egg territory. The very intimate interaction between a technology and its associated organizational forms should warn us against focusing too exclusively on the artefacts alone. But it is also clear that under certain conditions – though certainly not always (Macfarlane & Harrison, ch.7) – technologies can migrate between different organizational forms such as firms, industries and even national cultures – for example, the magnetic compass and China.[32] Indeed the 'transfer' of technology is an important element of modern business activity, although it is increasingly appreciated that successful transfer requires the associated transfer of skills, work practices and other supporting resources. This strongly suggests that such organizations as firms, industrial sectors or countries are too large as the key organizational entities in the evolution of technology. Organization is certainly important in technology evolution, but the crucial element is a more circumscribed sub-firm form, focused closely around the artefactual aspect of activity.

An exclusive focus on organization suffers the same drawbacks as a focus on ideas. It does not bring out the distinctive feature of technology, namely the real-world impact of artefacts. Thus, while a consideration of organization is necessary, again, by itself, it is far from sufficient for a convincing evolutionary analysis.

18.7   The artefact–activity couple

From what I have said above, not the artefact, nor the idea, nor the organization as a whole, is adequate by itself as the unit of analysis in a study of the detailed development (and hence evolution) of technology. 'Technology-in-use' is an amalgam of artefact, knowledge and organization.[33] Its basic unit is the *artefact–activity* couple. In effect, this is a dynamic ensemble of the artefact with the *immediate set* of human activities that sustain the use and development of the relevant technology. This ensemble, I argue, satisfies many of the requirements for a functional analogue of the biological gene.

The artefact–activity couple is the basic, minimal, necessary and sufficient element which can be derived from all the above accounts. In terms of the technology complex noted above, it comprises those terms most immediate to the artefact (see fig. 18.3). Its extent will vary, of course, with the complexity of the technology. For a very simple technology, such as a traditional carpenter's hand tools, it will include along with the actual tools the knowledge and operational skills required to use them, together with necessary maintenance activities such as sharpening saw teeth, or replacing worn out parts. In such simple cases the organizational aspect appears to be minimal, although in fact it is there, pre-assumed in the division of labour which isolates that particular trade as one discrete task among the many possible partitions of activity possible (Martin, ch.8; Turnbull, ch.9).

For slightly more complex technological entities (such as a piano, a motor car or an aeroplane) the operation, maintenance and production requirements are more extensive, and require different skills parcelled out among a group of people. In these cases the organizational aspects become more evident and marked. Even when individuals can operate the technology independently, as with playing a piano or driving a car, an important element of organization over time rather than space emerges in their need to learn the requisite skills, typically through some form of apprenticeship. But technological evolution involves production and maintenance, as well as operation. Without any one of these, there can be no development. They must all be included in the analysis.

And as technologies become yet more complex, their organizational aspects become more apparent. The relevant activities are performed by *teams* of people, each with a specialized part to play in an articulated organizational structure. A factory production line is a typical example. Here the artefactual components are precisely designed with explicit consideration of the interfaces with specialized segments of human skill. Interestingly, nearly identical artefactual components, such as machine tools and transport devices, may be reconfigured into slightly different layouts, with different assumptions about the human division of labour and responsibility, to yield remarkably divergent outcomes. This is very clearly illustrated by the shift from conventional assembly lines to the new, U-shaped, 'lean production' models promoted by the Japanese auto manufacturers and now widespread throughout the West.[34] Indeed, precisely because of the similarity of the artefactual components, some analysts use the term 'organizational innovation' to cover such cases. But on careful inspection, especially of the important physical *layout* characteristics, crucial differences are evident in the artefactual make-up of the technology, closely correlated with organizational distinctiveness. In such complicated cases, the artefact–activity couple is itself necessarily more intricate, and includes a larger chunk of the technology complex.

Technological complexity is fostered by the way that innovations can result from nested combinations of simpler technologies. This, indeed, is a much more important mode of variation and evolution in technology development than it is in biology, although it has ecological parallels. Nevertheless, in a very important sense, such complexes must have a definite functional integrity, with some overall clear purpose – that is, if they are to evolve as technologies, rather than as elaborate sculptural ventures.

18.8   Stable replication

The functional integrity required to address some broadly specified purpose, is the hallmark of a technology, as opposed to an *ad hoc* contrivance that just happens to be useful at any particular time. More crucially, the test of functional integrity lends remarkable long-term stability to an artefact–activity couple, enabling it to fulfil the part of an evolutionary 'replicator'. The artefact–activity couple comprises a real-world testing loop (Vincenti, ch.13; Constant, ch.16) which feeds back strongly into intentionality (e.g. redesign) to produce

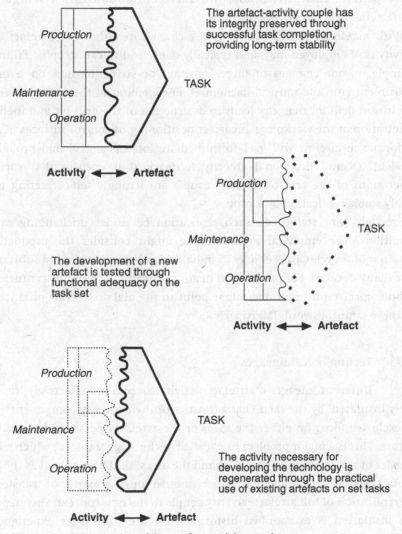

Figure 18.4 Reproduction of the artefact–activity couple.

successful outcomes. It should be noted that this functional-integrity feedback loop is not intrinsically a selectionist evolutionary mechanism – but it can play such a part when conditions change.

More importantly, the artefact-activity combination has a self-contained replicating capability. In effect, the activity element (which includes production and maintenance as well as operation) constitutes a template for the subsequent production of new artefacts, while the artefactual element provides a material template which 'filters' or 'recognizes' activities and ensures skills are kept tuned through practice (see fig. 18.4). As well as enabling replication over time then, the elements of the couple mutually condition each other to reinforce a stable pattern. This pattern holds together both knowledge (or skill) and the artefact, in some form of organization.

With a refined technology, a minor change in either the artefact or the activity/skill employed may, and typically does, lead to catastrophic failure. For example, minor changes of finger position or string tension on a musical instrument ruin the musical harmonies. Engagement of the wrong speed on a percussion drill for masonry leads to destruction of the drill bit and ineffective penetration of the workpiece. Incorrect positioning of control surfaces in a high performance aircraft will lead to immediate loss of control and probably an accident. Overall functional-integrity feedback thus ensures that successive generations of the artefact–activity couple are strongly self-correcting toward highly stable replication over time.

Precisely how stable can such replication be under unchanging external conditions? For empirical evidence, one might consider the production of certain motor cycles and vehicles in India, which have not changed substantially over many decades. For an extreme historical example of stable, centuries-long technological replication one might point to the elaborate craft production of Japanese samurai swords (Martin, ch.8).

18.9   Technological lineages

Different lineages of artefact–activity couples are effectively 'reproductively insulated' by the tacit characteristics of their activity components. Again an exclusive focus on either the ideas or the artefacts *per se* tends to miss this feature. This is a major problem in so-called 'technology transfer', which requires transfer of the production processes and the users' skill base as well as the actual artefact. Effective technology transfer is not so much a matter of transfer as of the replication of full artefact–activity couple in the new context. Such reproductive insulation is exemplified historically by the difficulties experienced in trying to 'transfer' machine tools from England to the United States in the nineteenth century.[35] Until personnel knowledgeable about the 'activity' aspects were also obtained, the artefacts remained unused and unusable.

Similar issues arise in the consideration of technology diffusion and development into third world countries. It is noticeable that many analyses emphasize the need for skills and 'quality of labour' – that is, the activity side of the couple.[36] It is clearly recognized that it is not sufficient merely to transfer the artefact. Such considerations are pre-eminently practical, and tend to be glossed over by a scholarly preoccupation with ideas.

Although an artefact-activity couple may be a 'hybrid' combination of several distinct technological lines of descent, the process of replication will strongly constrain its lineage. And on sufficiently detailed examination we can usually discern definable 'generations' (§3.5; §5.3). These generations are often obscured by a focus on associated generic ideas which change less, yet have more fluid boundaries. Essentially a technological generation corresponds to a 'made batch' of the artefact. For large, complex technologies such as power stations or space shuttles, each generation may be a singular instance of the artefact. For less complex technologies well into their life-cycles, each made batch may number thousands or, for very simple 'mass-produced' items, even millions. Moreover, as every car owner quickly learns, the precise generation is crucial when it comes to replacement of spare parts, and usually has to be accurately indicated. To replace an alternator, for example, specific model code, engine number, chassis number and part numbers will all be required. In principle, therefore, it is possible to define lineage trees indicating the relationship of all variants of a particular artefact–activity couple.

For engineers and production specialists who are well acquainted with the distinctiveness of each made batch, and know how each represents a particular setting of key parameters, the uniqueness of generations is very evident. Indeed, the opposite problem of exact matching between made generations is more of an issue, and elaborate procedures are undertaken to ensure sufficient standardization between batches. A major step in the evolution of manufacturing processes was made by Eli Whitney's system of 'interchangeable parts' in the manufacture of muskets for the US army in the late nineteenth century. A key ingredient here was the notion of 'tolerance', in which component parts were designed with sufficient slack to accommodate variations in manufacture. This was a dramatic shift from the ideal of a perfect fit which characterized the era of craft-produced weapons. There, each item was an individual with every component made to fit exactly into the unique ensemble comprising the entire musket. With such craft pieces, the failure of any part necessitated the manufacture of a new customized replacement. With Eli Whitney's system, a standard replacement part could be used – provided it had been manufactured within the

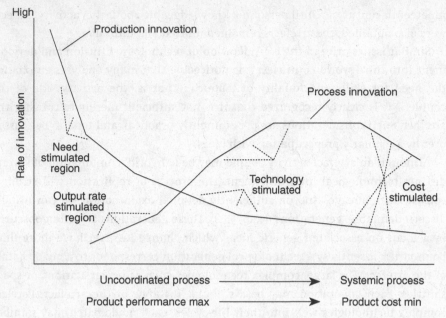

Figure 18.5 Abernathy and Utterback on the product/process relation.

appropriate tolerances. So successful has this system been that we now routinely think of modern manufactured artefacts as generic, and divorced from their origination – quite unlike, for example, works of art or craft pieces.

Generational differences are reflected in the close relation between artefacts and the processes through which they are made or produced. This product/process relation within the artefact–activity couple is explicitly recognized in the field of technology management and is an important factor in technology development. For each product, a characteristic process may be identified: at the same time, each process can only make certain products. What is more, distinct forms of innovation, stimulated by different factors, occur at different stages of a product/ process *life cycle*[37] (fig. 18.5).

In archaeology and anthropology the processes of production are difficult to 'read' directly from the artefacts themselves, but, as the example of the 'Iceman' shows,[38] much of interest can often be inferred. Indeed, a lot of effort is now being devoted to recreating empirically the production and use of archeological artefacts such as flint workpieces[39] and even such relatively 'high-tech' artefacts as the Greek trireme.[40] In such investigations, the production process and the forms of organization which created the original artefact are inferred, recreated and increasingly tested out for feasibility. To understand the artefact more fully, the whole artefact–activity couple has to be investigated and regenerated. For instance, reconstructive empirical research on the trireme has suggested that the Greeks made use of the sliding seat for rowing efficiency, an innovation which was lost until the nineteenth century.[41]

18.10   Variation and innovation

Although replication of the artefact-activity couple is stable, it is not ultra-stable. Indeed, if resource conditions or basic purposes change beyond certain bounds, the functional-integrity feedback loop will rapidly expose inadequacies, and the technology will become obsolescent. However, the dynamism of the artefact–activity couple induces and permits minor variations which are not disastrous, and may even be advantageous. So if the contextual changes are relatively minor, the artefact-activity couple may be able to track them to a new stable point (§12.5). Thus, with the advent of macadamized roads which were relatively smooth and dust-free, a new cycling artefact–activity couple became viable (§21.2), centred around the standard safety bicycle with smaller wheels and the use of gearing. This replaced the previous couple based on the 'penny farthing', in which the gearing effect was implicitly achieved by the large wheel and which required quite different sets of skills and maintenance activities.[42] This design change did not occur earlier because the dust of dirt tracks played havoc with the metals then available for gearing, while the large wheel of the penny farthing helped smooth the ride over large cart ruts.

Such changes, once proven in a given environment, may then remain stabilized in a new pattern because of artefactual failure if the supporting activity moves too far outside its initial bounds. Because of their dependency on particular patterns of activity, or 'routines',[43] such changes also tend to be restricted to particular organizations and are frequently quite difficult to diffuse successfully, especially in the early stages before 'entropic smoothing'[44] through standardization or simplification.

Hence a certain amount of reproduction noise or 'mutation' can play a part in technology development (see fig. 18.6). But of far greater importance are combinations of existing artefact–activity couples into larger complexes or systems (§17.9). A classic example is the development of electrical power generation and distribution.[45] Structured processes of standardization and the combination of components into configurations designed to address new emergent purposes are a driving force in this type of innovation.[46] Such purposes are usually offered by larger organizational entities.

Once a complex artefact–activity system is established, systematic scope for variation and innovation is subsequently afforded by adjusting the performance of specific component artefact–activity couples *within* the envelope provided by the overall system. Here the concepts of a 'reverse salient' or a 'technological

1	**Mutation**	changes via the reproductive process
2	**Simplification**	the stripping down of entities to the bare functional essentials
3	**integration**	the synthesis of previously separate functional systems into new more inclusive single systems
4	**Elaboration**	accretion of new features onto a basic functional entity
5	**Standardization**	the emergence of common interfaces or functional boundaries round technological entities
6	**Incrementation**	improvements of components within the bounds set by larger functional systems (leading to technological trajectories)
7	**Crystallization**	the emergence of new artefactual elements from working practices, procedures or software
8	**Configuration**	combinations (often nested) of new and old components to perform new tasks
9	**Innofusion**	innovation during diffusion – continued development during actual use, through various forms of 'learning'
10	**Resolution**	improvements in functional systems via the resolution of bottlenecks in development (what Hughes calls 'reverse salients')
11	**Cross-fertilization**	transfer of sub-components from within one functional system to another
12	**Implementation**	transfer of functional systems into new application contexts

Figure 18.6 Modes of technological variation/innovation.

trajectory'[47] are often relevant. A very complex activity system, involving a structured community[48] and very diverse forms of knowledge, as in aeronautical engineering,[49] provides considerable scope for progressive elaboration, and for various forms of 'cross-fertilization' among the constituent segments.

Yet another form of technological variation is provided when generic arte-fact–activity couples are 'transferred' into new particular application contexts. Recent work on the practical difficulties of *implementing* technology into new organizational contexts now recognizes the considerable scope for new developments and innovation in the 'transfer' process.[50] As Dorothy Leonard-Barton succinctly puts it: 'implementation *is* innovation'. In the struggle to get the artefactual components to work in a different context, substantive changes and even improvements may be made. By incorporating new elements into the activity side, constrained alterations to the overall artefact–activity couple result.

Here, we must not be tempted into the fallacy of making a distinction between the invention of a technology and its 'mere use'. The straightforward diffusion of a basic artefact into widespread use essentially as originally invented, is actually quite unusual. Increasingly, specialists of technology innovation reject such 'linear' models, and take note of continued development during use.[51] Von Hippel in particular has documented the important role of users in certain types of innovation.[52] These latter studies offer further evidence for the intrinsically evolutionary nature of technology development. But they do rather upset the ideological predilection of many popular histories for cataclysmic accounts of technological progress as the result of breakthroughs by heroic individuals.

18.11   Lamarckianism, Darwinism and neo-Darwinism in technological evolution

In this chapter I have shown that technological change can be represented as an evolutionary process whose basic unit – the artefact–activity couple – is an amalgam of artefact, knowledge and organization. I think that this approach avoids several theoretical obstacles to further exploration of the theme of this book.

For example, much is made (cf. §1.3; §3.1; §5.6; §9.2; §10.6; §12.6; §16.6) of the supposed distinction between 'Lamarckian' and 'Darwinian' modes of evolution. From the perspective outlined in this chapter, this distinction seems less important. The activity side of the artefact-activity couple inherently incorporates conscious decision-making. But at more comprehensive and inclusive levels of consideration, in terms both of time and space, the envelopes of Lamarckian processes are revealed to be clearly localized and therefore subject to broader selection constraints. Because no individual (or collectivity for that matter) has perfect foresight, 'blind' Darwinian selectionism will dominate beyond the horizon of bounded rationality.

Doubts have also been expressed (Ziman, ch.4) about whether it makes sense to apply a thorough-going neo-Darwinian model to technological change. But, despite its composite make-up, the artefact–activity couple operates dynamically as a unit in technology development, and is able to express both 'interactor' and 'replicator' functions. What is more important, such couples have evolutionary properties that shape the salient aspects of technological innovation. Thus, their historical traces delineate unambiguous patterns of descent, which would be indeterminate if one were to focus solely on the ideas, and whose reproductive stages would not be captured by focusing solely on the artefacts.

In this short account I can only indicate schematically how several distinct forms of technological innovation can be interpreted in this way as direct analogues of biological variation and selective retention. A more complete range of specifically technological mechanisms of change is set out in fig. 18.6. Having cleared away some of the conceptual obstacles, we can now see that there is a great deal of scope for comparing these with what we know empirically about their possible biological counterparts.

The organization of innovative enterprises

GERARD FAIRTLOUGH

19.1 Styles of organization and patterns of evolution

This chapter is about the ways in which innovation is organized, and how these ways of organizing change over time. Just as we can identify types of technology and patterns of technological evolution (Stankiewicz, ch.17), so we can distinguish types, or styles, of organization and patterns of evolution among these styles. Technological innovation takes place within a complex framework of individual knowledge and attitudes, organization routines, and structures in human society (Nelson, ch.6; Fleck, ch.18). An understanding of such frameworks, and of how they evolve, should help us to understand the evolution of technology.

The chapter starts with a typology of innovation, followed by a typology of organization styles, and then by a discussion of the relation between type of innovation and type of organization. Finally, it discusses how the evolution of organization styles relates to technological evolution. I draw on the experience I have gained from over forty years of work with technology-based businesses, mostly innovative ones.

19.2 A typology of innovation

During the course of his writing Schumpeter (§6.3) assumed two models of innovation.[1] In the Mark I model, innovation is the result of the constant formation of new entrepreneurial firms, each of which introduces some new product or process. Because of the competitive advantage these new products or processes provide, new firms drive older firms out of business, so technology advances not so much within a firm (after its formation) but within the whole population of firms. Schumpeter derived this model from the sort of innovation he saw as prevalent in the late nineteenth century.

In contrast, Mark II innovation emerges from the R&D laboratories of large corporations and was suggested to Schumpeter by the innovation typical of the first half of the twentieth century. Mark II innovation requires little formation of new firms, since innovative R&D within existing firms allows most of them to remain competitive over long periods.

We can generalize from Schumpeter's two models and classify innovation into two types – individualist and collaborative. Individualist innovation includes entrepreneurial firms, but also individual innovators and small academic groups, and even unauthorized 'skunk works' within large firms. Collaborative innovation includes corporate R&D laboratories, but also large government laboratories and perhaps a few large, well-coordinated academic research laboratories. Collaborative innovation benefits from sustained interaction within a group of people, who develop shared tacit knowledge, shared mental models and 'shared literacy' (Burns, ch.21).

We can put these pure individualist and pure collaborative types at either end of a spectrum, ranging from the individual investigator, who innovates alone, with only intermittent help from others, at one end and the largest and most coordinated research organizations at the other. In the middle there are small and medium-sized research-based firms and academic laboratories, together with various semi-permanent networks of research collaboration.

Stankiewicz (ch.17) proposes a classification of innovation as either design-driven or discovery-driven. Discovery-driven innovation depends in its pure form on the direct application of recent scientific advances, with new pharmaceuticals and novel materials being good examples. Design-driven innovation is often referred to as 'engineering' rather than as R&D. Klevorick *et al.* point out that most innovation uses all sorts of science as a tool kit for problem-solving, most of which is 'old' science rather than newly discovered science.[2] So we have another spectrum here, ranging from the pure discovery-driven type (e.g. when a newly cloned gene relevant to a human disease opens up the possibility of synthesizing a drug to treat that disease) to the pure design-driven type (e.g. the writing of computer software, which usually depends on a few well-known principles, applied as imaginatively as possible). Mixed types, in which some scientific experimentation is required, but which mainly rely on already established science and engineering principles, lie in the middle.

It should be noted that in discovery-driven innovation the relevant discoveries are often made outside the innovating organization. Spotting the usefulness of an external discovery may be just as valuable as actually making the discovery. The recent trend within very large technology-based corporations to disband

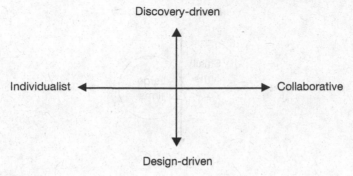

Figure 19.1

their internal laboratories devoted to basic research, replacing them with academic collaborations, recognizes this.

Combining the individualist/collaborative and discovery-driven/design-driven spectra gives us the typology matrix shown in fig.19.1.

At the top left of the matrix, the usually rejected linear model of technology transfer from academia to industry actually works – discoveries made by a individual scientist are picked up, usually from a published paper, and used by a firm. The bottom left corner is where we find lone inventors, those pictured in *New Yorker* cartoons sitting in the lobby of the patent office with strange-shaped parcels. At the top right is organized science and at the bottom right is organized engineering.

We can use the matrix to situate an industrial example. In the late 1970s, the pharmaceutical industry was concentrated in the top right quadrant of the matrix. A study by Malerba and Orsenigo[3] drawing on data for the years 1969–86 shows that the Schumpeter Mark II pattern then applied to this industry. Nearly all new drugs came from the R&D laboratories of thirty or so major corporations around the world. In these laboratories, chemists synthezised drug candidates and pharmacologists tested them to see if they worked. The barriers to entry to this highly profitable innovative pharmaceuticals industry were formidable, since entry usually meant building up a full-sized R&D organization and because of the lead-time of ten years or so for bringing an innovative pharmaceutical product to market. But the discoveries of modern biology, particularly in recent years the mapping of the human genome, changed this. Basic research in biology, usually in academia, now provides the basis for much of the pharmaceutical industry's innovation. A consequence is that new entrepreneurial companies, sometimes founded by academics and almost always well-attuned to using academic discoveries, are now where much of the early-stage R&D is done, while major corporations focus on later-stage development and marketing. This is the new pattern for the pharmaceutical industry (fig. 19.2).

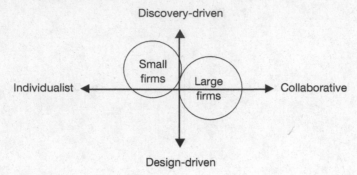

Figure 19.2

Abernathy and Clark introduced the concept of innovational *transilience*, meaning an innovation's capacity to destroy established organizational competences – to make existing production plant obsolete, or the present skills of workers redundant, or established links with customers irrelevant.[4] There is an analogy between the concept of transilience and the concept, drawn from biological evolution, of niche disturbance. A species may find a niche (cf. §2.5) in which it can thrive, undisturbed by competitors, for some time. But geological or other changes will sooner or later disturb the niche, and if these changes are sufficiently 'transilient', the species will be unable to compete in its new environment, becoming extinct as a result. Indeed, a high degree of adaptation to a particular niche makes a species, or a firm, particularly vulnerable to changes in its environment.

Modern biological discoveries have proved to be somewhat transilient as far as the pharmaceutical industry is concerned, since they broke the major firms' hold on drug discovery. But they have not been fully transilient, as they have not reduced major firms' competences in later-stage development or marketing. In the pharmaceutical field there is now extensive collaboration between small, entrepreneurial firms and major corporations.[5] As appropriate legal and organizational forms for such collaborations and strategic alliances get widely known, this marriage between large and small may become a general feature of discovery-driven innovation, since small firms generally display much greater creativity and speed in response to new discoveries than large firms. The view within conventional economics that competition is everything fails to take into account emerging industrial structures of this kind. Models from biology of coevolution and symbiosis (§2.5) are needed as well.

Transilient scientific discoveries plus new organizational forms are thus able to push innovation from a Mark II back to a Mark I type, which is counter to the trend towards the large scale that usually happens as a field matures. In the design-driven part of the matrix, transilience could derive from industrial advances outside the firm's own field. For example, cost reductions in fabrication equipment could reduce the advantages that already-installed fabrication plant gives to existing producers of a particular device.

Some further examples: the computer software industry would seem to be located in the bottom half of the matrix, much of it in the bottom right quadrant, with Microsoft right in that corner, and the aircraft industry may be similarly located, except for its use of new materials, which puts part of its innovation in the upper half of the matrix. The Malerba and Orsenigo study showed that much of the innovation in the mechanical industries, and in food, textiles and wood-based industry is of the Mark I type, while the oil, chemical, automobile and electric power industries had Mark II innovation.[6] Klevorick *et al.* ranked lines of business by their proximity to a particularly relevant area of science, which would indicate that innovation in these lines of business is heavily discovery-driven. In this study, pharmaceuticals, semiconductor devices, plastics materials and medical instruments emerged as the industries with the greatest proximity to a science.[7]

19.3   Types of organization

Nowadays, successful innovation seems to require 'loose–tight' organization structures, that provide both the freedom to create and the discipline to turn good ideas into commercial success. The loose–tight form has probably emerged through competitive pressures. Perhaps only these pressures could have forced people at the top of organizational hierarchies to become less controlling.

Discipline in loose–tight organizations should ideally be self-discipline, arising from widespread commitment to the purposes of the organization, to professional practice and to team-working, although many still think that some hierarchical, power-based control is, in practice, essential for business success. The innovative organization might well seek to emulate Japanese social organization, in which individuals find fulfilment through the progress of the various social units to which they belong – fulfilment that supports community self-discipline. In considering the combination of loose and tight in an organization, we can draw on European metaphors, looking for a combination of the classical values of order and clarity with the romantic values of freedom and creativity.

In an innovative organization, information is freely shared, there are high levels of trust, and everyone is bound together by a deep understanding of and commitment to organization goals. Leaders influence by the example of their own behaviour, by networking within their organization, by coaching, by articulating an inspiring vision. This type of organization is described as 'organic', in contrast to the 'bureaucratic' type.[8] In fact, innovative organizations, although primarily organic, do need some bureaucratic controls and information systems. The management systems of innovative organizations, such as those for project management, have to be limited in number, carefully chosen, well designed, usually tested on a pilot scale before general use and developed with widespread consultation. This is an example of loose–tight organization – tight in the selected areas covered by management systems, loose everywhere else.

So far, I have used the term 'organization' as if there were no problem about identifying what an organization is. But this taken-for-granted approach is not tenable. In the case of businesses, legal incorporation might seem to provide a suitable criterion, but subsidiary companies and joint ventures are legally separate, but usually not fully independent. Japan is famous for company spirit, but people working for a company nonetheless identify with various levels of organization. Companies like Sony and Honda work hard to build local loyalty within project teams charged with creating and developing new products. And there is commitment to the well-being of entities larger than the company – the groupings of companies called *zaibatsu*, and to Japan itself.

In Silicon Valley, firms form and reform quite rapidly, and someone's commitment to a network of collaborators spread over different firms is often as great as commitment to the firm where he or she currently works. Intercompany collaborations in the pharmaceutical industry can last for several years, often build their own loyalties and depend on good leadership and good management systems serving the collaboration rather than either of the participants. Malone uses the term 'radically decentralised organization' to describe institutions like the Internet, securities markets and scientific communities.[9] I will follow this terminology, using the term 'organization' to include not only separate legal entities such as firms, government bodies and not-for-profit bodies, but also networks and innovating communities of various kinds, some of them defined legally or professionally, others defined more by custom and practice.

19.4   Choosing the right management style

I will now return to the matrix of types of innovation developed above, and look at which organization style might be most suitable in each of the four quadrants of the matrix. If innovators in the individualist, discovery-driven top left quadrant (Carlson, ch.11) want their discoveries to lead to commercial success, they have to obtain resources to form and grow new companies, and they often have to attract other companies as collaborators. The organization skills they need are therefore those of influence across the boundaries of organizations like banks, venture capital firms and potential corporate partners – the skills of networking, boundary-spanning and technology-transfer. Capability in the protection of intellectual property is important too. The organization style best suited to this quadrant is thus quite a loose one, with 'tight' aspects being supplied by network loyalty, professional standards and the desire of most players to retain a reputation for honest and reliable dealing.

The individualist, design-driven bottom left quadrant may also require some boundary-spanning and intellectual property skills, but perhaps the characteristic most required is perseverance – more a matter of personal psychology than of organization. Compared with discovery-driven individualist innovators (whose discoveries will often be recognized as important even if the way to exploit them is not obvious) design-driven individualist innovators have a lonely struggle against not-invented-here attitudes in the rest of the world. Perseverance, combined with a sometimes fanatical belief in the value of the innovation, may be the only 'tight' element in this quadrant's natural organization style.

The organization style for the collaborative, design-driven bottom right quadrant is a blend of the organic and the bureaucratic. If the organization is well-managed and professionally orientated, most of the bureaucratic aspects, such as careful logging of engineering drawings, will be perceived as sound practice by those involved. Budgets and cost control may be less welcome, but good design of systems should make them tolerable. However, the organic, or loose, organization style will have to be well-represented in the blend, if the organization is to be a creative one. The challenge is to keep activities necessary for professional discipline from becoming mere tools of organizational power-seekers.

Lastly, the discovery-driven, collaborative top right quadrant is also an organic-bureaucratic blend, with laboratory notebooks (cf. Carlson, ch.11) in the place of engineering drawings. But usually less bureaucracy is required than in design-driven organizations, since experimentation provides a discipline of its own. Today this quadrant also needs some of the networking, boundary-spanning and technology interaction skills of the top left quadrant, since the largest discovery-driven firms and government laboratories now depend heavily on their contacts with academic researchers and small firms.

19.5   Features of innovative organizations

In this section I will expand on the following features of innovative organizations: focus on distinctive technological competences, attention to the multiple parentage of innovations, enhancement of group learning and building communication skills.

For all technology-based companies, the critical factor in strategy is the creation and constant enhancement of a firm's *distinctive technological competences*. This can be seen as continuing adaptation to a particular technological niche. For example, in a prospectus issued for a 1996 fund-raising, the biotechnology company Cantab Pharmaceuticals described the two 'technology platforms' on which its drug development programme is based. These are (1) specialist expertise in the clinical virology of Human Papilloma Virus and (2) a patented technology for disabling the replication of viruses, which allows disabled viruses to be given safely as medicines or vaccines. Using these two unique capabilities, a series of product candidates has been developed, which are now undergoing clinical testing. The main purpose of the fund-raising was to pay for this testing. The company sticks to products using these technologies, since further product development will, through learning-by-doing (Vincenti, ch.13; Constant, ch.16; Fleck, ch.18), enhance the value of its technology platforms. The strategy will be commercially successful if the platforms the company has chosen are fruitful ones and if its more general business and technological expertise is good enough to avoid mistakes in implementation.

There is therefore a hierarchical process of selection. Cantab and similar firms can survive if a few of their product candidates fail in clinical testing, but the loss of a technology platform (e.g. if the platform's key patents cannot be defended in litigation) would be serious. There is selection at the firm level – some biotech firms fail and their remaining assets are usually bought at a low value. And the biotech industry as a whole competes for resources (money and skilled people) with other industrial sectors.

Intellectual creativity often involves bringing together previously unconnected ideas, and innovative institutions may be successful because they know how to encourage these connections (Carlson, ch.11; Solomon ch.14). Thus innovations generally have multiple parentage (Mokyr, ch.5). In technology-based industries, the right layout in a laboratory building (including banal features like the location of coffee machines and toilets) is one that will foster informal interaction. An atmosphere of mutual trust and commitment to the success of the organization encourages uninhibited sharing of ideas. Factors like these are probably the reason why small organizations appear to be more creative than large ones – they provide a better environment for multi-parent innovation.

This inter-disciplinary interaction, most favourably within a close-knit community, is similar to the 'specialized activity of a closely co-operating group of well-trained people . . . .spread out across continents' that increasingly characterizes leading edge scientific research, but the need to integrate various scientific disciplines together with activities like finance and marketing, makes it harder to create a community in technology than in academic science.

The large majority of technological innovation depends less on leaps of the imagination (Perkins, ch.12) than on skilful choice and disciplined development of the right ideas. This is increasingly a social process, a process of group learning. During the 1980s at the biotechnology company Celltech, my colleagues and I developed a system for project selection that produced many creative proposals and allowed us to choose just a few of these in a way that did not demotivate people championing projects that were unsuccessful.[10] Thus anyone in the company could champion a proposal, and could have help in doing this from a group of experienced people whose task it was to support, not select. Within this group no one was allowed to make a negative comment on a proposal unless s/he had already made two positive comments. After a few weeks' work on the feasibility of a proposal, its champion(s) were encouraged to present it to a different group of company staff, who operated a 'selection gate', through which only about one in five proposals passed. Once through the gate, a proposal was given a formal allocation of resources (for patent searches, preliminary experiments and the like), following which the champion had to get the proposal through yet another gate. If it passed that selection, the proposal became a full project, with up to ten people working on it. The selection process was an open one, with the selectors obliged to give everyone involved the reasons for their selections. This openness not only demonstrated the fairness of the system, but also contributed powerfully to group learning.

In 1974, the year when the impact of the first 'oil shock' on the world economy became unpleasantly clear, I was involved in another memorable episode of organizational learning. At that time I had had twenty years of experience in the petrochemical industry – twenty years of more-or-less continuous growth, and of development of a worldwide industrial paradigm in which oil, natural gas and petrochemicals had a central role. This had given me and all my colleagues in the business a set of mental models (§11.4; §14.6) of the technical, economic and commercial forces at work in and on the industry that guided our decision-making. These models had become 'habits' in the sense that our actions were 'conditioned on beliefs that grow more and more difficult to cast off quickly' (David, ch.10).

A feature of the new world was much greater uncertainty – uncertainty about future costs of oil (our industry's raw material), about economic growth (that would determine the size of future markets) and about new industries (like IT) that could upset the dominance of the old ones (like petrochemicals). We came to see the period of the 1950s, 1960s and early 1970s as a golden age, where things were predictable and our old habits were useful. Those of us who worked in the Shell group were helped in making the transition to new ways of thinking by the practice of scenario planning. This uses multiple pictures of the next ten years or so (coherent, plausible, but pretty diverse pictures) against which one can test the desirability or otherwise of a particular research project or a particular investment in manufacturing capacity.

The experience of living through the 1973–4 discontinuity, and then of explicit modelling of alternative futures, has left me with an acute awareness of the effect of (often tacit) mental models on technological and commercial decision-making. Because of time and resource pressures, standard heuristic procedures are usually essential in innovation and the art of simplifying can make these procedures particularly effective. As we unpack the processes of learning and imagining on which innovation depends, I think we shall find that mental models (or mental habits) are deeply involved. Innovation usually depends on team work, and making team members' mental models explicit, and then shared, is a key feature of team building.

Innovation needs some combination of:

- basic scientific or engineering principles,
- either systematic experimentation or careful tinkering, and
- knowledge of markets or other contexts of application.

All this knowledge will seldom be available to a single person, and the process of knowledge-assembly will be an iterative one. So the innovating team has to have *good internal communication*. As teams get bigger, communication gets more difficult, and even greater skill in communication is essential. Japanese industry's attention to social interaction allowed it to work out development processes that were much faster and less resource-intensive than the West's. Internal communication is particularly important for innovation in the bottom right quadrant of the matrix.

Within particular scientific disciplines, communication seems to be easier, probably because of a lot of shared tacit knowledge. But confining communication to those 'in the know', may miss the opportunity for the problem-solving contributions of outsiders, who bring knowledge that removes the insiders' blinkers or who spark off some multi-parented innovation. A capability for promoting multi-disciplinary innovation may be one of the advantages of small, academically orientated, high-technology companies, since they do not have the disciplinary constraints to which career academics have to pay attention. Recent moves within academia to develop inter-disciplinary research units, usually with a bias towards industrial application, also seem to recognize the need for communication between disciplines. The traditional pure sciences and technological industry each have long-standing ways of allowing 'memes' to evolve by variation and selection. Academically orientated companies and industrially orientated academic centres may be seen as generating yet more variation.

Spanning the boundary between two communities, two organizations or two cultures is not easy. With determination it is possible to acquire skill in boundary-spanning, and this is helped by understanding the social psychological aspects of this task.[11] Another tricky area for communication is shared reflection on the outcome of actions when the outcome has been a bad one. By avoiding breast-beating or blaming, and by giving detailed attention to the steps that led to the failure, it is possible to learn by 'embracing error', as part of a recursive, reflective practice, as discussed in several other chapters in this volume.

19.6   Organization and technological evolution

Technologies and technologically innovative organizations must be expected to evolve together (Fleck, ch.18). Particular firms, or networks of firms, or other kinds of organization whose organization styles are less well adapted to the current state of their technology will fail to expand or may go out of business altogether. Thus, competitive pressures will co-select organization style plus technology. We might expect that organizations that are good at technological learning are also good at learning how to adapt their organization style to technological advances, at any rate when technological advances are gradual.

When coevolution of technologies and organization styles is a gradual process, the likely result is a trend towards the bottom right of the matrix, a trend to design-driven innovation and towards large, highly collaborative organizations. This might be the cause of the continuing reduction in the numbers of innovative organizations that is often observed in long-established industries – for example, in the commercial aircraft industry, which now has only two large players, worldwide.

But this trend can be reversed by unexpected discoveries in basic science, or drastic cost reductions in the products of other industries. Examples are the advances in biological science discussed above, and the effects of the decades-long reduction in the cost of information technology on the competitive structure of so many industries. Highly transilient innovation is likely to lead to punctuated, rather than gradual, evolution (§1.6; §2.6; §4.4) in the industries affected. The strategy of building on existing, distinctive competences may no longer work in highly transilient conditions. And when technology shifts towards individualist or discovery-driven types, organization styles may need drastic change as well. Existing organizations, especially those that have been the most successful under the old conditions, may find it hard to make the necessary changes in strategy and style. Schumpeter Mark I innovation, in which small is beautiful, will then get its chance once more – a chance to show that it is better at avoiding the evolutionary 'lock-in' that makes it hard to search for new fitness peaks (Perkins, ch.12) when old peaks have been eroded.

V
TECHNOLOGICAL CHANGE IN A WIDER
PERSPECTIVE

The evolution of war and technology

EDWARD CONSTANT

20.1   The partnership of Mars and Vulcan

Mars and Vulcan have been partners in the unhappy business of war since time immemorial. But who has been the managing director of the enterprise, and who has been the greater beneficiary? Much of the discussion of the coevolution of war and technology has been, at best, metaphorical. Some nineteenth-century Social Darwinists, and not a few twentieth-century fascists, believed that war in and of itself improved the 'race', by which they apparently meant not only the mean genetic quality of the species, but also its 'civilization' or its culture, including technology. But two World Wars and the advent of weapons of true mass destruction have refuted any general notion that advanced military technology in and of itself is a sign of advanced civilization.

The history of war does, however, raise serious scholarly questions[1] about the nature of the evolutionary linkage between military conflict and general technological change. On the one hand, it is argued that the exigencies of war (and, by extension, of military preparedness) reduced barriers to innovation and freed up resources for radical technological experimentation, thus promoting both technological and economic progress.[2] On the other hand, it is claimed that virtually all technological 'progress' associated with war or its preparations represented exploitation of already-developed civilian technologies.[3]

More recent historical scholarship makes more limited claims, in which the association between war and technological progress is less direct. William H. McNeill argues,[4] for example, that the conjunction of contingent historical circumstances and specific technological developments shaped the course of world history, and that broad technological progress was an unforeseen and unintended consequence of that process. Thus, after AD 1000, Europe had no imperial authority analogous to China's, which could and did suppress both military and commercial innovation. As a result the West set off on a thousand year exploration of what the combination of unhindered private entrepreneurship and innovation, and competing sovereign states, could engender. This system was dynamically unstable, but conducive to technological innovation. Even the awesome power of siege cannon, which threatened to restore imperial hegemony in the early sixteenth century, was frustrated by the *trace Italien* system of fortification, which assured continued political and economic fragmentation in Europe for several more centuries.

The end of the Cold War, and persisting concerns about Western economic competitiveness, have focused recent attention on whether a half-century's military preparedness 'distorted' technological and economic development, as well as scientific inquiry.[5] In its 'strong' form this 'distortionist hypothesis' claims not only that we know different *things* but also that, in a fundamental epistemological sense, we know *differently* than we would have without the Cold War. At a cognitive level, research into creative processes, which in general highlights the role of metaphor and analogy in creative thought, at least suggests that such might be the case.

More concretely, specific advances in technology and analytical techniques may have permitted new ways of knowing. For example, global climate models not only got their original impetus from military interest in meteorology, and the data accumulated in large-scale military aeronautical activity, but also depend both upon high-speed computerized data processing and upon advances in simulation techniques.[6] Even more spectacularly, Monte Carlo simulation techniques, which came originally directly out of nuclear weapons research, have created what Peter Galison so nicely terms an 'artificial reality' at the very foundations of modern microphysics.[7] Indeed, some observers have gone so far as to claim that such technologies have created a whole alternative culture, essentially different from either of the classical 'two cultures', humanities and science.[8] Such strong claims nevertheless beg two essentially unanswerable questions: whether the world thus known would have been 'discovered' anyway, or is simply a product of its cognitive and technical means of production, and what a counter-factual world to the one we claim to know would have been like without the effects of military enterprise.

In its 'weak' form the 'distortionist' thesis simply claims that military preparedness, military procurement, military-driven research and development, and, especially, the scientific and applied scientific research done in support of these endeavours, has resulted in our knowing different *things* than we likely would have known otherwise. This claim is relatively straightforward and not very controversial. Some of its corollaries, however, are more contentious – for example, the magnitude and importance of civilian 'spin-offs' from military-related technological development (computers, aircraft), or whether or not military acquisition policies and desiderata (e.g. mil-spec performance) have resulted in an economically dysfunctional style of research and development, ill-suited to consumer markets.

Some authors – especially McNeill – have portrayed the joint development of the state and its bureaucracy, military technology, military tactical and operational innovation, and civil technology metaphorically as 'coevolutionary' processes. But they have not used the metaphor as a *theory* about why and how those processes occurred. Thus, for our present purposes, many of the broad meta-claims about the relationship between the military environment and technological change are irrelevant.

Nevertheless, this relationship does highlight theoretically interesting aspects of the process of technological change. Although these do not seem to differ in essence from those observed in 'civilian' environments, they do perhaps appear in higher relief. Among the features that stand out are: the macro-coevolution of military power, technology and the state, which may have analogies in the macro-evolution of whole ecological systems; the coevolution of production practices, weapons, and operational and tactical innovations; the whole problem of selection for 'fitness', including its efficacy, multiple levels, and contingencies; and the analogues of 'directed mutation', 'vicarious selection' and 'memetic inheritance' in military institutions.

20.2   Macroevolution

War, the state and technology have always been, and remain, inextricably linked. As far back as history goes, only the state or its surrogates, with their comparatively immense resources, have been able to afford the most advanced technology. Technological change in turn, sometimes with state support, sometimes without, by transforming the basis of military and economic power, has often up-ended the state. In a short chapter, only a few examples will have to suffice to illustrate the general point.

In remote antiquity, war chariots were the first wheeled vehicles to have light-weight, spoked, precisely balanced wheels: up from the ox-cart, they were the elite high-technology of the Bronze Age. Cheap and plentiful iron weapons extinguished the elite basis of Bronze Age empire, and, combined with bureaucratic innovations both in the state tax-collecting apparatus and in the organization of armies themselves, ushered in the age of mass hoplite infantry and great agrarian empires.[9]

Again, the coming of European feudalism has been attributed to two specific technological innovations: the stirrup, which apparently diffused from China through the Middle East to Europe, and the heavy mold-board plough, which permitted cultivation of rich North-European soils.[10] The stirrup, together with a heavy saddle, for the first time welded horse and rider into a single shock weapon, whilst the mold-board plough, together with the three-field system and alfalfa, is said to have enhanced agricultural productivity sufficiently to support the great war horses necessary to carry an armoured knight. Together, these innovations may have shifted the locus of military power in Europe to locally based elites for nearly half a millennium.

The rise of the nation-state, itself a product of military innovations, both technological and tactical (such as siege cannon, arquebuses, and mass pike formations), led to direct state support for technological development, much of it either purely military or 'dual-use'. In the seventeenth century, for example, the British navy offered a substantial prize for an accurate marine chronometer, to determine longitude at sea. Since neither pendulum escapements nor weight drives would work on a tossing sailing ship, this was a non-trivial technical challenge with far-reaching consequences for world commerce and empire.

Again, in the late eighteenth century, the French army supported what is possibly the first modern, in-house military research and development pro-gramme (cf. §17.5). Jean Bapiste Vacquette de Gribeauval systematically analysed and improved field artillery pieces, and reorganized and rationalized the field artillery itself, thus laying the *ancien regime* foundations for Napoleon's 'grand battery'.[11] In like manner, in the first half of the nineteenth century, the young United States government promoted in its own arsenals, and among its contrac-tors, what became known as 'armory practice' or what a British parliamentary inquiry in 1851 dubbed 'the American system of manufacturing': production by special-purpose, water-powered machine tools of weapons with uniform, inter-changeable parts.[12] Critically, until the very end of the nineteenth century, only governments could afford the expensive luxury of true interchangeability: civilian spin-off 'mass production' depended upon looser tolerances and minimal fitting (§18.9).

Not all effects of the military on technological change are confined to military hardware, or even just to hardware. In America, during the French and Indian War (Seven Years' War, 1756–1763), British desire to strike at French interests in Canada could only be fulfilled by exploiting the many lakes and rivers of northern New York and New England, which required a massive number of small boats. Boatwrights from all over the American east coast were assembled and set to work building large numbers of a simple, very cheap, sturdy design – what became the classic American dory. After the war, the boatwrights carried the design up and down the east coast: the dory transformed coastal, then the

Grand Banks and North Atlantic fisheries, and diffused ultimately to the Baltic, the Atlantic coast of Europe, and into the Mediterranean.[13]

Military-related 'software' innovations – institutional, financial and organizational – may well have been at least as important as military-induced changes in hardware. The Bank of England, founded in 1694 during the war of William III, provided 'an efficient centralized credit mechanism for financing war . . .'.[14] For the first time, new commercial and proto-industrial wealth could be fully and effectively mobilized, far beyond what specie reserves or the real-time extractions of the Exchequer could pay for. Such deficit financing remains the major resource for war-fighting, as well as for fiscal support of state policy more generally. In addition, this highly developed system of credit, institutionalized for state purposes, had the unintended and unforeseen consequence of providing the foundation for the system of commercial and industrial finance necessary to support large-scale industrialization.

That large-scale industrial or commercial enterprise could be managed also derived from military innovation. Following the generally dismal performance of American armies in the war of 1812, the War Department initiated a series of bureaucratic reforms intended to rationalize and standardize army operations and their reporting and budgeting. Particularly challenging was the widely scattered and very expensive fortification construction programme, under the direction of the Army Corps of Engineers.[15] The organizational innovations pioneered by the Army, and especially by the Corps of Engineers, diffused, usually via personnel, first to early American railroads, and from there to virtually all large-scale industrial and commercial concerns.

The twentieth century, of course, saw these macro-coevolutionary processes multiplied and magnified: the century of total war, of the military-industrial (and academic) complex, of the 'national security state', had the profound interaction of technology and military power as perhaps its defining characteristic. Much of the fabric of contemporary life owes its warp and weft specifically to the exigencies of the military's technological interests.

Again, just a few examples will have to suffice. Following the First World War, and its own laggard adoption of radio, the United States Navy brokered creation of the Radio Corporation of America (RCA) in order to assure American control of national-security-critical technology, and explicitly to exclude British Marconi from any influence. RCA initially comprised a patent pool including General Electric and AT&T. When RCA was formed, radio was still officially conceived to be a point-to-point medium, analogous to the telegraph or the telephone. Quite to the astonishment of its official sponsors, rogue Westinghouse engineers, and an underground community of enthusiasts, began radio *broadcasts*. Westinghouse was soon added to the RCA cartel, which quickly gave birth to the National

Broadcasting Company (NBC), who, of course, would distribute its nationally syndicated programming to local broadcast stations over AT&T's telephone network. The Federal Radio Commission (FRC, later the Federal Communications Commission, FCC) quickly assumed the tasks of allocating both civilian and military-reserved frequencies, of setting technical standards (especially maximum transmitter power), and of issuing and regulating broadcast licences.[16] In brief, the contemporary structure of telecommunications, radio and television, right down to the half-hour 'sit-com' format, traces its origins – not just its technology but its institutional structure – to the military's intervention and interests.

Military interests also have profoundly shaped, sometimes to their detriment, other technologies that are the crown jewels of twentieth-century progress: aircraft, microelectronics and computation, to name just three. Both the Royal Aircraft Establishment (RAE) and the National Advisory Committee for Aeronautics (NACA), created just before and during the First World War respectively, were responses to the possible military utility of aeroplanes. In the inter-war period, NACA especially attended to military-relevant problems; indeed, NACA's research agenda was so closely tied to military and commercial development issues that it largely neglected fundamental high-speed research.[17] Nevertheless, during the 1930s, NACA research did underpin creation of the world's most advanced commercial aviation industry.

The efficacy of NACA's research contribution, however, was contingent on other government policies. During the 1930s, the US postal service issued airmail contracts to chosen airlines on a 'space available' basis (the space was paid for regardless of the actual volume of air-mail carried). This policy encouraged the airlines to purchase new, larger, more advanced aircraft – and thus provided an indirect subsidy to American aircraft manufacturers. The extraordinary quality of the resulting American twin and four-engined commercial airliners in turn reinforced the ideological preference within the US Army Air Corps for strategic precision bombardment, which was not only a military strategy, but also a bureaucratic strategem for distancing the Air Corps from regular Army interests, for distinguishing the role of the Air Corps from that of the Navy, and for asserting the need for a separate air force, along the lines of the Royal Air Force.[18]

During and after the Second World War, the Army Air Force, then its successor service, the US Air Force, blamed NACA for largely having missed the turbojet revolution, and not only pushed NACA into the forefront of high-speed, transonic research,[19] but also developed its own research capability in its Air Materiel Command laboratories. NACA, for its part, now freed from the developmental burden imposed during the 1930s, eagerly and effectively pursued its new 'fundamental' research mission. The resulting rapid development of aerodynamics in general, and of turbojet engines in particular (Constant, ch.16), combined with the manufacturing pre-eminence of American airframe manufacturers established in the 1930s and enhanced during the War, allowed American engine and airframe manufacturers to dominate world airliner production until the advent of the heavily subsidized Airbus Industries. This occurred despite the fact that Britain had a substantial development lead in turbojet aircraft in the immediate post-war period, and despite American airlines' protracted resistance to adopting jet aircraft. What emerges in this story is a very complex interaction between military ideology (serving as a not very good vicarious selector), military strategy and operational requirements, military agenda-setting, and contingent historical factors, such as the air-mail subsidy.

The history of micro-electronics is equally complex. Although point-contact transistors first were invented at Bell Labs, the early development of practical transistors was largely dominated by the US military, especially the Army. The US Army's electronics needs in the early 1950s placed a premium on miniaturization, robustness and moderate power consumption, for such applications as walkie-talkies, proximity fuses, and the radar-guided NIKE anti-aircraft missile system, then in development. As a result, military requirements drove the direction of not only the transistor's development (the Army wanted transistors that could amplify high-frequency signals, of little use to civilian users), but also the development of production techniques and facilities, many of which were directly funded by the Army.[20] Similarly, the whole history of very large-scale integrated circuits, computation, computer science, artificial intelligence, encryption, and even the Internet and World Wide Web is driven by the Defense Advanced Research Projects Agency (DARPA), and assuredly would not have evolved in the directions it did without the military's interests and lavish support.

Quite apart from its direct effect on technology, military support and intervention undeniably has had a profound effect on the way both scientific research and technological research and development are conceived and conducted. Vannevar Bush's Second World War Office of Scientific Research and Development, together with the legacy of the Manhattan Project, defined a style and an ideology (basic science breeds advanced technology) of research and research management that arguably continues to pervade the Office of Naval Research (ONR), the civilian National Science Foundation (NSF), The National Institutes of Health (NIH), the National Cancer Institute (NCI), and even the Human Genome Project. Likewise, the bureaucratic apparatus put in place in the 1960s to manage large-scale, multi-vendor military research, development and procurement projects altered the competitive structure of defence and space-related industries, and likely affected consumer-oriented firms and American economic competitiveness as well. The sheer complexity of the Planning, Programming and Budgeting System (PPBS), and the review processes supporting it, imposed such horrendous paperwork burdens and overhead costs that not only were the total costs of defence development and procurement inflated, but also competition for military contracts was limited to only a few very large, experienced (and likely well-connected) firms.[21]

Clearly then, in a multiplicity of ways, the military has directly and indirectly impacted both technology itself and its macro-ecology. Ecologies of innovation, whether in technology or in biology, is a topic not well defined or explored in either evolutionary theory or in ecology. Here, as in other contexts (Jablonka, ch.3; David, ch.10), it remains problematic whether or not the cumulative adaptation and differentiation of species (or technologies) creates contingent, time and path-dependent ecological systems in which further development of both its constitutive species (or technologies) and the ecological system as a whole is more probable in one direction or a small set of directions than in others of all the directions that are apparently open to it.

20.3   Coevolution and complementarities

Although the role of the military in the secular macroevolution of technology (and society) may be historically unique and contingent, the microprocesses governing the evolution of military technology are no different in principle from the microprocesses governing any other technology. Military and military-related technologies are affected profoundly by systemic coevolution and complementarities. Coevolution (§1.4; §2.5; §3.2; §6.4) occurs when two species (or technologies) constitute a paramount feature of each other's environment, that is, when they are strongly interdependent, and when they evolve *vis à vis* the rest of their ecology virtually as a linked unit.

For a given technology, complementarities are changes in other technologies, or in production or organizational practices, without which the technology is not viable. Almost all discrete innovations in military technology require corresponding systemic adaptations in other elements of military technology, with which they coevolve over time. Most, if not all, discrete innovations depend upon complementary innovations in production technology (Fleck, ch.18; Fairtlough, ch.19). Virtually all require or initiate complementary organizational, tactical and operational adaptations, without which the discrete innovation is still-born. Again, for brevity's sake, a small number of illustrative examples will suffice.

From the beginning of the seventeenth century until the middle of the nineteenth century, the dominant infantry weapon was the muzzle-loading, smooth-bore musket, with ring bayonet. It had an effective range of perhaps 100 yards, and a useful range somewhat less than that. Yet combined with innovations in drill and discipline introduced by Maurice of Nassau at the end of the sixteenth century, the smooth-bore musket made the infantry battalion, in line or square, the 'queen of battle'. Although rifles were used by skirmishers, their slow rate of fire (the ball had to be rammed down the rifled barrel) made them useless for mass formations.

The introduction of the Minié ball, or conoidal bullet, in the mid-nineteenth century revolutionized warfare, without commanders or soldiers quite realizing it. The Minié ball was a trivially simple innovation, and in principle could have been invented any time after 1690: it was simply a minimally sub-calibre, elongated lead bullet with its bottom slightly indented or hollowed out; it could be dropped easily down a rifled bore, but when the gun fired, the pressure of the charge distorted the slight flange formed by the indentation in the base outward to engage the rifling, thereby spinning and stabilizing the bullet.[22] The result was a musket lethal out to a thousand yards and effective out to 500, a range greater than or equal to that of contemporary smooth-bore field artillery. Thus began the precipitous increase in the advantages offered by defensive firepower that reached its apogee during the First World War.

Ironically, and tragically, at least through the American Civil War, military leaders do not seem to have comprehended the way such firepower altered the nature of combat, and did not formally redefine tactical doctrine: formations thinned out and took cover irregularly, and on a purely *ad hoc* basis.[23] As a result, not only did casualties go up, but also, in the Civil War, bullets accounted for some 86% of all combat casualties, the highest percentage for any war in history: sad testimony to a case in which complementary tactical and organizational adaptations to a new technology were *not* made.

In contrast, the introduction, in the 1860s, of the first practical breech-loading rifle, the Prussian bolt-action 'needle gun' (for its elongated firing pin), *was* accompanied by complementary tactical innovations. Although the needle gun was still a single-shot weapon using a paper cartridge, the Prussian General Staff realized that its introduction would likely have two serious consequences. First, because it could be fired repeatedly from a sitting or prone position, without standing to reload, the needle gun would permit soldiers to disperse and take cover. The problem was to maintain control, and get them moving again in offensive situations. Second, the needle gun would permit substantially higher rates of sustained fire, thus presenting an insuperable logistics problem without much greater fire discipline – which of course would be much more difficult given the dispersion the weapon encouraged. The Prussian solution, successful as it turned out, was a complete retraining of its Army (cf. §18.7), concentrating on the junior and non-commissioned officers on whom the heaviest burden of decentralized control would fall.

Although tactically challenging, breech-loading rifles did not have a profound effect on volume of fire until a series of complementary innovations appeared: most importantly, brass cartridges (which depended in turn upon developments in industrial extrusion processes), and the advent of smokeless powders, which reduced fouling and burned slower, thereby increasing muzzle-velocity in longer-barreled rifles. These complementarities permitted development of reliable, powerful and accurate bolt-action repeating magazine rifles, as well as, very quickly, magazine and belt-fed machine guns.

Similar coevolutionary processes, both among the component parts of the weapon as system, and between the weapon system and its industrial and materials basis, dominated the evolution of quick-firing field artillery. Industrial complementarities included development of high-strength steels, smokeless powders, and mass-production of reliable fuse mechanisms. Weapons-specific technological innovations included fabrication techniques for built-up tubular barrels (longer, stronger and lighter in weight than single-piece cast barrels), quick opening and closing breech mechanisms, and, perhaps most importantly for tactical efficacy, precisely balanced buffer-recoil systems which permitted a field gun, once registered, to fire repeatedly and rapidly at the same target without being re-sighted. The combination of rapid fire with precision fuses and steel-cased shells containing high-explosives permitted precisely placed 'air bursts', which produced thousands of jagged pieces of shell casing, or highly lethal shrapnel.[24]

The cumulative effect of these innovations – magazine rifles, machine guns, and quick-firing field artillery – was that, by the eve of the First World War, open-field manoeuvre by masses of infantry had become all but impossible. But tactical and operational doctrine hardly acknowledged the new reality, however soon it was to be discovered empirically.[25] Organizational adaptation and operational doctrine lagged far behind the capabilities of the new weapons. Modern breech-loading artillery by 1890 was fully capable of indirect fire: that is, shooting rapidly and accurately at targets beyond the line-of-sight of the gunners. Yet it was 1917 before the majority of leaders in either the German or the Allied armies fully understood indirect artillery fire, and had developed the required precise maps and trained forward observers and fire-control personnel necessary to its effective execution, as well as begun the task of integrating it into standard operational doctrine.[26] It was also 1917 before the first effective antidotes to the defensive-firepower-imposed stalemate of trench warfare came to be applied: use of tanks *en masse* by the British, 'Hutier' or infiltration tactics by the Germans, set-piece indirect artillery support by both sides.[27]

The industrial and logistics burden of the new weapons came as much of a shock to military staffs as their unforeseen tactical and operational consequences. No army began the First World War with what would turn out to be adequate ammunition reserves and, despite forty years of planning, even the vaunted German General Staff, with its precise railway time-tables, quickly faced insuperable logistics snarls.[28] Although by 1918 the problem of 'breaking through' the trench lines was solved in at least three different ways, no army in the Great War really had solved the riddle of logistical support very far forward of the railheads, or was ever capable of supplying a sustained advance. Inter-war advocates of mechanized warfare, especially J.F.C. Fuller, Charles De Gaulle, and Heinz Guderian, fully understood the critical importance of complementarity in military systems. And yet, despite the success of the Blitzkrieg, the *panzers* still had a logistical tail of horse-drawn wagons. It took the vast resources of the US motor and oil industries finally to equip an army – the Allied Expeditionary Force in Normandy – for a fully mechanized campaign.[29]

What emerges from these examples is a complex pattern of interdependencies, complementarities, and coevolution. Intra-system, subsystems, components and production processes coevolve: high-strength steels, gun-tube fabrication, and smokeless powders. Systems depend upon complementary innovations: tanks and mechanized infantry and logistics, communications. Virtually no discrete innovation or system makes much difference without complementary or coevolutionary innovations in organization and doctrine (Fleck, ch.18). In very few cases, if any at all, are the multiplicity of interconnections, consequences, and contingencies susceptible to analytical solution *ex ante*. In the evolution of military technology and techniques, as for all other technologies, and as for biological evolution, trial-and-error discovery, and sometimes learning, is much more common than prescient anticipation – a topic to which we return briefly below.

20.4   'Fitness' and the problem of selection

On the face of it, armed combat ought to provide about as rigorous a selection environment as there is – 'nature red in tooth and claw' not excepted. Usually in war there are winners and losers, certainly the quick and the dead. But as in biological evolution, the problem of precisely what is being selected in combat, at what level, and on what grounds, remains problematic. This is apparent even in historical episodes for which there is a superabundance of evidence. Two examples, one at an intermediate level, the battle of Jutland, one at a microlevel, Supermarine Spitfires versus Messerschmitt Bf 109s in the Battle of Britain, will serve to illustrate the general point.

Of the battle of Jutland in 1916, a dour and underwhelmed Basil Henry Liddell Hart (Royal Army, Ret.) wrote in 1930: 'No battle in all history has spilled so much – ink'.[30] Given that for the only time in history two great fleets of Dreadnought battleships locked in combat, it ought to be possible to determine which ships were superior, or at least which fleet was better, that is, more fit. Not so. Eighty years of analysis have not produced agreement on who even 'won' the battle, much less why.

In the great naval race leading up to the First World War, and ultimately Jutland, Britain and Germany had pursued well-defined, carefully thought-out strategies. Germany, under the inspiration of Admiral Alfred Tirpitz, embraced 'risk theory'. Forever doomed to numerical inferiority at sea, Germany still could challenge British naval hegemony, not by matching the Royal Navy ship for ship, but simply by building a High Seas fleet sufficiently large and powerful to put that hegemony at risk, at least in the North Sea. The Royal Navy countered with three moves: concentrating ever more of its strength in the Home (Grand) Fleet; shifting the Grand Fleet's principal anchorages to Scapa Flow and Rosyth, from whence it more easily could dominate the North Sea and its exits to the great waters of the Atlantic; and, beginning with construction of the first all big-gun battleship, *HMS Dreadnought*, in 1905, pursuing not only numerical superiority in capital ships, but also a sustained development lead in their qualitative design and construction.

The technical characteristics of German and British capital ships reflected the strategic choices of the respective navies. Class-for-class, German capital ships were somewhat smaller in displacement, had smaller guns, and were slightly slower. They were also substantially better armoured, had greater internal compartmentation, and were greater in beam, all of which reflected German calculation that it would be quicker and cheaper to repair a severely damaged ship than to build a new one. This was consistent with the German hope of slowly whittling the Grand Fleet down to size.[31] German optics and range finders also proved to be superior.

The larger, more heavily gunned British capital ships, by contrast, were less beamy, but offered greater range and better sea-keeping capabilities, especially in the North Atlantic and beyond, and provided better habitability for their crews (all of which reflected Britain's worldwide imperial commitments). Under Admiral Sir John Fisher, British design deliberately sacrificed protection, both armour and internal subdivision, to gain speed and gun power. Because of bureaucratic ineptitude, if not worse, British fire-control provisions were also inferior to those of the Germans.[32]

So which ships, or which fleet, was 'better fit'? In the 'cash transaction of battle' at Jutland, among modern capital ships (excluding pre-*Dreadnought* designs), the British lost three battle cruisers (*Invincible, Indefatigable*, and *Queen Mary*) to the German's one (*Lützow*). Casualties were similarly disproportionate. No battleships were sunk on either side. Indeed, although designed to dominate the high seas, *no Dreadnought*-type battleship *ever* sank another, even though such battleships did sink from virtually every other cause: torpedoes (launched from torpedo boats, submarines, and aircraft), bombs, mines and internal explosions. Yet, despite the disparity in losses, John Keegan's strategic assessment of Jutland is compelling:

> The Battlecruiser and Grand Fleets, with their accompanying shoals of destroyers and cruisers, had returned to Scapa Flow and Rosyth by 2 June. At 9:45 that evening Jellicoe reported to the Admiralty that his warships were ready to steam at four hours' notice. That signal writes the strategic verdict on Jutland. Britain's navy remained fit for renewed action, however soon it should come. Germany's did not.[33]

But again, so what? Other commentators suggest that it didn't matter anyway, that as a species battleships never accomplished anything of military value for any navy, that the Germans would have been better off devoting the resources that went into the High Seas Fleet to submarines, and that the British were correspondingly derelict in providing anti-submarine protection for the sea-borne commerce so essential to their war effort.[34] These criticisms, are, of course, profoundly ahistorical, projecting a counter-factual from 1917, when the war truly had become 'total', back to the pre-war period, when unrestricted submarine warfare was still unthinkable for a civilized nation. Even so, that such arguments are seriously entertained illustrates just how slippery imputation of 'fitness' to a military technology can be.

Just on narrowly technical criteria, it is virtually impossible to make well-founded judgements about the relative merits of British and German capital ship design: first, the ships themselves are a nested hierarchy of interdependent systems and subsystems, each of which is the material enactment of a set of arguments, negotiations and compromises. As noted, all capital ship designs, like all designs, as well as all operating procedures, represent a set of explicit, as well as sometimes tacit or unrecognized, trade-offs, between, say, gun power and protection, or magazine security and rate of fire. There is currently no simple metric for judging the global optimality of any single set of design decisions, or for comparing two designs. Second, not only is a given ship itself a sociotechnical system, but it is also embedded within concentrically larger sociotechnical systems (formations, fleets, logistic and support organizations, planning staffs): the performance of a ship, or even of its subsystems (such as its optics), cannot easily be separated from its whole sociotechnical context. Third, a capital ship is a product of historical processes, from deliberate strategic decisions, such as 'risk theory', to pure historical contingency and accident: a lucky hit on one of the after turrets of the German battlecruiser *Seydlitz* at Dogger Bank in 1915 revealed the danger of ready ammunition, held in the turret, flashing down the magazine trunks. The Germans modified their turret arrangements, but the British, not having been made aware of the severity of the problem, did not. All three of the British battlecruisers sunk at Jutland suffered magazine explosions, likely set off by turret hits.[35]

So, how should all these considerations be taken into account? How should the vicarious selectors (§4.1; §5.3; §12.3; §13.3; §16.4) of the two rival admiralties have weighed them in a hypothetical index of 'fitness' (§15.5)?

Similar considerations bedevil the evaluation of *any* military technology. The Supermarine Spitfire and the Messerschmitt Bf 109 are arguably the most famous, and probably the most analysed, fighter aircraft of all time. Certainly the Battle of Britain, from August to October 1940, is one of the most momentous battles in history: the *Luftwaffe*'s failure to suppress the RAF precluded German invasion of England. That the British won, however, hardly entails that the Spitfire was unequivocally superior to the Bf 109.

Indeed, there is no *the* Spitfire and no *the* Bf 109; both aeroplanes were in production throughout the war, and went through multiple versions and iterations. In the Battle of Britain, the predominant version of the Spitfire was the Spitfire I; of the Bf 109, the Bf 109 E-3. The Spitfire I was slightly faster, and considerably more manoeuvrable, but the Bf 109 E-3 climbed and dived better, had a higher service ceiling and a more powerful and longer-ranging armament.[36] Both fighters used liquid-cooled V-12 engines. The specific output of the Spitfire's Rolls Royce Merlin was superior, but, because it was a carburetted engine, it tended to cut out in negative-g manoeuvres, which gave the Bf 109, with its larger, direct fuel-injected Daimler-Benz DB 601A inverted-V engine, a tactical advantage.

Just as for capital ships at Jutland, any inferences about which aircraft was superior, or 'more fit', based on combat experience in the Battle of Britain, are muddled by context and contingency. First, British Fighter Command deliberately tried to avoid challenging German fighters, preferring instead to allocate its scarce resources to knocking down bombers. This strategem was made possible by Chain Home radar operators, who, by the time the Battle of Britain began in earnest, had learnt to distinguish fighter from bomber formations on their relatively crude scopes. Second, the *Luftwaffe* had mistakenly thought the larger (and much less maneouvrable) twin-engined Messerschmitt Bf 110 'Zerstörer' could provide long-range bomber escort, and had not anticipated operating Bf 109s over England. Because of this strategic (and tactical) miscalculation, the Bf 109s were not equipped with long-range drop tanks: on internal fuel, their combat time over southern England was limited to about ten minutes. The defending fighters, alerted by radar and vectored direct to their targets by fighter command using radar tracking, suffered no such disadvantage.[37] Thus not much can be inferred about the relative 'fitness' of the two aircraft, other than to say that the Spitfire and the Messerschmitt Bf 109 were 'about evenly matched'.

Nevertheless, this inability to determine relative 'fitness' in this specific instance does not refute the concept itself. Virtually all authorities agree that the Bf 109 E was substantially superior to the Hawker Hurricane I in the Battle of Britain, and even more superior to the French Morane-Saulnier M. S. 406 in the Battle of France in the spring of 1940.[38] Thus, for a technological system (and, one suspects, for a biological system as well), relative 'fitness value'. despite the term, should not be considered a single discrete value, or the proverbial Euclidean dimensionless point. Rather, 'fitness value' might be better construed as a notional quantity in an amorphous, multi-dimensional space defined by a multiplicity of performance parameters, perhaps only loosely 'centred about' some specific value calculated by arbitrarily weighting the various dimensional vectors. Within some unspecified domain or overlapping space around this value, the differences between two competing systems cannot be distinguished: a kind of poorly specified Heisenberg uncertainty principle for technology. But for differences larger than that domain of uncertainty, some systems are more fit than others, and will be selected for quite unequivocally.

Unfortunately for analysis, even when such unequivocal judgements about technological fitness can be made, they say nothing about the fitness, or survival value, of the larger sociotechnical systems (§17.5), or ecology, within which a given 'superior' technology is embedded: Messerschmitt Me 262 turbojets were superior to any Allied fighter aircraft in the Second World War, but did not enter service quickly enough, or in sufficient numbers, to have any effect on the outcome of the war. German Panther V, Tiger I and Tiger II tanks, and even up-gunned Panzer IVs, were markedly better than American M4 Shermans. But, given constraints on trans-Atlantic shipping capacity, the United States Army deliberately chose to have a much larger number of lighter, inferior tanks. In the context of US Army doctrine and resources, especially its abundance of artillery, it is not at all clear that that was a bad decision.

In general, then, inferences about technological fitness run into the same difficulties regarding selection criteria and levels of selection, not to mention simple historical contingency, as imputations of fitness to biological characters.

20.5   Directed mutation, vicarious selection and institutional memory

Like John Milton's protagonist in *Paradise Lost*, who thought 'Better to reign in Hell than serve in Heaven', our species seems unable to resist the temptation to plan. For technology, this planning commonly goes under the acronym 'R&D', or research and development (Perkins, ch.12; Stankiewicz, ch.17; Fleck, ch.18). It represents an attempt to direct the 'mutation' of technology along certain thought-to-be-desirable paths. Organizations and institutions, in this case military, deliberately choose the direction of this mutation, and vicariously select which variants survive and reproduce, and which go extinct. Moreover, among all organizations, the military has perhaps the best developed institutional memory – what ought to represent accumulated wisdom – to guide its technological choices.

But, as the foregoing discussion indicates, even direct technological selection, if it is presumed to be based on some sort of 'fitness', is an iffy business at best. Whether or not the military's directed mutation and vicarious selection are efficacious is even more problematic. Planning runs amok in at least three different ways. Innovations are fraught with unintended and unanticipated consequences. Consequences that were expected or intended often fail to materialize. Rejected alternatives, variants that all the best minds consign to extinction or oblivion somehow not only survive but also turn out to be superior.

We have already noted a number of cases of unintended consequences: the adoption of stirrups; the development of infantry rifles, or quick-firing, breech-loading field artillery; 'risk theory'. Likewise disappointed expectations: Sir John Fisher's all big-gun *Dreadnoughts* arguably never attained that absolute dominance of the high seas he and their other advocates foresaw for them. Furthermore, even the best, most quantitatively rigorous benefit/cost analyses, done in real time with minimal speculation or uncertainty, can go ludicrously awry. In 1969, *Project Hindsight* carefully compared the performance and cost advantages of the new turbojet Lockheed C-141 Starlifter, just then entering service, with the old fashioned and obsolete turboprop aircraft it was to replace, the Lockheed C-130 Hercules, which first flew in 1952.[39] In 1999 the C-141 had reached the end of its service life; the C-130 was still in service – and in production. The Hercules' robustness and short- and rough-field capabilities proved invaluable, while the C-141 suffered a series of embarrassing structural deficiencies.

The military's adventures in vicarious selection and directed mutation often fare no better. In the late 1940s, the United States Navy was convinced that the future of fleet air defence lay with guided missiles, both ship and air launched. It seemed obvious that the powerful radar capability required to detect targets at long range should be used either to control the missile directly or to 'paint' the target with electromagnetic energy that a passive sensor in the missile could lock on to. But a small group at the Naval Ordnance Test Station (NOTS) at China Lake, California, thought otherwise. Although not even officially tasked with missile guidance system development, they proposed a fully self-contained, infrared homing missile. The Navy Bureau of Ordnance repeatedly rejected their proposals. The NOTS team simply bootlegged R&D resources for two years, until they had an innovative design that was robust and cheap enough to persuade the Bureau of Ordnance to support prototype development. A little over two years later, in January of 1954, the Sidewinder infrared-homing anti-aircraft missile was first successfully tested.[40] It has become perhaps the most widely deployed air-defence missile in the world, with derivatives still in first-line service. In effect, weapons evolve best in institutional environments with sufficient organizational slack and surplus resources (cf. Fairtlough, ch.19) to afford experimentation, and its inevitable concomitant, failure.

Finally, at least since the formation of permanent general staffs during the Napoleonic Wars, most modern military establishments have historical sections whose function it is not only to keep minutely detailed historical records, but also to mine those records for tactical, operational and strategic insights into contemporary military problems. While these activities very likely do develop officers' tactical and operational intuition and judgement, there is no evidence that they render the military any more prescient about the future than other formal organizations. Having a mother lode of memetic material appears not to systematically promote adaptation, much less pre-adaptation: as we have seen, the military is not noticeably better at choosing technology or predicting its consequences than any other institution.

Similarly, since the Napoleonic Wars, there has been a tradition of scholarship about things military which claims that there exist immutable and veritable 'principles of war'.[41] Perhaps: but like purported 'rules of discovery' in science (§16.5), the devil is always in the details. As normally stated, the 'principles' are little more than common sense: 'economy of force', for example, translates into 'don't waste resources'. The question is, 'how?'. The trick with victory, as with discovery, is to reify the 'principles' to fit local, real situations.[42]

In short, despite commanding immense material, intellectual and historical resources, military institutions have proven no better at directing technological mutation, vicariously selecting technologies, or otherwise anticipating the future, than any other group. The military too gropes along via a process not demonstrably superior to blind variation and selective retention.

Military endeavour, combat and preparation for it has had an undeniable and profound impact on the content of technological and scientific knowledge, and on the direction of technological development. The effects of military technologies on the contingent course of world history are no less profound. Nevertheless, there is no reason to believe, and virtually no evidence to suggest, that the underlying processes of technological evolution for military technologies are in any way different from those for any other technology. Military technology too is an inductive achievement, conjectural, corrigible and hypothetical, like all other instances of enhanced fit of noumena to phenomena (Constant, ch.16).

21

Learning about technology in society: developing liberating literacy

JANET DAVIES BURNS

21.1   Understanding technological change

The relationship between technology and society has largely been neglected in formal technology education, which is mainly concerned with learning technological problem-solving, particularly in relation to recent technologies. The aim is to train future technologists for research and development and ultimately economic growth within nation-states. While such training is important for the development of technical inventiveness, on its own it also constrains future technology within established value and goal boundaries. In effect, the evolution of technological trajectories is limited to paths already envisaged within the existing paradigms.

In this chapter I discuss the extension of technology education, formal and informal, to include the history, philosophy and sociology of technology. I argue that this reform would provide for improved understanding of the nature of technological change and thus for reflection and critique of the process itself. In particular, education *about* technology contributes to the 'Lamarckian' efficacy of design in technological evolution by increasing the reflexivity and social diversity of actual and potential contributors to it. This broader technology education is especially significant in relation to contemporary technological change where new technologies have greater power to inflict harm than ever before.

21.2   The role of social groups in technological development

Recent historical work has improved our understanding of the relationship between technology and society. It is now recognized that technology is

299

culturally located (Mokyr, ch.5; Nelson, ch.6) and does not follow a pre-determinable line of 'advance'. Cultural values, social organization and technical knowledge are all significant in technology practice.[1] Technological products are 'hardened history, frozen fragments of human and social endeavour'.[2]

The adaptive relationship between a developing technology and its social and cultural environment shows up particularly clearly in cases of failure. All technologies must eventually become extinct when they get out of step with their cultural context.[3] For example, the original design of the Fordist assembly plant was influenced not only by the technical constraints of machine tool production (§18.7), but also by the prevailing view of immigrant workers as in need of paternalistic help. But rigidities developed with the appearance of a 'maintenance constituency' of people who had come to depend on the automobile industry and related industries for employment. As a result of the promotion of multiskilling and flexibility in the workforce, the assembly plant concept eventually became obsolete – with serious consequences for the 'impact constituency'.

But the Fordist assembly line may be considered only a stage in the evolution of a more general technology of mass production, giving way today to teamwork on the factory floor. Looking at technological change over a longer timeframe, Basalla sees it as a continuous process.[4] He traces the development of particular technologies, like the stone axe, and finds a process of incremental change indicative of an evolutionary mechanism. He thus rejects discontinuous 'revolutionary' accounts which depend heavily on the identification of 'heroic' inventors, motivated and supported by patents (cf. Carlson, ch.11). Basalla describes technological evolution as occurring within technical and cultural boundaries, but still subscribes to the notion of technological 'progress' within these boundaries.

The evolution of a particular technology[5] can take very different directions in association with different social groups. In the early years of the evolution of the bicycle (§18.10), for example, the high-wheeled 'Ordinary' bicycle was promoted for middle-class males as a sporting pastime. The danger associated with the difficulty of control, mainly because of the rider's height above the road and the absence of brakes, gave it an added fillip. When other social groups, including women, saw the bicycle less as a risky sport than as a means of transport, its structure was gradually changed to reduce the height and to add brakes. Bijker explains this in terms of the 'interpretative flexibility' of the product and thus the different meanings which may be attached to it by 'relevant social groups'. He goes further and suggests that each relevant social group establishes a 'technological frame' of goals for the technology and of ideas and 'tools for action' in its achievement. The frame both enables and constrains product development, and thus explains the 'stabilization' of a product as well as its change.

The involvement of different social groups, then, provides openings for innovation in technological development and for changes in the direction of technological evolution. Yet the generally accepted history of Western technology suggests the disproportionate involvement of white, middle-class men. What is counted as technology, as historically accredited in patents, clearly contributes to the construction of this view.[6] Thus, methods of food gathering and food preparation which women have been developing since pre-history, have not been recognized as technologies comparable with, say, hunting. Where women have been responsible for or contributed to accredited technological development in the past, patent regulations made it impossible for them to register as the legally recognized inventors. In any event, it was considered unseemly for women to draw attention to themselves by claiming responsibility for technological development.

The involvement of women in technological development has always been hindered by lack of access to education, and exclusion from what were originally all-male scientific institutions, such as the Royal Society. To obtain access to technical information they have had to rely on the patronage of male family members. The organization of technology today still impedes the entry of women through educational and social filters. Cockburn and Furst-Dilic[7] report that in a five-year study of a series of technological developments in Europe in the early 1990s women's involvement was negligible. Indeed almost no women were to be found seriously involved in the critical field of design engineering. It is true that all-male design teams contrived to introduce women through the familiar buyer behaviour surveys, 'imagining' women's preferences, asking their wives to try out products and employing women technicians and assembly line workers to 'stand in for the housewife' and advise on details of the final product. But this involvement was little more than token, since major design decisions, such as the development of a front-loading rather than a top-loading washing machine had already been made. Economic considerations took precedence over, for example, the problems for women who have to bend and stoop to fill the machine. The authors conclude that there is a mutual shaping of new technologies and gender relations.

Women have also been notably absent in the development of information technology, whose language is dominated by male and military metaphors.[8] 'Execute', 'kill' and 'abort', for example, have been used rather than other equally effective but less aggressive words such as 'run', 'cancel' and 'stop'. The approach to programming is designed for hierarchical and abstract thought, ways of working which are not preferred by girls and women. The language and structure of computer technology serve to discourage uptake by social groups who feel out of place in such an environment. By contrast, an alternative approach, characterized by negotiation, association and closeness to the materials, has developed among a small number of participants, including young women.

Until recently, Western sponsored technological programmes in developing countries have failed to involve the indigenous peoples whom the technology was supposed to assist, beyond the identification of outcomes and possible solutions. As a result, products were introduced that were inappropriate or failed in the local environment (cf. Macfarlane & Harrison, ch.7). Thus, Appleton[9] describes the biogas plant installed in a village in India where farming practices did not produce enough dung to operate it. She goes on to describe the re-evaluation of their roles by those working for sponsored technological programmes. Through 'participatory technology development' they learned to involve local people in decision-making regarding the particular technologies and skills required. Local knowledge as well as Western technological knowledge is now directed towards longer-term goals such as management and sustainability. Participatory technology development is not aimed just at developing appropriate technology, but also at enhancing the problem-solving capabilities of the local people so that they can take over more of the responsibility for technological change for themselves and thus escape from the trap of poverty and dependence.

21.3   Liberating literacy

While it is clear that the involvement of different social groups changes the direction of development of a technology, it is not so clear how their involvement comes about. How does a group, or an individual from a group, external to that concerned with an ongoing technological development, become involved to an extent that will allow change in the development?

Everybody is familiar with the exclusive character of professional talk in particular fields, such as architecture, engineering, medicine and music.[10] Goals and problem-solving methods differ significantly from field to field, and even within fields from site to site, reflecting historically constructed values and beliefs that are embodied in diverse theories and languages.[11] According to Gee,[12] the language used within a social group is part of a distinctive Discourse (with a capital 'D') – a combination of appropriate behaviour, social role and apparent values constituting a 'way of being in the world'. Discourses, he claims, cannot be learned in the conventional sense but have to be acquired through enculturation into social practices, primarily through personal apprenticeship.

We acquire a Discourse more or less fluently depending on how far we subscribe to its values and beliefs and are admitted to its institutions. Thus, individuals within the relevant social group of a particular technological frame have acquired the Discourse of that frame (§17.7).

Complete fluency within a single Discourse usually stabilizes a technology by internalizing uncritical acceptance of its practices and compliance with its values and beliefs. This is exemplified by the iron-age technology of scythe manufacture in present-day Nepal, as described by Macfarlane, Harrison and Martin.[13] Scythes are made by outcasts of the community who are literally untouchable and thus have little opportunity to acquire other Discourses through which to critique their ironwork.

In less constricted cultures, however, technologists often find that their Discourse conflicts to some degree with other Discourses they have acquired. Through such conflicts they are able to obtain meta-knowledge of their Discourse and to develop what Gee calls 'liberating literacy' – that is, the ability to challenge accepted practice and envisage novel variants worth testing. In effect, liberating literacy reveals the ideologies underlying social practices,[14] and opens the way to evolutionary change. In industrialized countries, technological Discourses belong to a dominant category, protected by their association with social prestige and other goods. Nevertheless, women who engaged with the technology of the Ordinary bicycle,[15] and would thus have been apprenticed to that Discourse, were able to critique it through the Discourses of their everyday lives.

Indeed, liberating literacy often develops out of the clash of two different technological Discourses. Baekeland, for example, worked in electrochemistry as well as in celluloid chemistry.[16] The meta-knowledge gained through participation in the electrochemistry Discourse enabled him to critique practices within the technological frame of celluloid chemistry and thus develop a novel material, 'baekelite'. Again, Edison's development of a telephone,[17] although primarily within an electrical frame (Carlson, ch.11), was significantly influenced by his access to the Discourses of acoustics (including interestingly a non-professional Discourse) and human biology. In evolutionary terms, an established technological Discourse delineates a circumscribed region of the 'search space' for conceivable innovations. But the most successful inventions are often those that combine features drawn from the search spaces of several Discourses.

Again, there are contradictions between contemporary feminist Discourses and the traditional scientific Discourses developed historically by mainly male scientists.[18] These contradictions enhance the ability of feminist scientists to critique scientific Discourses and contribute to their ongoing development. I would argue that a similar form of liberating literacy holds for feminists and other social groups whose values and beliefs conflict with traditional technological Discourses. Indeed, even what counts as a rational argument may differ from one Discourse to another. Thus the mediating influence of background assumptions is as significant in the identification and solution of technological problems as it is in science.[19] Technology professionals belonging to different Discourses often arrive quite rationally at different interpretations of what is needed and how these needs should be met. There is nothing 'illogical' about the way that social groups construct different technological frames and are responsible for the development of quite different technological products.[20]

As Walter Vincenti emphasizes (ch.13) working engineers are well aware of the mediating effects of social considerations on technical rationality, although maybe less so of its effects on their tacit knowledge and intuitive thinking. One of the main themes in evolutionary thinking about cultural entities (Nelson, ch.6) is the long-running interaction between 'material' and 'social' factors in the production, selection and adaptive survival of novel variants. It is clear that innovations that can satisfice the very diverse requirements – both technical and volitional (not to be understood in psychological terms, but rather as commitment to choice based on disposition with respect to values and beliefs)[21] – of the range of lifeworld Discourses[22] (where the 'lifeworld' embodies human-technology relations in both immediate practical and broader cultural contexts) are only likely to emerge from a 'search space' (Perkins, ch.12; Stankiewicz, ch.17) spanned by equally diverse selection criteria at all stages of development.

But such extended spaces do not develop automatically – especially in an advanced capitalist culture where specialized research and development teams are created by market competition and intellectual secrecy. Longino[23] argues for the deliberate promotion of more 'collective' conceptions of rationality and objectivity. Thus 'rationality in a social context' would identify the background assumptions that are invisible to professionals who are only fluent in a particular Discourse and expose them in the wider community, where they become open to criticism, and can be defended, modified or abandoned. Clearly, this alternative Discourse would require support by new social institutions embodying recognized avenues for criticism, shared standards, responsiveness to the whole community, and equality of intellectual authority of qualified practitioners. But the outcome would be a socio-technological ecology favouring innovations that had been better evaluated and better fitted for their purpose than at present.

21.4   Liberating literacy in a risk society

Post-modern scepticism about the possibility of achieving a universally enlightened social order is no argument against the need to promote 'rationality in a social context'. Indeed, with globalization and the rise of international intellectual elites, this need is becoming ever more acute. No educated person nowadays can be unaware of the long-term dangers posed by such techno-scientific developments as the manufacture of nuclear weapons, the diffusion of ecotoxins into the environment, the highly geared social organization of cities and the engineering of genetic systems for commercial profit.[24] There is a widespread feeling that we need to develop worldwide ethical control, by legislation or otherwise, of emerging technologies before they inflict their ethics on us.[25]

These threats are summed up in the concept of a 'risk society'. According to Beck,[26] this is an industrialized society in which products of technological change have damaging effects on future generations of human beings, animals and plants as well as on the physical environment. These effects often spread far outside the nation-state in which the technology is produced, and cannot be avoided even by people of great wealth and social power. In a risk society, therefore, science and technology have to be concerned not only with nature but with the products of their own success. In effect, *modernity* has to become *reflexive* in its pursuit of 'progress'.

Perceptions of escalating risk are increasing public critique of science and technology. Citizens' groups challenge local developments such as storage of nuclear waste, and transnational groups tackle global issues such as deforestation. Parents of children whose health has been affected by smog, and farmers whose livelihoods have been taken away by radioactive fallout, challenge the bases on which risk is calculated, often demonstrating the value of local knowledge[27] derived from their lifeworld Discourses.

More generally, many of the values and beliefs underpinning traditional science and technology are coming under public scrutiny. In particular, one of the central values of technoscience has been the control of nature. This is linked to anthropocentrism and, through individualism, to the typically reductive rationale of science. It is driven on by 'economic' competition between nation-states,[28] supported by the social contract between individuals and the state in which decisions are made on the basis of what is good for the state, regardless of what is good for the planet as a whole.

But feminist and indigenous groups have long supported an alternative view[29] where cooperation and identity with nature are particularly valued. Thus, the interdependence of human beings, other living organisms and the physical components of the earth embodied in the concept of Gaia[30] requires that what counts as 'economic' should fully include social and environmental costs – through a new technological Discourse that mimics the ecosystem in maintaining energy flows, and recycling all its waste products.[31] The mutual dependence of human beings supports emotional fellowship, not intelligence, as the key to humanity.[32]

But how should this reflexive response to the threats of a risk society be put into effect? At present technology is run by technicians and business, and operates 'in the grey zone between law and politics'.[33] It is possible to think of alternative modes of organization that would give lower priority to crude economic and military priorities, such as the public funding of technological innovation and the establishment of new institutions, such as 'technology courts', to manage it.[34] But such developments within nation-states will have little effect on the globalized economy, where the ability to move capital between states will mean that technology is relocated to states where there is least constraint. There is thus a need for new international institutions, with authority above that of nation-states.[35]

Opposition to such institutional change by supporters of 'free enterprise' suggests that constraint will prevent the generation and selection of the variants necessary for technological innovation in a modern industrial society, and thus stifle technological evolution.[36] What such opposition ignores is the constraint which already exists, albeit by largely tacit, self-interested agreement, in the prioritization of narrow economic goals, and the reflexiveness of markets.[37] Markets are created; they do not exist outside human values, they respond to human perceptions and inflict their consequences on us in a two-way relationship. Doubts about our ability to anticipate threats of technological change are challenged by advances in risk research which recognize social as well as scientific criteria.[38]

Anticipation and constraint of long-term and widespread risk foreshadows change in the social organization of technological change. The threats of a risk society bring into sharper relief the influence of cultural values and organizational structures on the direction of technological change, and thus the 'Lamarckian' character of technological evolution. Changing values drive the liberating literacy evident among socially and environmentally conscious citizens and, more recently, multinational organizations and governments. Selection does not take place with respect to any random generation of variants.

In practical political terms, as people become aware of their exclusion from the Discourses where action is shaped, they demand the increased openness of public debate that is characteristic of reflexive modernity.[39] Already we see self-help groups on, for example, local environmental issues and social movements like feminism opening up spaces for public dialogue and developing new global solidarities. The evolutionary riposte to the risk society is not necessarily an uneasy balance of power between countervailing institutions. It may well be the emergence of a 'dialogic' or 'deliberative' democracy,[40] largely operating outside the formal political sphere and achieving its goal of conflict resolution through public discussion of policy in which all points of view can be heard. Instead of seeking 'the correct answer', the aim might be to develop and legitimate procedures and criteria for policy assessment in a particular sphere, thus broadening the search space for generally acceptable outcomes.

21.5   Education for technology

Technology, conceptualized holistically through technological problem-solving, has recently appeared in the school curricula of most Western and many developing countries (Solomon, ch.14). It is often a compulsory replacement for optional technical subjects and has usually been introduced to meet governments' perceptions of the need to foster entrepreneurial activity for economic success. Though expressed in a variety of forms that to a greater[41] or lesser[42] degree recognize the significance of the social context for technological problem-solving, it seldom consciously enables the internal critique that would permit the development of new Discourses. Even the recognition of social contexts is difficult if the focus is on technological problem-solving within one's own society. Underlying values and beliefs show up much more clearly against the background of other historical times or cultural locations.[43]

Evolutionary accounts of technological change are sometimes enlisted in support of the notion that technological education need not go beyond this policy of training people to perform technological tasks 'blindly', without regard to their social and moral implications.[44] But the idea that technological development merely requires the technical training of a certain percentage of inventive individuals[45] is too narrow even by its own criteria, since it fails to recognize the way that the contributions from individuals with different ideas, experiences and beliefs combine to enhance collective creativity. While it is in the immediate interests of governments to protect current goals and practices in technological problem-solving, it puts off the day when the hazards of a risk society must be met by new goals and practices. As we saw above, better technologies for society can only come through the emergence of Discourses that enable critique of current technological Discourses. The citizens of a 'dialogic' democracy clearly require various types of liberating technological literacy (depending on their membership of other Discourses) where these distinct elements are combined. In other words, as well as being encultured to some extent into particular technological Discourses, they need to acquire more general understanding of the nature of technological change and its relationship with society.

This meta-knowledge is accessible through the medium of technology studies – the history, philosophy, sociology and psychology of technology. Some education in these subjects must now be considered a primary responsibility for schools, colleges and universities. But it may require new educational practices. Helping students 'to acquire . . . Discourses that lead to effectiveness in their society . . . and to imagine better and more socially just ways of being in the world'[46] is not the same as enculturing them, through apprenticeship or through formal education,[47] into a conventional technological Discourse. The relationship between these two very different elements of liberating technological literacy is obviously critical, and still needs to be thought through by enlightened educationalists.

What of potential recipients of this education? The surge of interest in promoting technology for research and development and economic growth has stimulated many studies of public attitudes towards, and understanding of, science and technology. People in the industrialized world evidently have very mixed feelings about technology. While they are interested in new technologies, they are concerned about their possible effects. In New Zealand,[48] for example, it was found that adults frequently watched television programmes and read newspaper articles about science and technology, but struggled with doubts about the trustworthiness of scientists and technologists. Junior secondary school students were keen to study and to look for careers in technology, but were concerned about the potential of new technologies for harm.

Overall positive public attitudes toward technology thus tend to mask perceptions of problematic issues. Any proposed technology education needs to recognize the complexity of such feelings. Access to Discourses in technology studies would provide individuals with the means for resolving conflicting views – means that are not available from uncritical training in the application and development of current technological capabilities. Media education which provides no potential threat to government positions in democracies incorporates such internal critique. Although it is still possible in school technology for students to develop the meta-knowledge with which to critique technological practice, this is largely reliant on individual teachers.

In universities and colleges, technology is taught through the specialisms of, for example, engineering, information technology, food technology and bio-technology. Curricula are traditionally concerned with the relevant body of knowledge and with practical processes of experimentation and problem-solving. The philosophical underpinnings of such disciplines are not normally included. In an innovatory programme in Sweden, however, Benckert and Staberg[49] addressed the gendered nature of science and technology through a course on feminist perspectives on science and technology. This course was intended for undergraduate engineering students, with the aim of strengthening women's position in the field. It required both male and female students to read material presenting various positions and to adopt a position of their own – an approach which these engineering students, used to material for which there was always a right answer, found difficult, but which clearly introduced the need for critical reflection.

Non-formal education, unconstrained by formal examination systems, offers more scope for the development of innovatory programmes in technology. Courses on environmental issues, alternative medicines and organic gardening, for example, provide new ways of living in the world and new ways of viewing the world. Many 'new start' courses for women, and especially for women from minority ethnic groups, are founded on feminist philosophies which assist women to review their roles in society and set new directions. There are also courses that introduce women to technology, but few as yet that combine the two approaches. But another Swedish programme at a provincial city technology centre[50] offers women courses based on local industry and home technology. These courses demonstrate the value of 'knowledge in context'[51] by giving participants the opportunity to develop knowledge of value to them in their lives by integrating professional Discourses with their existing knowledge.

Informal technology education occurs widely through the media. According to Nelkin,[52] however, the media have developed a special relationship with science and technology that is absent in all other areas of reporting. Whereas media professionals exercise critical reflection in reporting other matters, they connive with scientists and technologists to present science and technology as 'progress' and 'a good thing'.

In the case of television, however, there is evidence of change. Though most TV programmes are uncritically supportive of new technologies, a small but increasing number do provide opportunities for reflection. Bennett[53] has traced the development of the British Broadcasting Corporation's *Horizon* programmes since the 1960s. Awe at the wonders of new technology gave way in the 1970s to suspicion, especially with regard to ecological issues. This sceptical attitude was followed in the 1980s by concern about the politics of science itself. In the present decade, programmers are responding to public interest by bringing science 'closer to home', thus developing a healthier relationship between the media and scientists.

In a similar vein, museums of science and technology have begun to take a more critical attitude in their presentations. Some interactive science centres,[54] which would normally prioritize current scientific and technological knowledge through hands-on – though not always minds-on – activities, now contextualize activities. It may be that these and other initiatives for 'public understanding of science and technology' mainly appeal to the small sector of society which already has such interests.[55] Nevertheless, people who would not otherwise engage seriously with science and technology are becoming more aware of its values and beliefs, especially when personally motivated by, for example, the illness of a family member or the construction of a nuclear power station in their locality. The first step towards liberating literacy is the confidence to seek access to scientific and technological information and the acquisition of the knowledge required to do so.

21.6   Conclusion

What I have tried to do in this chapter is to demonstrate the profound educational implications of seeing technological change as a purposive evolutionary process. It is often thought that evolution is a completely blind process, pre-determined perhaps by some intangible force for 'progress',[56] but as inscrutable and as implacable as Fate. But that assumes that all evolutionary processes are as narrowly 'Darwinian' as they mostly are in biology. The evolution of cultural entities, such as technological artefacts and systems, can to some extent be steered through such 'Lamarckian' factors as learning from experience and design for intended use. These factors are shaped by changing cultural values and social organization and appear in the factors from which selection is made.

What is more, it is a mistake to underestimate the role of the selective environment in the evolutionary paradigm. As we have shown throughout this book, the selective environment for technological products and practices – the ecosystem that supports such a diversity of technological Discourses – is society at large. And as Giambattista Vico observed, nearly three centuries ago,[57] this is more intelligible than the God-given natural world, because we make it ourselves. In other words, by systematically pressing for change in general values and beliefs, we slowly alter the social 'climate' to which technology must adapt.

The threats of a risk society impel us towards a climate of values that gives much greater weight to the longer-term, wider consequences of each and every technological innovation. For professional educators, for their unofficial colleagues in the media and for enlightened scientists and technologists, this is a challenge that should rule all their work. But formal and informal means of education are not the only channels for systematic pressures of this kind. The direction of technological evolution can be altered by institutional changes that bring the knowledge and beliefs of different social groups into the variation-selection-replication cycle.

A wider perspective thus shows technological innovation as only one aspect of the evolutionary transformation of our whole culture. This transformation goes beyond opening up the creation and selection of technology to a diversity of

Discourses. It involves the installation of an ethic of reflexive modernity, practised through dialogic democracy. This would seem to be the selective environment required for the emergence of imaginative technologies chosen to meet the needs of the planet and all its inhabitants.

An end-word

BY ALL CONTRIBUTORS

This book is already too long, so these final remarks must be correspondingly brief. We started out with a simple question: could the obvious analogy between technological innovation and biological evolution be developed from a 'metaphor' into a 'model'? Where have we got to? Certainly not to a 'model' in the mechanistic sense. That was not to be expected. Technological change is, above all, a social phenomenon. As such, its categories and their interactions are too imprecise and contextual to be represented realistically by a computable algorithm. Indeed, the same holds for any evolutionary system. A mathematical simulation of its contingent, path-dependent behaviour can never be true to life. It too is a metaphor in its representation of a real system with complex unquantifiable structural relationships between its elements.

What is more, a closer study of the structural analogies between technological and biological systems did not make them seem less problematic. The notion that an artefact is 'just like' an organism began to seem a little strained. We became more aware, for example, of the distinction between biological self-reproduction and replicative manufacture, and of the multiplicity of their forms, from cell-division to placental birth in the one case, from flint-knapping to computer-aided engineering in the other. It seems presumptuous to try to cover such a diversity of processes under a single conceptual umbrella.

But it would be a mistake to water it down into an abstract metaphysical tautology. Darwin's naturalistic account of temporal change in the living world has a logical coherence and proven explanatory power which are hard to match. We have no wish to set it up as a comprehensive theory of cultural change, in competition with social constructivism, historical determinism or self-organizing complexity. Indeed, we were divided amongst ourselves on the fundamental issue of whether or not it is proper to see cultural evolution as a human extension of biological evolution, or as an entirely different type of process.[1]

But we have come to see the evolutionary perspective as an indispensable tool of thought, highlighting a vital aspect of all historical processes that is easily overlooked in particular cases. In real-world stories of technological innovation, it may well be very difficult to dissect out the separate components of a classical 'Darwinian' process – variation, selection, replication, etc. Nevertheless, these operations can often be identified as the fundamental functional elements in the dynamics of change. Whether or not such an approach can be considered sufficiently rigorous to count as a 'model-building', it certainly enriches, facilitates and partially shapes our understanding of what is going on.

Our contributions to this book are not, then, jigsaw pieces that are supposed to fit together overall into a coherent picture. What they do, rather, is illustrate the explanatory power of 'evolutionary reasoning' in a very wide variety of contexts. They show the effectiveness of 'selectionism' as a unifying 'paradigm of rationality',[2] that can provide a conceptual vocabulary for the particular interpretations that it evokes in otherwise very diverse circumstances. Although it deals with an extraordinarily heterogeneous human activity, this book was made and can be read as a single work. Each chapter is on a different topic, yet they are all interwoven with these common concepts.

Some of the old issues scarcely figure on this new tapestry. For example, it does not seem to matter now whether technological evolution should be considered 'Darwinian', 'neo-Darwinian' or 'Lamarckian'. These were grand, if ill-defined, theoretical schemes in the history of biology, but their names serve nowadays mainly as coded references to particular mechanisms of heredity, such as strictly genic replication, or the inheritance of 'acquired' characteristics. What we now realize is that these are only a few of the many mechanisms by which populations of evolving entities – biological as well as cultural – actually maintain and reconstitute themselves, generation by generation. Systematic inheritance of selectable traits is essential for evolution to take place at all. The task is to discover the relevant mechanism in each case, not to squeeze it into one or another of these preconceived categories.

Similarly, we no longer feel impelled to find technological analogues for the most familiar evolutionary concept in biology – the notion of a *species*. There can be little doubt that evolutionary change almost always involves a considerable amount of categorial diversification. The history of technology is full of radical advances, differentiations and divergences that could be described as speciation events. But it also contains convergences and mergers. Material artefacts and the concepts they embody do not always mimic living organisms by separating into 'natural kinds'. Inventive institutions exhibit features both of individual organisms and of whole populations. What is more, speciation is not an essential feature of evolution, since many evolving organisms, such as bacteria, cannot be differentiated rigorously into mutually exclusive species.

Another much-debated question – whether or not to use the term 'meme' – also seems very academic. In any case, we have no right, or duty, to try to impose a systematic nomenclature on a subject that is still so attractively protean. But in talking about thinking, one cannot avoid referring to mental entities such as ideas, concepts, notions, beliefs, designs, theories (etc.). These all have the property of being able to persist and be transmitted, more or less unchanged, from mind to mind. They can also 'evolve' – that is change by 'variation' and 'selection' along the way. These properties undoubtedly play a major part in all forms of cultural change. Evolutionary reasoning about technological innovation thus needs a broad generic term for all such entities, indicating that they typically have these capabilities. But selectionism does not require that 'memes' should also be atomic, fecund, faithfully replicable, long-lived, binary-codable and so on. The fact that 'meme' rhymes (evocatively, if imperfectly!) with 'gene' does not mean that they must be as like as two peas. On this understanding, therefore, we now feel quite free to use this term wherever it helps the argument.

What we have found, indeed, is that 'memes' are major players – perhaps *the* major players – in technological change. From an evolutionary point of view, material artefacts cannot be considered in isolation from their cognitive and social correlates. Their making and use are closely linked with various bodies of human knowledge and with the collective activities of various human groups – craft guilds, factories, industrial corporations, commercial firms, research institutes, shops, armies, and so on. As the artefact changes, so does the cloud of ideas and social activities that surround it. An anthropologist's 'thick description' of what evolves would necessarily include numerous entities in these other domains. Indeed, we now realize that the evolutionary biologist must similarly include the epigenetic and behavioural inheritance mechanisms that run in parallel with the genic system.

But the situation in technology is complicated by the fact that both ideas and organizations can change in ways that do not directly involve the artefacts with which they are connected. It is characteristic of advanced societies that 'techno-memes' and scientific concepts can be transmitted, stored, mutated, recombined, accepted and rejected by a variety of processes, such as publication, theoretical analysis, experimental refutation, critical debate, and so on, as well as by direct appeals to the crucial test of use. Similarly, organizations such as research groups, industrial firms and military units may prosper or become extinct for a variety of other reasons than the competitive quality of their ideas, 'routines' or technological capabilities. Thus, the various other elements in the evolving artefact complex are often geared into other evolutionary cycles in their larger cultural and/or intellectual domains. In other words, from some points of view technological innovation can be considered an 'ecological' process, where a number of entities of different types – material, social and cognitive – mutually interact and *coevolve* as they adapt to one another.

Whatever way we look at it, however, the knowledge element in technology gives it some peculiar dynamical properties. These are often summed up in the word 'design'. Thus, the inventor, by the manipulation of intangible mental models, is able to design – or 'breed', or 'artificially select' – far more promising variants than would turn up by chance. But the inclusion of design does not entirely bar an evolutionary approach. What it signifies, in our more general perspective, is that imaginative linkages with selective functions are established between different points along the historical trajectory of the system. The range of feasible variants at a given moment is not limited solely by *present* circumstances, such as the materials and tools currently available: it is also conditioned by memories of *past* circumstances, such as unsuccessful configurations and ideas, and by mental images of *future* circumstances, such as of a hypothetical device in action.

Our evolutionary perspective thus helps us to understand the involvement of both *learning* and *imagination* in invention. Technological innovation has long made use of 'vicarious selection', based on past experience, to model the adaptive environment and thus reduce wasted effort on variants that are almost certain to fail in use. In recent years, this has been extended into a whole system of RDD&D (Research, Development, Design and Demonstration) where 'virtual artefacts' – for example, preliminary designs – are subjected to theoretical analysis, computer simulations and practical tests before being put on the market. Complex, multi-level search processes have thus emerged, where nested hierarchies of variation and selection accumulate and use vast quantities of experienced and inferred information. Indeed, just as in the inner workings of individual inventive minds, the basic 'BVSR' operations – Blind Variation and Selective Retention – intermingle and alternate so rapidly that they merge into a continuous creative activity resembling 'directed variation'.

The long history of technological innovation thus exemplifies the most macro of evolutionary phenomena: the emergence of new modes of change. The collectivization and institutionalization of invention in the twentieth century is only the latest in a sequence of transitions to new technological regimes, where artefacts, knowledge, organizations and individuals take on new configurations and where even the basic evolutionary processes identified by Donald Campbell enter into new relations. Whatever the underlying explanation for any such event, it obviously has far-reaching implications for evolutionary change throughout the cultural domain.

It is worth noting, moreover, that there is an entirely analogous 'evolution of evolution' in the biological world. This shows clearly in the succession of major transitions to new 'evolutionary regimes' identified by Maynard Smith and Szathmáry.[3] Despite our divisions of opinion on this point, it is evidently possible to view the whole scene through the wide-angled lens of evolutionary epistemology, where the various phases of technological evolution appear as merely the most recent extensions of the adaptive knowledge-handling capabilities of life itself.

But at a much more mundane level, an evolutionary perspective – whether we call it an 'analogy', a 'metaphor' or a 'model' – is clearly a very fruitful way of looking at the actual business of making and doing. As we have seen at various points throughout this book, it poses practical questions and suggests useful answers for designers, technology managers, policy makers and others in industry, government and academia. These insights are not only valuable in themselves: they also help us to develop and refine our understanding of the underlying principles of technological change.

Notes

CHAPTER 1

[1] Collins English Dictionary

[2] Gould 1996

[3] Toulmin 1972; Campbell 1974, 1977; Oeser 1984; Wuketits 1984; Mokyr 1990

[4] Lorenz 1962; Campbell 1974; Riedl 1984; Munz 1993

[5] Basalla 1988; Mokyr 1990

[6] Maynard Smith 1972; Ridley 1996

[7] Dawkins 1976

[8] Ziman 1995

[9] Vincenti 1990

[10] Dennett 1995

[11] Edelman 1992; Plotkin 1994

[12] Campbell 1974

[13] Boden 1994

[14] Collins English Dictionary

[15] Ziman 1996b

[16] Nelson 1995

[17] For example, Boyd & Richerson 1985; Durham 1991; Barkow et al. 1992

[18] Diamond 1997

[19] Searle 1995

[20] Boulding 1981; Nelson & Winter 1982; Saviotti 1996

[21] For example, Hull 1988

[22] For example, Jablonka & Lamb 1995

[23] Campbell 1977

[24] For example, Levy 1992; Langton 1995

[25] Cohen & Stewart 1994; Kauffman 1995

[26] Plotkin 1994; Munz 1993; Cziko 1995

CHAPTER 2

[1] Cziko 1995

[2] Maynard Smith 1986

[3] Mayr 1982

[4] Maynard Smith & Szathmáry 1995

[5] Edelman 1992

[6] Wilson & Sober 1994

[7] Heyes & Galef 1996

[8] Brown 1979

[9] Gould & Vrba 1982

[10] Lewontin 1978

[11] Laland et al 1996

[12] Bush 1994

[13] Ruse 1996

[14] Gould 1996

[15] Bonner 1988

[16] Jablonka & Lamb 1995; Jablonka & Szathmáry 1995; Maynard Smith & Szathmáry 1995

CHAPTER 3

[1] Johanssen 1911

[2] Sapp 1987; Jablonka & Lamb 1995

[3] Waddington 1957

[4] Mayr 1982, p.828

[5] Maynard Smith 1986

[6] Maynard Smith & Szathmáry 1995; Jablonka & Szathmáry 1995

[7] Holliday 1987; Jablonka & Lamb 1989, 1995

[8] Heyes & Galef 1996

[9] Sniegowski & Lenski 1995

[10] Brenner 1992

[11] Cavalli-Sforza & Feldman 1981; Boyd & Richerson 1985

[12] Durham 1991; Feldman & Laland 1996

[13] Hull 1988

[14] Jablonka & Lamb 1995

[15] Blau 1992

[16] Grimes & Aufderheide 1991

[17] Grimes & Aufderheide 1991

[18] Penrose 1959
[19] Grimes & Aufderheide 1991
[20] Patino *et al.* 1996
[21] Ettinger & Doljanski 1992
[22] Cavalier-Smith 1996
[23] Jablonka & Lamb 1989
[24] Holliday 1990
[25] Moehrle & Paro 1994
[26] Wolffe 1994
[27] Meyer *et al.* 1992
[28] Fedoroff 1989
[29] Mikula 1995
[30] Das & Messing 1994
[31] Jablonka & Lamb 1995; Jablonka *et al.* 1995
[32] Roemer *et al.* 1997
[33] Buss 1987; Edelman 1987; Klekowski 1988; Sachs 1988; Jablonka & Lamb 1995
[34] Alberts *et al.* 1989
[35] Edelman 1987
[36] Klekowski 1988
[37] Sachs 1988
[38] Krakauer & Pagel 1996
[39] Dawkins 1976
[40] Griesemer 1999
[41] Griffiths & Gray 1994
[42] Griffiths & Gray 1994; p.296

CHAPTER 4
[1] Kauffman 1993, 1995
[2] Campbell 1960, 1965, 1974
[3] Plotkin 1994
[4] Dawkins 1976; Hull 1988; Plotkin 1994
[5] Munz 1993
[6] Campbell 1974
[7] Cziko 1995
[8] Edelman 1992
[9] Craggs 1989
[10] Campbell 1977; Cziko 1995
[11] Csikszentmihalyi 1993; Munz 1993
[12] Plotkin 1994
[13] Arthur 1994

[14] Lovelock 1979
[15] Edelman 1992
[16] Edelman 1992
[17] Campbell 1966, 1990; Callebaut 1993
[18] Boden 1991; Donald 1991; Dennett 1991, 1995; Plotkin 1994; Cziko 1995
[19] Cavalli-Sforza & Feldman 1981; Boyd & Richerson 1985; Durham 1991
[20] Pinker 1994
[21] Flam 1994
[22] Holland 1975
[23] Levy 1992; Kelly 1994; Langton 1995
[24] Langton 1984, 1989
[25] Rasmussen *et al.* 1990
[26] Ray 1992
[27] Wilson 1986
[28] Hillis 1990
[29] Collins & Jefferson 1992
[30] Boyd & Richerson 1985
[31] Kauffman 1993
[32] Kauffman 1993, 1995; Gell-Mann 1994; Cohen & Stewart 1994

CHAPTER 5
[1] Marshall 1930 [1890]; Thomas 1991; Hahn 1991
[2] Ramstad 1994; p.82; Penrose 1952
[3] Campbell 1974
[4] Dawkins 1976
[5] Dennett 1995
[6] Mokyr 1996
[7] Basalla 1988; Petroski 1993; Farrell 1993; Kauffman 1995
[8] Nelson & Winter 1982
[9] Dawkins 1982; Lewontin 1992
[10] Dawkins 1976
[11] Vincenti 1990

[12] Campbell 1974
[13] Crouch 1989; Vincenti 1990; Bagley 1990
[14] Stebbins 1982
[15] Maynard Smith 1976
[16] Kimura 1992
[17] Cziko 1995
[18] Stebbins 1982
[19] White 1978; Mokyr 1996
[20] Gould 1997
[21] Gould & Vrba 1982
[22] Ohlman 1990
[23] Campbell 1974
[24] Cavalli-Sforza & Feldman 1981
[25] Eldredge 1989
[26] Kauffman 1995
[27] Futuyma 1986
[28] Cavalli-Sforza & Feldman 1981
[29] Cavalli-Sforza & Feldman 1981
[30] Dawkins 1976
[31] Vincenti 1990
[32] Nelson & Winter 1982
[33] Evans & Rydén 1996
[34] Polanyi 1958
[35] Hull 1988; Lloyd 1992
[36] Dawkins 1982; Eldredge 1995
[37] Penrose 1952
[38] Dawkins 1982
[39] Eldredge 1995
[40] Ruse 1996
[41] Bijker 1995
[42] Sen 1993
[43] Mokyr 1997
[44] Dennett 1995, p.111
[45] Ridley 1985
[46] Ruse 1996

CHAPTER 6
[1] Nelson 1995
[2] Lumsden & Wilson 1981
[3] Cavalli-Sforza & Feldman

1981; Boyd & Richerson 1985; Durham 1991

4 Campbell 1960, 1974

5 Plotkin 1982; Hull 1988

6 Popper 1968; Kuhn 1970

7 Landes & Posner 1987; Demsetz 1997

8 Chandler 1962, 1990

9 Nelson & Winter 1982

10 Nelson 1995

11 Vincenti 1994

12 Abrahamson 1996

13 Nelson 1962

CHAPTER 7

1 Hayami 1987

2 Campbell 1960

3 Mokyr 1990, p.35

4 Mokyr 1990, p.38

5 Mokyr 1990, p.34

6 Mumford 1947, p.115

7 Mokyr 1990, pp.44–5

8 Birdsall and Cipolla 1980, p.86

9 Mumford 1947, p.118

10 Bird 1984 [1880], p.49

11 Bird 1984 [1880], p.128

12 Bird 1984 [1880], p.121

13 Bird 1984 [1880], p.176

14 Purchas 1938

15 Kaempfer 1906 [1727], vol.1, pp.194–6

16 Kaempfer 1906 [1727], vol.2, p.376

17 Kaempfer 1906 [1727], vol.1, p.194

18 Thunberg 1796 [1793], iv, p.95

19 King 1911, pp.237–99

20 Chamberlain 1990 [1904], p.20

21 Kaempfer 1906 [1727], vol.3, p.202

22 Thunberg 1796 [1793], iii, p.108

23 Thunberg 1796 [1793], iii, p.134

24 Oliphant 1859, p.139

25 Alcock 1863, p.477

26 Morse 1936, Day i, p.425

27 Morse 1936, Day i, p.347

28 Jansen & Rozman 1988, p.463

29 Morse 1936, Day 1, p.66

30 Morse 1936, Day 1, pp.345–6

31 Morse 1936, Day 1, p.46

32 Morse 1936, Day 1, p.9

33 Morse 1936, Day ii, p.51

34 Morse 1936, Day ii, p.284

35 Morse 1936, Day i, pp.46–7

36 Morse 1936, Day ii, p.271

37 Purchas 1938, p.147

38 Kaempfer 1906 [1727], vol.3, p.202

39 Thunberg 1796 [1793], iii, p.137

40 Alcock 1863, i, p.295

41 Morse 1936, Day ii, p.139

42 Morse 1936, Day ii, p.332

43 Morse 1936, Day ii, p.326

44 Thunberg 1796 [1793], iv, pp.94–5

45 Hayami 1987, p.37

46 King 1911, p.363

47 Morse 1936, Day i, pp.55–6

48 Regamey 1892, p.185

49 Thunberg 1796 [1793], iii, p.149; cf. iv, p.85

50 Farris 1985, pp.97–8; 115

51 Morse 1936, Day ii, p.344

52 Morse 1936, Day ii, pp.285–6

53 Alcock 1863, vol.2, p.283

54 Kaempfer 1906 [1727], vol.1, p.185

55 Kaempfer 1906 [1727], vol.3, p.314

56 Thunberg 1796 [1793], iii, p.257

57 Oliphant 1859, p.186

58 Thunberg 1796 [1793], iv, p.81

59 Thunberg 1796 [1793], iv, p.94

60 King 1911, p.135

61 Beardsley et al. 1959, p.177

62 Boserup 1981, p.49

63 Embree 1946, p.31

64 Mokyr 1990, p.161; cf. Durham 1991, pp.226–85

65 Goody 1971, p.26

66 Bairoch 1988, p.377

67 Fairbank 1992, p.172

68 Alcock 1863, vol.1, p.319

69 Alcock 1863, vol.1, p.296

70 King 1911, p.284

71 King 1911, p.239

72 Geertz 1968

CHAPTER 8

1 Those interested in the science and technology of historically important artefacts including detailed laboratory examination of the Japanese sword will gain much from Smith 1981.

2 Smith 1981

3 Much more complete descriptions of the processes may be found in Kapp, Kapp & Yoshihara 1990: this is unusually informative, being a book about a craft by a man who actually practises it.

4 For the modern theory of the structure of materials at an atomic level, with fundamental explanation of hardness, brittleness and ductility see Callister 1997, especially pp.148–53.

CHAPTER 9

1 Gould & Lewontin 1979
2 Dennett 1995
3 Gould & Lewontin 1979, p.147
4 Gould & Lewontin 1979, p.583
5 Gould & Lewontin 1979, p.584
6 Dennett 1995; p.273
7 Gould & Lewontin 1979, p.156
8 Margulis & Sagan 1987, p.34
9 Latour 1986a, 1986b, 1987a; Law 1987, 1991
10 Staudenmeir 1995
11 Dennett 1995, p.228
12 Gould 1991, pp.65–6
13 Rouse 1987, pp.21–3
14 Turnbull 1993a
15 Turnbull 1993b; Watson-Verran & Turnbull 1995
16 Turnbull 1996a, 1996b
17 Steadman 1979
18 Sapp 1994
19 Turnbull 1993b, 1996a, 1996b
20 Suchman 1987
21 Steadman 1979, p.72
22 James 1982, p.9
23 James 1982, p.123
24 Harvey 1974, pp.32–3
25 Turnbull 1993a
26 Mark & Clark 1984
27 Toker 1985
28 Watkins 1990
29 Harvey 1972, pp.119,174
30 James 1979, p.543; James 1982, p.34; Harvey 1974, pp.119, 174
31 James 1989, p.2; see also Shelby 1971
32 Shelby 1976
33 Mark 1982, p.56
34 Shelby 1970
35 Shelby 1970, pp.18, 48
36 Shelby 1981
37 Mark 1990, pp.169–70
38 Long 1985, p.267
39 Burford 1972; Geoghegan 1945, p.229
40 O'Neill 1991, p.32
41 Hoch 1990
42 Strong 1996
43 Steadman 1979, p.225
44 Steadman 1979, p.231

CHAPTER 10

1 David 1975, chs. 2–3; 6
2 Arrow 1962
3 Hounshell 1996, p.20
4 David & Sanderson 1997; David 1998
5 Suppes & Atkinson 1960; Atkinson et al. 1965
6 David 1993a, 1993b
7 David & Sanderson 1986
8 Kahneman, Slovic & Tversky 1982
9 Simon 1957
10 Campbell 1960

CHAPTER 11

1 Carlson 1994, pp.171–3
2 Edison Papers (Rosenberg et al. 1991), Vol. 2, pp.313–14, 372–5, 378–80; Israel 1992, pp.138, 147
3 Edison Evidence (TAEM 1985–, Reel 11) Vol. 1, p.4
4 Helmholtz 1875; pp.174–82; 604–7
5 Baile 1872
6 Edison 1873
7 Edison Evidence, Vol. 1, pp.4–5; Edison Testimony, Vol. 2, pp.600–2; Israel 1992, pp.138–40
8 Thompson 1883, p.5
9 Thompson 1883, pp.50–66
10 Edison Evidence, Vol. 2, pp 509 ff.; Thompson 1883, pp.70–8, 182–3
11 Edison Evidence, Vol. 1, pp.6–9
12 Carlson & Gorman 1990, pp.405–6
13 cf. Weber & Perkins 1989
14 Edison Evidence, Vol. 1, p.224
15 Edison Evidence, Vol. 1, p.9
16 Carlson & Gorman 1992, pp.65–72
17 Batchelor diary, entry for 30 July 1877 (TAEM 1985–)
18 Prescott 1884, p.224
19 Batchelor diary, entry for 9 Nov. 1877 (TAEM 1985–)

CHAPTER 12

1 Ziman 1996a, 1996b
2 Dawkins 1987
3 Campbell 1960
4 Campbell 1960, p.398
5 Boden 1991, p.40
6 Langley, Simon, Bradshaw & Zytkow 1987
7 Wright 1932
8 Newell & Simon 1972
9 Perkins 1981; Langley, Simon, Bradshaw & Zytkow 1987; Boden 1991
10 Langton 1989; Langton, Taylor, Farmer & Rasmussen 1992; Waldrop 1992; Kauffman 1993; Gell-Mann 1994
11 Dawkins 1987
12 Perkins 1992, 1994, 1995
13 Sternberg & Davidson 1995
14 Koestler 1964; Gruber 1974; Perkins 1981
15 Gould 1980
16 Perkins 1992, 1994, 1995

[17] Carlson & Gorman 1992
[18] Campbell 1992
[19] Goodwin 1994
[20] Ziman 1996a, 1996b
[21] Boden 1991
[22] Langton 1989; Langton, Taylor, Farmer & Rasmussen 1992
[23] Wesson 1991
[24] Wesson 1991
[25] Weber 1992; Weber & Dixon 1989
[26] Langley, Simon, Bradshaw & Zytkow 1987
[27] Weber & Perkins 1992
[28] Weber & Perkins 1992
[29] Langley, Simon, Bradshaw & Zytkow 1987
[30] Campbell 1960
[31] Boden 1991

CHAPTER 13
[1] Petroski 1994, 1995
[2] Petroski 1994, p.135
[3] Petroski 1995, p.302
[4] Petroski 1995, p.301
[5] Rosenberg and Vincenti 1978
[6] Clark 1850, p.144
[7] Jevons 1865, p.97
[8] Vincenti 1990, pp.245–8
[9] Vincenti 1994
[10] Vincenti 1994, pp.26–8

CHAPTER 14
[1] Mead and Métraux 1957
[2] For example, Brickhouse 1989; Solomon et al. 1996
[3] Meadows 1996
[4] Solomon 1987
[5] Pacey 1983
[6] Hall 1996, p.37
[7] Vygotsky 1978
[8] Staudenmaier 1985, p.40

[9] Garner 1990
[10] Light & Simmons 1983; Solomon & Hall in press
[11] Piaget & Inhelder 1958
[12] Meadows 1996
[13] Schon 1983
[14] For example, Biggs 1988
[15] Burns et al. 1991
[16] Polanyi 1958
[17] Johnson-Laird 1983
[18] Koestler 1964
[19] Csikszentmihalyi 1990

CHAPTER 15
[1] See Lewontin 1974
[2] Fogel, Owens & Walsh 1966
[3] Holland 1975
[4] Rechenberg 1973; Schwefel 1977
[5] Grefenstette 1985
[6] Goldberg 1989
[7] Koza 1992
[8] Goldberg 1989; Davis 1991; Mitchell 1996
[9] Holland 1975
[10] Miller, Todd, & Hegde 1989
[11] Cliff & Miller 1996
[12] Cliff & Miller 1996
[13] Campbell 1974
[14] Cliff & Miller 1996
[15] Miller 1994
[16] Kelly 1994
[17] Koza 1992
[18] Miller, Todd & Hedge 1989; Cliff & Miller 1996
[19] Cliff & Miller 1996
[20] Husbands, Harvey, Cliff & Miller 1994
[21] Service 1997
[22] Dawkins 1986
[23] Sims 1991; Todd & Latham 1992
[24] Miller 1994; Miller & Todd 1995
[25] Diamond 1997

[26] Kelly 1994

CHAPTER 16
[1] Bijker 1995; Pickering 1995
[2] Law 1997; Akrich & Latour 1992
[3] Mackenzie 1990; Hughes 1983
[4] Campbell 1966, 1974
[5] Campbell et al. 1974; Lorenz 1962
[6] Campbell & Fiske 1959; Campbell 1969
[7] Hull 1988
[8] Bucciarelli 1994; especially pp.55–65
[9] Vincenti 1982
[10] Popper 1963, 1978
[11] Duhem 1902; Quine 1953
[12] Galison 1987, 1997
[13] Constant 1980, chs 3 and 4
[14] Ziman 1978; Hull 1988
[15] Hesse 1974; Howson & Urbach 1989; Glymour 1992, p.249; Kantorovich 1993; Kitcher 1993
[16] Glymour 1992
[17] Glymour 1980; Ziman 2000
[18] Giere 1988
[19] Hesse 1974; Howson & Urbach 1989; Kantorovich 1993; Kitcher 1993
[20] Cipra 1996
[21] MacKenzie 1990
[22] Glymour 1980, p.239
[23] Ziman 2000
[24] Constant 1980
[25] McNeill 1982
[26] Ziman 1978
[27] Pursell 1995; Kline 1995
[28] Kline 1985; Hughes 1989
[29] Simon 1969
[30] Pickering 1995
[31] Dawes 1993; Mark 1990
[32] Vincenti (in preparation)

33 Shapiro 1995; Galitski & Roth 1995

34 Mayr 1963

35 Vincenti, private communication

36 Campbell 1974

37 Poincaré 1921; Hadamard 1945

38 Ziman 1984, p.51

39 Lorenz 1962

CHAPTER 17

1 Basalla 1988

2 Constant 1984; Bucciarelli 1994

3 Layton 1974; Vincenti 1990; Constant 1980; Petroski 1993

4 Simon 1969

5 For example: Dosk 1982; Pavitt 1984; Kline 1991

6 Constant 1980; Dosi 1982; Laudun 1984

7 Dosi 1982

8 Ferguson 1992

9 Lindgren 1996

10 Bucciarelli 1994

11 Vincenti 1990; Simon 1969

12 Gibbs 1994; Gabriel 1996

13 Alexander 1964, 1977, 1979

14 Kodama 1992

15 Drexler 1981, 1986, 1991

16 Pinker 1995; Hofstadter 1995; Hull 1988

17 Carlsson & Stankiewicz 1991

CHAPTER 18

1 Rosenberg 1982

2 Basalla 1988; Nelson & Winter 1982; Ziman 1996a, 1996b

3 Ziman 1996a, 1996b

4 Basalla 1988

5 Ziman 1996a, 1996b

6 Nelson & Winter 1982

7 Georghiou et al. 1986

8 Hull 1988

9 Hull 1988

10 Basalla 1988, p.210

11 Vincenti 1990

12 Polanyi 1967; Collins 1974

13 Pacey 1983, p.4

14 Porter & Miller 1985

15 Boddy & Buchanan 1987

16 Harrington 1984

17 Basalla 1988, p.30

18 Ibid. p.137

19 Georghiou et al. 1986

20 Prahalad & Hamel 1990

21 Georghiou et al. 1986, pp.31–2

22 Plotkin 1994

23 Dawkins 1976

24 Georghiou et al. 1986, p.44

25 Ibid. p.34

26 Nelson and Winter 1982, p.134

27 Ibid. p.259

28 MacDonald et al. 1983

29 Woodward 1980

30 Freeman et al. 1982

31 Bessant 1991

32 Basalla 1988

33 Ziman 1996a, 1996b

34 Womack et al. 1990

35 Basalla 1988

36 UNICTAD 1978

37 Utterback and Abernathy 1975

38 Spindler 1994

39 Schick & Toth 1993

40 Hale 1996

41 Morrison & Coates 1986

42 Bijker et al. 1987

43 Nelson & Winter 1982

44 Ravetz 1971

45 Hughes 1983

46 Fleck 1995

47 Dosi 1982

48 Constant 1984

49 Vincenti 1990

50 Leonard-Barton 1988; Swanson 1988; Voss 1988; Fleck 1993

51 Fleck 1994; von Hippel and Tyre 1996

52 Von Hippel 1988

CHAPTER 19

1 Nelson & Winter 1982

2 Klevorick et al. 1995

3 Malerba & Orsenigo 1995

4 Abernathy & Clark 1985

5 Fairtlough 1996

6 Malerba & Orsenigo 1995

7 Klevorick et al. 1995

8 Burns & Stalker 1966; for a recent review, see Fairtlough 1994a

9 Malone 1997

10 Dodgson 1991; Fairtlough 1994b

11 Michael 1973

CHAPTER 20

1 Smith 1985a; Forman 1995; Kay 1993; Leslie & Kargon 1994

2 Sombart 1913

3 Nef 1950

4 McNeill 1982

5 Foreman 1993

6 Edwards 1996

7 Galison 1997; esp. ch.8

8 Kelly 1998

9 McNeill 1982

10 White 1962

11 McNeill 1982

12 Hounshell 1984; Smith 1985b

13 Gardner 1987

14 McNeill 1982, p.178

15 O'Connell 1985

[16] Douglas 1985; Jennifer Bannister, seminar paper CMU
[17] Constant 1980
[18] McFarland 1995
[19] Becker 1980; Dawson 1991; Vincenti 1997
[20] Misa 1985
[21] Allison 1985
[22] McNeill 1982
[23] Dupuy 1984
[24] McNeill 1982; Dupuy 1984
[25] Travers 1979
[26] Brown 1998
[27] Liddell Hart 1964 [1930]; Dupuy 1984
[28] Van Creveld 1977; McNeill 1982
[29] Van Crefeld 1977
[30] Liddell Hart 1964 [1930]
[31] Keegan 1989
[32] Keegan 1989, Sumida 1993
[33] Keegan 1989, pp.150–1
[34] Garcia y Robertson 1987; O'Connell 1991
[35] McMahon 1978; Keegan 1989; Lambert 1998
[36] Green 1967 [1960], 1968 [1961]
[37] Wood & Dempster 1969 [1961]; Hough & Richards 1989
[38] Green 1967; de Seversky 1942[39] Isenson 1998
[40] Allison 1985; Friedman 1985
[41] Dupuy 1984; Keegan 1987
[42] Pickering 1995

CHAPTER 21
[1] Pacey 1983
[2] Noble 1984, p.xiii
[3] Staudenmaier 1985
[4] Basalla 1988
[5] Bijker 1995
[6] Rothschild 1982; Alic 1986; Schiebinger 1989
[7] Cockburn and Furst-Dilic 1994
[8] Turkle and Papert 1990
[9] Appleton 1994
[10] Schon 1983
[11] Longino 1990
[12] Gee 1989
[13] Private communication
[14] Street 1984
[15] Bijker 1995
[16] Bijker 1995
[17] Carlson 1996
[18] Gee 1996, p.190
[19] Longino 1990
[20] Bijker 1995
[21] Lindblom 1990
[22] Ihde 1990
[23] Longino 1990
[24] Giddens 1994
[25] Monbiot 1998
[26] Beck 1992
[27] Wynne 1991
[28] Midgley 1996
[29] Braidotti et al. 1994
[30] Lovelock 1991
[31] Daly & Cobb 1989
[32] Midgley 1996
[33] Beck 1994, p.28
[34] Beck 1994
[35] Giddens 1994

[36] Basalla 1988; Mokyr 1990; McNeill 1982
[37] Soros & Giddens 1997
[38] Beck 1992; Morgall 1993
[39] Giddens 1994
[40] Giddens 1994
[41] Ministry of Education 1995
[42] Department for Education 1995
[43] Burns 1997
[44] Feinberg and Horowitz 1990; Hyland 1993
[45] Ogburn, cited by Basalla 1988; p.22
[46] Gee 1996; p.190
[47] Vygotsky 1978; Bruner 1985; Ramsden 1992
[48] Burns 1990a, 1990b
[49] Benckert & Staberg 1993
[50] Israelsson 1993
[51] Wynne 1991
[52] Nelkin 1995
[53] Bennett 1996
[54] Rennie and McClafferty 1996
[55] Levinson & Thomas 1997
[56] Rusc 1996
[57] Vico 1725 (Fisch & Bergin 1944)

CHAPTER 22
[1] Nelson 1995
[2] Kantorovich 1993
[3] Maynard Smith & Szathmáry 1995

Bibliography

The numbers in square brackets at the end of each entry indicate the chapter(s) and end-note number(s) where the reference is cited. Thus 'Bagley, J.A. 1990 … [**5**, *13*]' is cited in note 13 of chapter 5.

Abernathy, W. & Clark, K. 1985. Innovation: mapping the winds of creative destruction. *Research Policy* **14**:3–22 [**19**, *4*]

Abrahamson, E. 1996. Management Fashion. *Academy of Management Review*:254–85 [**6**, *12*]

Akrich, M. & Latour, B. 1992. A summary of a convenient vocabulary for the semiotics of human and non-human assemblies. In *Shaping Technology/Building Society: Studies in Sociotechnical Change*, edited by Bijker, W. & Law, J. (Cambridge MA: MIT Press) pp.259–64 [**16**, *2*]

Alberts, B., Bray D., Lewis, J., Raff, M., Roberts, K. & Watson, J.D. 1989. *Molecular Biology of the Cell* (New York NY: Garland) [**3**, *34*]

Alcock, S.R. 1863. *The Capital of the Tycoon: A Narrative of a Three Years' Residence in Japan* (London) [**7**, *25*, *40*, *53*, *68*, *69*]

Alexander, C. 1964. *Notes on the Synthesis of Form* (Cambridge MA: Harvard University Press) [**17**, *13*]

Alexander, C. 1977. *A Pattern Language* (New York NY: Oxford University Press) [**17**, *13*]

Alexander, C. 1979. *The Timeless Way of Building* (New York NY: Oxford University Press) [**17**, *13*]

Alic, M. 1986. *Hypatia's Heritage: A History of Women in Science from Antiquity to the Late Nineteenth Century* (London: The Women's Press) [**21**, *6*]

Allison, D.K. 1985.U. S. Navy Research and Development since World War II. In *Military Enterprise and Technological Change: Perspectives on the American Experience*, edited by Smith, M.R. (Cambridge MA: MIT Press), pp.289–328 [**20**, *21*, *40*]

Appleton, H. 1994. Ownership through participation. *Appropriate Technology* **21** (1):1–4 [**21**, *9*]

Arrow, K.J. 1962. The economics of learning by doing. *Review of Economic Studies* [**10**, *2*]

Arthur, W.B. 1994. *Increasing Returns and Path Dependency in the Economy* (Ann Arbour MI: University of Michigan Press) [**4**, *13*]

Atkinson, R.C., Bower, G. & Crothers, E.J. 1965. *Introduction to Mathematical Learning* (New York NY: Wiley & Sons) [**10**, *5*]

Bagley, J.A. 1990. 'Aeronautics' in *An Encylopedia of the History of Technology* edited by MacNeil, I. (London: Routledge) [**5**, *13*]

Baile, J. 1872. *The Wonders of Electricity*, translated by Armstrong, J.W. (New York NY: Charles Scribner) [**11**, *5*]

Bairoch, P. 1988. *Cities and Economic Development from the Dawn of History to the Present* (London) [**7**, *66*]

Barkow, J.H., Cosmides, L. & Tooby, J. eds. 1992. *The Adapted Mind: Evolutionary Psychology and the Generation of Culture* (New York NY: Oxford University Press) [**1**, *17*]

Basalla, G. 1988. *The Evolution of Technology* (Cambridge: Cambridge University Press) [**1**, *5*; **5**, *7*; **17**, *1*; **18**, *2, 4, 10, 17, 18, 32, 35*; **21**, *4, 36, 45*]

Beardsley, R., K., Hall, J.W. & Ward, R.E. 1959. *Village Japan* (Chicago IL) [**7**, *61*]

Beck, U. 1992. *Risk Society: Towards a New Modernity* (London: Sage) [**21**, *26, 38*]

Beck, U. 1994. The reinvention of politics: Towards a theory of reflexive modernisation. In *Reflexive Modernization: Politics, Tradition and Aesthetics in the Modern Social Order*, edited by Beck, U., Giddens, A. & Lash, S. (Cambridge: Polity Press) pp.1–55 [**21**, *33, 34*]

Becker, J.V. 1980. *The High-Speed Frontier: Case Histories of four NACA Programs, 1920–1950. NASA SP-445* (Washington DC: NASA) [**20**, *19*]

Benckert, S. & Staberg, E.M. 1993. Feminist critiques of science and technology in engineering education: Challenges and possibilities. *GASAT* 7:874–82 [**21**, *49*]

Bennett, J. 1996. Do the media give a utopian view of contemporary science and technology? Paper read at Here and Now: Improving the Presentation of Contemporary Science and Technology in Museums and Science Centres, at The Science Museum, London [**21**, *53*]

Bessant, J. 1991. *Manufacturing Advanced Manufacturing Technology: The Challenge of the Fifth Wave* (Oxford: NCC Blackwell) [**18**, *31*]

Biggs, J. 1988. The role of enhanced learning in metacognition. *Australian Journal of Education* **32** (2):127–38 [**14**, *14*]

Bijker, W.E. 1995. *Of Bicycles, Bakelite and Bulbs: Towards a Theory of Sociotechnical Change* (Cambridge MA: MIT Press) [**5**, *41*; **16**, *1*; **21**, *5, 15, 16, 20*]

Bijker, W.E., Hughes, T.P. & Pinch, T.J., eds. 1987. *The Social Construction of Technological Systems* (Cambridge MA: MIT Press) [**18**, *42*]

Bird, I. 1984 (1880). *Unbeaten Tracks in Japan* (London) [**7**, *10, 11, 12, 13*]

Birdsall, D. & Cipolla, C. 1980. *The Technology of Man* (London) [**7**, *8*]

Blau, H.M. 1992. Differentiation requires continuous active control. *Annual Review of Biochemistry* **61**:1213–30 [**3**, *15*]

Boddy, D. & Buchanan, D.A. 1987. *Management of Technology: The Technical Change Audit* (London: Manpower Services Commission) [**18**, *15*]

Boden, M.A. 1991. *The Creative Mind: Myths and Mechanisms* (New York NY: Basic Books) [**4**, *18*; **12**, *5, 9, 21, 31*]

Boden, M.A. 1994. *Dimensions of Creativity* (Cambridge MA: The MIT Press) [**1**, *13*]

Bonner, J.T. 1988. *The Evolution of Complexity by Means of Natural Selection* (Princeton NJ: Princeton University Press) [**2**, *15*]

Boulding, K. 1981. *Evolutionary Economics* (Beverley Hills CA: Sage) [**1**, *20*]

Boserup, E. 1981. *Population and Technology* (London) [**7**, *62*]

Boyd, R. & Richerson, P.J. 1985. *Culture and the Evolutionary Process* (Chicago IL: University of Chicago Press) [**1**, *17*; **3**, *11*; **4**, *19*, *30*; **6**, *3*]

Braidotti, R., Charkiewicz, E., Hasler, S. & Wieringer, S., eds. 1994. *Women, the Environment and Sustainable Development: Towards a Theoretical Synthesis* (London: Zed Books) [**21**, *29*]

Brenner, S. 1992. Dicing with Darwin. *Current Biology* 2:167–8 [**3**, *10*]

Brickhouse, N. 1989. The teaching of the philosophy of science in secondary classrooms: case studies of teachers' personal theories. *International Journal of Science Education* 11 (4):437–49 [**14**, *2*]

Brown, E.D. 1979. The song of the common crow, *Corvus brachyrhynchos*. Master's thesis, University of Maryland, College Park MD [**2**, *8*]

Brown, I.M. 1998. *British Logistics on the Western Front, 1914–1919* (Westport CN: Praeger) [**20**, *26*]

Bruner, J. 1985. Vygotsky: A historical and conceptual perspective. In *Culture, Communication and Cognition; Vygotskian Perspectives*, edited by Wertsch, J. (Cambridge: Cambridge University Press) pp.21–34 [**21**, *47*]

Bucciarelli, L.L. 1994. *Designing Engineers* (Cambridge MA: MIT Press) [**16**, *8*; **17**, *2*, *10*]

Burford, A. 1972. *Craftsmen in Greek and Roman Society* (Ithaca NY: Cornell University Press) [**9**, *39*]

Burns, J. 1990a. Public attitudes towards and understanding of science and technology in New Zealand: Implications relating to women. Wellington: Ministry of Research and Technology. [**21**, *48*]

Burns, J. 1990b. Students' attitudes towards and concepts of technology. Wellington: Ministry of Education. [**21**, *48*]

Burns, J. 1997. Access to technology education: The role of historical and cultural studies. In *Technology in the New Zealand Curriculum: Perspectives on Practice*, edited by Burns, J. (Palmerston North: Dunmore Press) [**21**, *43*]

Burns, J., Clift, J. & Duncan, J. 1991. Understanding understanding: Implications for learning and teaching. *British Journal of Educational Psychology* 61:276–89 [**14**, *15*]

Burns, T. & Stalker, G. 1966. *The Management of Innovation* (Lonon: Tavistock) [**19**, *8*]

Bush, G.L. 1994. Sympatric speciation in animals: new wine in old bottles. *Trends in Ecology and Evolution* 9:285– [**2**, *12*]

Buss, L.W. 1987. *The Evolution of Individuality* (Princeton NJ: Princeton University Press) [**3**, *33*]

Callebaut, W. 1993. *Taking the Naturalistic Turn: How Real Philosophy of Science is Done* (Chicago IL: University of Chicago Press) [**4**, *17*]

Callister, W.D. 1997. *Material Science and Engineering* (New York NY: John Wiley) [**8**, *4*]

Campbell, D.T. 1960. Blind variation and selective retention in creative thought as in other knowledge processes. *Psychological Review* 67:380–400 [**4**, *2*; **6**, *4*; **7**, *2*; **10**, *10*; **12**, *3*, *4*, *30*]

Campbell, D.T. 1965. Variation and selective retention in socio-cultural evolution. In *Social Change in Developing Areas: A Re-interpretation of Evolutionary Theory*, edited by

Barringer, H.R., Blanksten, G.I. & Mack, R.W. (Cambridge MA: Schenkman) pp.19–49 [4, 2]

Campbell, D.T. 1966. Pattern matching as an essential in distal knowing. In *The Psychology of Egon Brunswil*, edited by Hammond, K.R. (New York NY: Holt, Rinehart & Winston), pp.81–106 [4, 17; 16, 4]

Campbell, D.T. 1969. A phenomenology of the other one: corrigible, hypothetical and critical. In *Human Action: Conceptual and Empirical Issues*, edited by Mischel, T. (New York NY: Academic Press) pp.41–69 [16, 6]

Campbell, D.T. 1974. Evolutionary Epistemology. In *The Philosophy of Karl Popper*, edited by Schilpp, P.A. (La Salle IL: Open Court), pp.413–63 [1, 3, 4, 12; 4, 2, 6; 5, 3, 12, 23; 6, 4; 15, 13; 16, 4, 36]

Campbell, D.T. 1977. Descriptive epistemology: Psychological, sociological, evolutionary. In Secondary Title: Unpublished draft of William James Lectures, Harvard [1, 3, 23; 4, 10]

Campbell, D.T. 1990. Epistemological roles for selection theory. In *Evolution, Cognition and Realism: Studies in Evolutionary Epistemology*, edited by Rescher, M. (Lanham MD: University Press of America) [16, 5]

Campbell, D.T., Riecken, H.W., Boruch, R.F., Caplan, N., Glenman, T.K., Pratt, J. & Williams, W. 1974. Quasi-experimental designs. In *Social Experimentation: A Method for Planning and Evaluating Social Interventions*, edited by Riecken, H.W. & Boruch, R.F. (New York NY: Academic Press) pp.87–116 [4, 17]

Campbell, W.C. 1992. The genesis of the antiparasitic drug Ivermectin. In *Inventive Minds: Creativity in Technology*, edited by Weber, R.J. & Perkins, D.N. (New York NY: Oxford University Press) pp.194–214 [12, 18]

Carlson, W.B. 1994. Entrepreneurship in the early development of the telephone: How did William Orton and Gardiner Hubbard conceptualize this new technology? *Business and Economic History* 23 (Winter):161–92 [11, 1]

Carlson, W.B. 1996. Invention as re-representation: The case of Edison's sketches of the telephone. *History and Technology* [21, 17]

Carlson, W. & Gorman, M.E. 1990. Understanding invention as a cognitive process: The case of Thomas Edison and early motion pictures, 1888–91. *Social Studies in Science* 20:387–430

Carlson, W.B. & Gorman, M. 1992. A cognitive framework to understand technological creativity: Bell, Edison, and the telephone. In *Inventive Minds: Creativity in Technology*, edited by Weber, R.J. & Perkins, D.N. (New York NY: Oxford University Press) pp.48–79 [11, 16; 12, 17]

Carlson, B. & Stankiewicz, R. 1991. On the nature, function and composition of technological systems. *Journal of Evolutionary Economics* 1:93–118 [17, 17]

Cavalier-Smith, T. 1996. The Origin and Diversification of Cells. Paper read at ICESB V [3, 22]

Cavalli-Sforza, L.L. & Feldman, M.W. 1981. *Cultural Transmission and Evolution: A Quantitative Approach* (Princeton NJ: Princeton University Press) [3, 11; 4, 19; 5, 24, 28, 29; 6, 3]

Chamberlain, B.H. 1990 (1904). *Japanese Things, Being Notes on Various Subjects Connected with Japan* (Tokyo) [7, 20]

Chandler, A.D. 1962. *Strategy and Structure: Chapters in the History of Industrial Enterprise,* (Cambridge MA: Harvard University Press) [**6**, *8*]

Chandler, A.D. 1990. *Scale and Scope: The Dynamics of Industrial Capitalism* (Cambridge MA: Harvard University Press) [**6**, *8*]

Clark, E. 1850. *The Britannia and Conway Tubular Bridges* (London: Day & Son) [*13*, *6*]

Cliff, D. & Miller, G.F. 1996. Co-evolution of pursuit and evasion II: Simulation methods and results. In *From Animals to Animats 4 (SAB96)*, edited by Maes, P. *et al.* (Cambridge MA: MIT Press), pp.608–17 [**15**, *11*, *12*, *14*, *18*, *19*]

Cockburn, C. & Furst-Dilic, R., eds. 1994. *Bringing Technology Home: Gender and Technology in a Changing Europe.* (Buckingham: Open University Press) [**21**, *7*]

Cohen, J. & Stewart, I. 1994. *The Collapse of Chaos: Discovering Simplicity in a Complex World* (Harmondsworth: Penguin) [**1**, *25*; **4**, *32*]

Collins, H.M. 1974. The TEA set: Tacit knowledge and scientific networks. *Science Studies* 4:165–86 [**18**, *12*]

Collins, R.J. & Jefferson, D.R. 1992. AntFarm: Towards simulated evolution. In *Artificial Life II*, edited by Langton, C.G., Taylor, C., Farmer, J.D. & Rasmussen, S. (Reading MA: Addison-Wesley), pp.579–602 [**4**, *29*]

Constant, E.W. 1980. *The Origins of the Turbojet Revolution* (Baltimore MD: Johns Hopkins University Press) [**16**, *13*; *24*; **17**, *3*, *6*; **18**, *48*; **20**, *17*]

Constant, E.W. 1984. Communities and hierarchies: structure in the practice of science and technology. In *The Nature of Technological Knowledge*, edited by Laudan, R. (Dordrecht: Reidel), pp.27–46 [**17**, *2*]

Craggs, C.B. 1989. Evolution of the steam engine. In *Issues in Evolutionary Epistemology*, edited by Hahlweg, K. & Hooker, C.A. (Albany NY: SUNY Press), pp.313–56 [**4**, *9*]

Crouch, T. 1989. *The Bishop's Boys: A Life of Wilbur and Orville Wright* (New York NY: Norton) [**5**, *13*]

Csikszentmihalyi, M. 1990. The domain of creativity. In *Ecology and Culture*, edited by Runco, M. & Albert, R. (Beverly Hills CA: Sage), pp.190–212 [**14**, *19*]

Csikszentmihalyi, M. 1993. *The Evolving Self: A Psychology for the Third Millennium* (New York NY: HarperCollins) [**4**, *11*]

Cziko, G. 1995. *Without Miracles: Universal Selection Theory and the Second Darwinian Revolution* (Cambridge MA: MIT Press) [**1**, *26*; **2**, *1*; **4**, *7*, *10*, *18*; **5**, *17*]

Daly, H.E. & Cobb, J.B. 1989. *For the Common Good: Redirecting the Economy Toward Community, the Environment, and a Sustainable Future.* 2nd edn (Boston MA: Beacon Press) [**21**, *31*]

Das, O.P. & Messing, J. 1994. Variegated phenotype and developmental methylation changes of a maize allele originating from epimutation. *Genetics* 136:1121–41 [**3**, *30*]

David, P.A. 1975. *Technical Choice, Innovation and Economic Growth* (Cambridge: Cambridge University Press) [**10**, *1*]

David, P.A. 1993a. Historical economics in the long run: Some implications of path dependence. In *Historical Analysis in Economics*, edited by Snooks, G.D. (London: Routledge) pp. [**10**, *6*]

David, P.A. 1993b. Path dependence and predictability in dynamic systems with local network externalities: A paradigm for historical economics. In *Technology and the Wealth of Nations*, edited by Foray, D. & Freeman, C. (London: Pinter) Ch.10, pp.208–31 [**10**, *6*]

David, P.A. 1998. Path dependent learning, and the evolution of beliefs and behaviours: Implications of Bayesian adaptation under computationally bounded rationality, In *The Evolution of Economic Diversity*, edited by Pagano, U. & Nicita, A. (London: Routledge) [**10**, *4*]

David, P.A. & Sanderson, W.C. 1986. Rudimentary contraceptive methods and the American transition to fertility control, 1855–1915. In *Long-Term Factors in American Economic Growth*, edited by Engerman, S.L. & Gallman, R.E. (Chicago: The University of Chicago Press, for the NBER) pp.307–90 [**10**, *7*]

David, P.A. & Sanderson, W.C. 1997. Making use of treacherous advice: Cognitive learning, Bayesian adaptation and the tenacity of unreliable knowledge. In *The Frontiers of Institutional Economics*, edited by Nye, J.V.C. & Drobak, J.N. (San Diego CA: Academic Press), [**10**, *4*]

Davis, L. 1991. *Handbook of Genetic Algorithms* (Reinhold: Van Nostrand) [**15**, *8*]

Dawes, R.M. 1993. Prediction of the future versus an understanding of the past: A basic asymmetry. *American Journal of Psychology* **106**:1–24 [**16**, *31*]

Dawkins, R. 1976. *The Selfish Gene* (Oxford: Oxford University Press) [**1**, *7*; **3**, *39*; **4**, *4*; **5**, *4*, *9*, *30*; **18**, *23*]

Dawkins, R. 1982. *The Extended Phenotype* (San Francisco CA W.H. Freeman) [**5**, *9*, *36*, *38*]

Dawkins, R. 1986. *The Blind Watchmaker* (London: Longman) [**12**, *2*, *11*; **15**, *22*]

Dawson, V.P. 1991. *Engines and Innovation: Lewis Laboratory and American Propulsion Technology*. Vol. NASA SP-4306, *The NASA History Series* (Washington DC: NASA) [**20**, *19*]

Demsetz, H. 1967. Toward a Theory of Property Rights. *American Economics Review* **57** (2):347–59 [**6**, *7*]

Dennett, D.C. 1991. *Consciousness Explained* (London: Penguin) [**4**, *18*]

Dennett, D.C. 1995. *Darwin's Dangerous Idea* (London: Penguin) [**1**, *10*; **4**, *18*; **5**, *5*, *44*; **9**, *2*, *6*, *11*]

Department for Education 1995. Design and Technology in the National Curriculum. London: HMSO [**21**, *42*]

Diamond, J. 1997. *Guns, Germs and Steel: The Fates of Human Societies* (London: Jonathan Cape) [**1**, *18*; **15**, *25*]

Dodgson, M. 1991. *The Management of Technological Learning: Lessons from a Biotechnology Company* (Berlin: de Gruyter) [**19**, *10*]

Donald, M. 1991. *Origins of the Modern Mind: Three Stages in the Evolution of Culture and Cognition* (Cambridge MA: Harvard University Press) [**4**, *18*]

Dosi, G. 1982. Technological paradigms and technological trajectories: a suggested interpretation of the determinants and directions of technical change. *Research Policy* **11**:147–62 [**17**, *5*, *6*, *7*; **18**, *47*]

Douglas, S.J. 1985. The Navy adopts the radio, 1899–1919. In *Military Enterprise and Technological Change: Perspectives on the American Experience*, edited by Smith, M.R. (Cambridge MA: MIT Press), pp.117–74 [**20**, *16*]

Drexler, K.E. 1981. Molecular engineering: An approach to the development of general capabilities for molecular manipulation. *Proceedings of the National Academy of Science USA* **78** (9):5275–8 [**17**, *15*]

Drexler, K.E. 1986. *Engines of Creation* (New York NY: Anchor Books, Doubleday) [**17**, *15*]

Drexler, K.E. 1991. *Unbounding the Future. The Nanotechnology Revolution* (New York NY: Quill, William Morrow) [**17**, *15*]

Duhem, P. 1902 (1996). *Pierre Duhem, Essays in the History and Philosophy of Science*, translated and edited by Ariew, R. & Barker, P. (Indianapolis IN: Hackett) [**16**, *11*]

Dupuy, T.N. 1984. *The Evolution of Weapons and Warfare* (Fairfax VA: Hero Books) [**20**, *23, 24, 27, 41*]

Durham, W.H. 1991. *Coevolution: Genes, Culture and Human Diversity* (Stanford CA: Stanford University Press) [**1**, *17*; **3**, *12*; **4**, *19*; **6**, *3*; **7**, *64*]

Edelman, G.M. 1987. *Neural Darwinism: the Theory of Neuronal Group Selection* (New York: Basic Books) [**3**, *33, 35*]

Edelman, G.M. 1992. *Bright Air, Brilliant Fire: On the Matter of the Mind* (London: Penguin) [**1**, *11*; **2**, *5*; **4**, *8, 15, 16*]

Edison, T.A. 1873. *Relay Magnets.* [**11**, *6*]

Edwards, P. 1996. *The Closed World* (Cambridge MA: MIT Press) [**20**, *6*]

Eldredge, N. 1989. *Macroevolutionary Dynamics* (New York NY: McGraw Hill) [**5**, *25*]

Eldredge, N. 1995. *Reinventing Darwin: The Great Evolutionary Debate* (London: Weidenfeld & Nicholson) [**5**, *36, 39*]

Embree, J.F. 1946. *A Japanese Village, Suye Mura* (London) [**7**, *63*]

Ettinger, L. & Doljanski, F. 1992. On the generation of form by the continuous interactions between cells and their extracellular matrix. *Biological Reviews* **67**:459–89 [**3**, *21*]

Evans, C. & Rydén, G. 1996. Recruitment, kinship, and the distribution of skill: Bar iron production in Britain and Sweden, 1500–1860. Paper read at Technological Revolutions in Europe, 1760–1860, at Oslo [**5**, *33*]

Fairbank, J.K. 1992. The paradox of growth without development. In *China: A New History* (Cambridge MA: Harvard University Press), pp. [**7**, *67*]

Fairtlough, G. 1994a. Innovation and organisation. In *The Handbook of Industrial Innovation*, edited by Dodgson, M. & Rothwell, R. (Aldershot: Edward Elgar), pp. 325–36 [**19**, *8*]

Fairtlough, G. 1994b. *Creative Compartments: A Design for Furture Organisation* (London: Adamantine Press) [**19**, *10*]

Fairtlough, G. 1996. A marriage of large and small: R&D for healthcare products. *Business Strategy Review* **7** (2):14–22 [**19**, *5*]

Farrell, C.J. 1993. A theory of technological progress. *Technological Forecasting and Social Changes* **44**:161–78 [**5**, *7*]

Farris, W.W. 1985. *Population, Disease and Land in Early Japan* (Cambridge MA: Harvard University Press) [**7**, *50*]

Fedoroff, N.V. 1989. About maize transposable elements and development. *Cell* **56**:181–91 [**3**, *28*]

Feinberg, W. & Horowitz, B. 1990. Vocational education and equality of opportunity. *Journal of Curriculum Studies* 22:188–92 [**21**, *44*]

Feldman, M. & Laland, K.N. 1996. Gene-culture coevolutionary theory. *Trends in Ecology and Evolution* 11:453–7 [*3*, *12*]

Ferguson, E.S. 1992. *Engineering and the Mind's Eye* (Cambridge MA: MIT Press) [**17**, *8*]

Fisch, M.H. & Bergin, T.H. 1994. *The Autobiography of Giambattista Vico* (Ithaca NY: Cornell University Press) [**21**, *57*]

Flam, Faye 1994. Co-opting a blind watchmaker. *Science* **265**: 1032–3 [**4**, *21*]

Fleck, J. 1993. Configurations: Crystallizing contingency. *International Journal on Human Factors in Manufacturing* 3 (1):15–36 [**18**, *50*]

Fleck, J. 1994. Learning by trying: The implementation of configurational technology. *Research Policy* 23:637–52 [**18**, *51*]

Fleck, J. 1995. Configurations and standardization. In *Soziale und okonomische Konflicte in Standardisierungsprozessen*, edited by Esser, J., Fleischmann, G. & Heimer, T. (Franfurt: Campus Verlag) pp.38–65 [**18**, *46*]

Fogel, L.J., Owens, A.J. & Walsh, M.J. 1996. *Artificial Intelligence through Simulated Evolution* (New York NY: John Wiley) [**15**, *2*]

Foreman, S.W.L. 1993. *The Cold War and American Science: The Military-Industrial-Academic Complex at MIT and Stanford* (New York NY: Columbia University Press) [**20**, *5*]

Forman, P. 1995. 'Swords into ploughshares': Breaking new ground with radar hardware and technique in physical research after World War. *Reviews in Modern Physics* 67:397 [**20**, *1*]

Freeman, C., Clark, J. & Soete, L. 1982. *Unemployment and Technical Innovation: A Study of Long Waves and Economic Development* (London: Pinter) [**18**, *30*]

Friedman, N. 1985. *US Naval Weapons* (Annapolis MD: Navy Institute Press) [**20**, *40*]

Futuyma, D.J. 1986. *Evolutionary Biology* (Sunderland MA: Sinauer Publishers) [**5**, *27*]

Gabriel, R.P. 1996. *Patterns of Software* (New York NY: Oxford University Press) [**17**, *12*]

Galison, P. 1987. *How Experiments End* (Chicago IL: University of Chicago Press) [**16**, *12*]

Galison, P. 1997. *Image and Logic: A Material culture of Microphysics* (Chicago IL: University of Chicago Press) [**16**, *12*; **20**, *7*]

Galitski, T. & Roth, J.R. 1995. Evidence that F plasmid transfer replication underlies apparent adaptive mutation. *Science* **268**:421–3 [**16**, *33*]

Garcia y Robertson, R. 1987. The failure of the heavy gun at sea, 1898–1922. *Technology and Culture* 28 (July):539–57 [**20**, *34*]

Gardner, J. 1987 *The Dory Book* (Mystic, Connecticut: Mystic Seaport Museum) [**20**, *13*]

Garner, S. 1990. Drawing and designing: The case for reappraisal. *Journal of Art and Design* **9**: 39–55 [**14**, *9*]

Gee, J.P. 1989. Literacy, discourse, and linguistics: Introduction. *Journal of Education* **171** (1):5–17 [**21**, *12*]

Gee, J.P. 1996. *Social Linguistics and Literacies: Ideology in Discourse* (2nd edn) (London: Taylor and Francis) [**21**, *18*, *46*]

Geertz, C. 1968. *Agricultural Involution* (University of California Press) [**7**, *72*]

Gell-Mann, M. 1994. *The Quark and the Jaguar: Adventures in the simple and the complex* (London: Little, Brown & Co.) [**4**, *32*; **12**, *10*]

Geoghegan, A.T. 1945. *The Attitude Towards Labour in Early Christianity and Ancient Culture* (Washington DC: Catholic University of America Press) [**9**, *39*]

Georghiou, L., Metcalfe, J.S., Gibbons, M., Ray, T. & J., E. 1986. *Post Innovation Performance* (Basingstoke: Macmillan) [**18**, *7, 19, 21, 24, 25*]

Gibbs, W.W. 1994. Software's chronic crisis. *Scientific American* **271** (3):72–81 [**17**, *12*]

Giddens, A. 1994. *Beyond Left and Right: The Future of Radical Politics* (Cambridge: Polity Press) [**21**, *24, 35, 39, 40*]

Giere, R.N. 1988. *Explaining Science: A Cognitive Approach* (Chicago IL: University of Chicago Press) [**16**, *18*]

Glymour, C. 1980. *Theory and Evidence* (Princeton NJ: Princeton University Press) [**16**, *17, 22*]

Glymour, C. & Kelly, K. 1992. *Logic, Computation and Discovery* (New York NY: Cambridge University Press) [**16**, *15, 16*]

Goldberg, D.E. 1989. *Genetic Algorithms in Search, Optimization, and Machine Learning* (Reading MA: Addison Wesley) [**15**, *6, 8*]

Goodwin, B. 1994. *How the Leopard Changed its Spots: The Evolution of Complexity* (New York NY: Simon & Schuster) [**12**, *19*]

Goody, J. 1971. *Technology, Tradition and the State in Africa* (Oxford: Oxford University Press) [**7**, *65*]

Gould, S.J. 1980. *The Panda's Thumb: More Reflections in Natural History*. (New York NY: Norton) [**12**, *15*]

Gould, S.J. 1991. *Bully for Brontosaurus: Reflections in Natural History* (London: Hutchinson Radius) [**9**, *12*]

Gould, S.J. 1996. *Full House* (New York NY: Harmony Books) [**1**, *2*; **2**, *14*]

Gould, S.J. 1997. Evolution: The pleasures of pluralism. *New York Review of Books* (26 June 1997) [**5**, *20*]

Gould, S.J. & Lewontin, R. 1979. The spandrels of San Marco and the Panglossian paradigm: A critique of the adaptationist paradigm. *Proceedings of the Royal Society, London* **B205**:147–65, 581–98 [**9**, *1, 3, 4, 5, 7*]

Gould, S.J. & Vrba, E.S. 1982. Exaptation – a missing term in the science of form. *Palaeobiology* **8** (1):4–15 [**2**, *9*; **5**, *21*]

Green, W. 1967 [1960], 1968 [1961]. *Warplanes of the Second World War: Fighters, Volumes 1 and 2* (Garden City NY: Doubleday & Co.) [**20**, *36, 38*]

Grefenstette, J.J., ed. 1985. *Proceedings of an International Conference on Genetic Algorithms and Their Applications* (Hillsdale, NJ: Lawrence Erlbaum) [**15**, *5*]

Griesemer, J. 1999. *Reproduction in the evolutionary process* (in preparation) [**3**, *40*]

Griffiths, P. & Gray, R.D. 1994. Developmental systems and evolutionary explanations. *Journal of Philosophy* **91**:277–304 [**3**, *41, 42*]

Grimes, G.W. & Aufderheide, K.J. 1991. *Cellular Aspects of Pattern Formation: the Problem of Assembly*. Vol. 22, *Monographs in Developmental Biology* (Basel: Karger) [**3**, *16, 17, 19*]

Gruber, H. 1974. *Darwin on Man: A Psychological Study of Scientific Creativity* (New York NY: E.P. Dutton) [**12**, *14*]

Hadamard, J. 1945. *An Essay on the Psychology of Invention in the Mathematical Field* (Princeton NJ: Princeton University Press) [**16**, *37*]

Hahn, F. 1991. The next hundred years. *Economic Journal* **101**:47–50 [**5**, *1*]

Hale, J.R. 1996. The Lost Technology of Ancient Greek Rowing. *Scientific American* (May):66–71 [**18**, *40*]

Hall, S. 1996. Helping primary children to understand the relationship between needs, wants, and technology. *Journal of Design and Technology* **1** (1):37 [**14**, *6*]

Harrington, J.J. 1984. *Understanding the Manufacturing Process: Key to Successful CAD/CAM Implementation* (New York NY: Marcel Dekker) [**18**, *16*]

Harvey, J. 1972. *The Mediaeval Architect* (London: Wayland Press) [**9**, *29*]

Harvey, J. 1974. *Cathedrals of England and Wales* (London: B.T. Batsford) [**9**, *24*, *30*]

Hayami, A. 1987. Population growth in pre-industrial Japan. In *Evolution Agraire et Croissance Demographique* (Liege: Ed. Antoinette Fauve-Chamoux) [**7**, *1*, *45*]

Helmholtz, H.L.F. 1875. *On the Sensations of Tone as a Physiological Basis for the Theory of Music*, translated by Ellis, A.J. (London: Longmans Green) [**11**, *4*]

Hesse, M. 1974. Changing concepts and stable order. *Social Studies of Science* **16**:714–26 [**16**, *15*, *19*]

Heyes, C.M. & Galef, B.G., eds. 1996. *Social Learning in Animals: The Roots of Culture* (San Diego CA: Academic Press) [**2**, *7*; **3**, *8*]

Hillis, W.D. 1990. Co-evolving parasites improve simulated evolution as an optimization procedure. *Physica D* **42**:228–34 [**4**, *28*]

von Hippel, E. 1988. *The Sources of Innovation* (Oxford: Oxford University Press) [**18**, *51*, *52*]

von Hippel, E. & Tyre, M. 1996. The mechanics of learning by doing: Problem discovery during process machine use. *Technology and Culture* **37** (2):312–29

Hoch, P. 1990. Institutional mobility and the management of technology and science. *Technology Analysis and Strategic Management* **2**:341–56 [**9**, *41*]

Hofstadter, D.R. 1995. *Fluid Concepts and Creative Analogies* (New York NY: Basic Books) [**17**, *16*]

Holland, J.H. 1975. *Adaptation in Natural and Artificial Systems*, 2nd edn (Cambridge MA: MIT Press), 1st edn 1992. (Ann Arbor MI: University of Michigan Press) [**4**, *22*; **15**, *9*]

Holliday 1987. The inheritance of epigenetic defects. *Science* **238**:163–70 [**3**, *7*]

Holliday, R. 1990. Mechanisms for the control of gene activity during development. *Biological Review* **65**:431–71 [**3**, *24*]

Hough, R. & Richards, D. 1989. *The Battle of Britain: The Greatest Air Battle of World War II* (New York NY: W.W. Norton) [**20**, *37*]

Hounshell, D.A. 1984. *From the American System to Mass Production, 1800–1932: The Development of Manufacturing Technology in the United States* (Baltimore MD: Johns Hopkins University Press) [**20**, *12*]

Hounshell, D.A. 1996. The medium is the message, or how context matters: The RAND Corporation builds an economics of innovation, 1946–1962. Paper read at Spread of the Systems Approach, 3–5 May 1996, at MIT (Dibner Institute) [**10**, *3*]

Howson, C. & Urbach, P. 1989. *Scientific Reasoning: The Bayesian Appraoch* (La Salle IN: Open Court) [**16**, *15*, *19*]

Hughes, T.P. 1983. *Networks of Power: Electrification in Western Society, 1880–1930* (Baltimore MD: Johns Hopkins University Press) [**16**, *3*; **18**, *45*]

Hughes, T.P. 1989. *American Genesis: A Century of Invention and Technological Enthusiasm* (New York NY: Viking) [**16**, *28*]

Hull, D.L. 1988. *Science as a Process: An Evolutionary Account of the Social and Conceptual Development of Science* (Chicago IL: University of Chicago Press) [**1**, *21*; **3**, *13*; **4**, *4*; **5**, *35*; **6**, *5*; **16**, *7*, *14*; **17**, *16*; **18**, *8*, *9*]

Husbands, P., Harvey, I., Cliff, D. & Miller, G.F. 1994. The use of genetic algorithms for the development of sensorimotor control systems. In *Proceedings of the International Workshop from Perception to Action (PerAc94)*, edited by Gaussier, P. & Nicoud, J.D. (Los Alamitos CA: IEEE Computer Society Press), pp.100–21 [**15**, *20*]

Hyland, T. 1993. Vocational reconstruction and Dewey's instrumentalism. *Oxford Review of Education* **19**:98–100 [**21**, *44*]

Ihde, D. 1990. *Technology and the Lifeworld: From Garden to Earth* (Bloomington IN: Indiana University Press) [**21**, *22*]

Isenson, R.S. 1998. C-130. *Flying* **125**:31 [**20**, *39*]

Israel, P. 1992. *From Machine Shop to Industrial Laboratory: Telegraphy and the Changing Context of American Invention, 1830–1920* (Baltimore MD: Johns Hopkins University Press) [**11**, *12*]

Israelsson, A.-M. 1993. A science centre to serve the missing half. *GASAT* **7**:940–9 [**21**, *50*]

Jablonka, E. & Lamb, M.J. 1989. The inheritance of acquired epigenetic variations. *Journal of Theoretical Biology* **139**:69–83 [**3**, *7*]

Jablonka, E. & Lamb, M.J. 1995. *Epigenetic Inheritance and Evolution: the Lamarckian Dimension* (Oxford: Oxford University Press) [**1**, *22*; **2**, *16*; **3**, *2*, *7*, *14*, *23*, *31*, *33*]

Jablonka, E., Oborny, B., Molnár, I., Kisdi, E., Hofbauer, J. & Czárán, T. 1995. The adaptive advantage of phenotypic memory in changing environments. *Philosophical Transactions of the Royal Society London* B **350**:133–41 [**3**, *31*]

Jablonka, E. & Szathmáry, E. 1995. The evolution of information storage and heredity. *Trends in Ecology Evolution* **10**:206–11 [**2**, *16*; **3**, *6*]

James, J. 1979. *The Contractors of Chartres*. Vol. 11 (Wyong: Mandorla Publications) [**9**, *30*]

James, J. 1982. *Chartres: The Masons Who Built a Legend* (London: Routledge and Kegan Paul) [**9**, *22*, *23*, *30*]

James, J. 1989. *The Template-Makers of the Paris Basin* (Leura: West Grinstead Nominees) [**9**, *31*]

Jansen, M.B. & Rozman, G., eds. 1988. *Japan in Transition from Tokugawa to Meiji* (Princeton NJ: Princeton University Press) [**7**, *28*]

Jevons, W.S. 1865. *The Coal Question* (London: Macmillan) [**13**, *7*]

Johannsen, W. 1911. The genotype conception of heredity. *American Naturalist* **45**:129–59 [**3**, *1*]

Johnson-Laird, P. 1983. *Mental Models* (Cambridge: Cambridge University Press) [**14**, *17*]

Kaempfer, E. 1906 [1727]. *The History of Japan, together with a Description of the Kingdom of Siam, 1690–1692*, translated by Scheuchzer, J.G. (London) [**7**, *15*, *16*, *17*, *21*, *38*, *54*, *55*]

Kahneman, D., Slovic, P. & Tversky, A., eds. 1982. *Judgement under Uncertainty: Heuristics and Biases* (Cambridge: Cambridge University Press) [**10**, *8*]

Kantorovich, A. 1993. *Scientific Discovery: Logic and Tinkering* (Albany NY: SUNY Press) [**16**, *15, 19*; **22**, *2*]

Kapp, L., Kapp, H. & Yoshihara, Y. 1990. *The Craft of the Japanese Sword* (Tokyo & New York NY: Kodansha International) [**8**, *3*]

Kauffman, S.A. 1993. *The Origins of Order: Self-Organization and Selection in Evolution* (Oxford: Oxford University Press) [**4**, *1, 31, 32*; **12**, *10*]

Kauffman, S.A. 1995. *At Home in the Universe: The Search for Laws of Complexity* (London: Viking Press) [**1**, *25*; **4**, *1*; **5**, *7, 26*]

Kay, L.E. 1993. *The Molecular Vision of Life: Caltech, the Rockefeller Foundation, and the Rise of the New Biology* (New York NY: Oxford University Press) [**20**, *1*]

Keegan, J. 1987. *The Mask of Command* (New York NY: Penguin) [**20**, *41*]

Keegan, J. 1989. *The Price of Admiralty: The Evolution of Naval Warfare* (New York NY: Viking) [**20**, *31, 32, 33, 35*]

Kelly, K. 1994. *Out of Control: The New Biology of Machines* (London: Fourth Estate) [**4**, *23*; **15**, *16, 26*]

Kelly, K. 1998. The Third Culture. *Science* **279**:992–3 [**20**, *8*]

Kimura, M. 1992. Neutralism. In *Keywords in Evolutionary Biology*, edited by Keller, E.F. & Lloyd, E.A. (Cambridge MA: Harvard University Press), pp.225–30 [**5**, *16*]

King, F.H. 1911. *Farmers of Forty Centuries, or permanent agriculture in China, Korea and Japan* (London) [**7**, *19, 46, 60, 70, 71,*]

Kitcher, P. 1993. *The Advancement of Science* (Oxford: Oxford University Press) [**16**, *15, 19*]

Klekowski, E.J. 1988. *Mutation, Developmental Selection, and Plant Evolution* (New York NY: Columbia University Press) [**3**, *33, 36*]

Klevorick, A.K., Levin, R.C., Nelson, R.R. & Winter, S.G. 1995. On the sources and significance of interindustry differences in technological opportunities. *Research Policy* **24**:185–205 [**19**, *2, 7*]

Kline, S.J. 1985. Innovation is not a linear process. *Research Management* **28**:36–45 [**16**, *28*]

Kline, S.J. 1991. Styles of innovation and their cultural bases. *CHEMTECH* **21**(4):472–80 [**17**, *5*]

Kline, R. 1995. Construing 'Technology' as 'Applied Science'. *ISIS* **86**:194–221 [**16**, *27*]

Kodama, F. 1992. Technology fusion and the new R&D. *Harvard Business Review* (July–August):70–8 [**17**, *14*]

Koestler, A. 1964. *The Act of Creation* (New York NY: Dell) [**12**, *14*; **14**, *18*]

Koza, J.R. 1992. *Genetic Programming: On Programming Computers by Means of Natural Selection* (Cambridge MA: MIT Press) [**15**, *7, 17*]

Krakauer, D.C. & Pagel, M. 1996. Selection by somatic signals: the advertisement of phenotypic state through costly intercellular signals. *Philosophical Transactions of the Royal Society London* B **351**:647–58 [**3**, *38*]

Kuhn, T.S. 1970 [1962]. *The Structure of Scientific Revolutions* (Chicago IL: University of Chicago Press) [**6**, *6*]

Laland, K.N., Odling-Smee, F.J. & Feldman, M.W. 1996. The evolutionary consequences of niche construction: a theoretical investigation using two-locus theory. *Journal of Evolutionary Biology* **9**:293–316 [*2*, *11*]

Lambert, N.A. 1998. 'Our bloody ships' or 'Our bloody systems'? Jutland and the loss of the battle cruisers, 1916. *Journal of Military History* **62**:29–56 [*20*, *35*]

Landes, W.M. & Posner, R.A. 1987. *The Economic Structure of Tort Law* (Cambridge MA: Harvard University Press) [*6*, *7*]

Langley, P., Simon, H.A., Bradshaw, G.L. & Zytkow, J.M. 1987. *Scientific Discovery: Computational Explorations of the Creative Processes* (Cambridge MA: MIT Press) [*12*, *6*, *9*, *26*, *29*]

Langton, C.G. 1984. Self-reproduction in cellular automata. *Physica D* **10**:135–44 [*4*, *24*]

Langton, C.G., ed. 1989. *Artificial Life* (Redwood City CA: Addison-Wesley) [*4*, *24*; *12*, *10*, *22*]

Langton, C.G., 1995. *Artificial Life: An Overview* (Cambridge MA: MIT Press) [*1*, *24*; *4*, *23*]

Langton, C.G., Tayor, C., Farmer, J.D. & Rasmussen, S., eds. 1992. *Artificial Life II* (Redwood City CA: Addison-Wesley) [*12*, *10*, *22*]

Latour, B. 1986a. The powers of association. In *Power, Action and Belief: A New Sociology of Knowledge?*, edited by Law, J. (London: Routledge & Kegan Paul) [*9*, *9*]

Latour, B. 1986b. Visualisation and cognition: Thinking with eyes and hands. *Knowledge and Society* **6**:1–40 [*9*, *9*]

Latour, B. 1987. *Science In Action* (Milton Keynes: Open University Press) [*9*, *9*]

Laudun, R., ed. 1984. *The Nature of Technological Knowledge: Are Models of Scientific Change Relevant?* (Dordrecht: D. Reidel) [*17*, *6*]

Law, J. 1987a. On the social explanation of technical change: The case of the Portuguese maritime expansion. *Technology and Culture* **28**:227–53 [*9*, *9*]

Law, J. 1987b. Technology and heterogeneous engineering: The case of Portugese expansion. In *The Social Construction of Technological Systems; New Directions in the Sociology and History of Technology*, edited by Bijker, W., Hughes, T. & Pinch, T. (Cambridge MA: MIT Press) pp.111–34 [*9*, *9*]

Law, J., ed. 1991. *A Sociology of Monsters: Essays on Power, Technology and Domination*. Vol. 38, *Sociological Review Monograph* (London: Routledge) [*9*, *9*]

Law, J. 1997. On the social explanation of technical change: The case of the Portugese maritime expansion. *Technology and Culture* **28**:227–52 [**16**, *2*]

Layton, E. 1974. Technology as knowledge. *Technology and Culture* **15**:31–41 [**17**, *3*]

Leonard-Barton, D. 1988. Implementation as mutual adaptation of technology and organization. *Research Policy* **17**:251–67 [**18**, *50*]

Leslie, S.W. & Kargon, R.H. 1994. Electronics and the geography of innovation in post-war America. *History and Technology* **11**:217–32 [**20**, *1*]

Levinson, R. & Thomas, J.N. 1997. *Science Today: Problem or Crisis?* (New York NY: Routledge) [**21**, *55*]

Levy, S. 1992. *Artificial Life: The Quest for a New Creation* (London: Penguin) [*1*, *24*; *4*, *23*]

Lewontin, R.C. 1974. *The Genetic Basis of Evolutionary Change* (New York NY: Columbia University Press) [**15**, *1*]

Lewontin, R. 1992. Genotype and phenotype. In *Keywords in Evolutionary Biology*, edited by Keller, E.F. & Lloyd, E.A. (Cambridge MA: Harvard University Press) pp.137–44 [*5*, *9*]

Lewontin, R.C. 1978. Adaptation. *Scientific American* **239**:156–69 [**2**, *10*]

Liddell Hart, B.H. 1964 [1930]. *The Real War, 1914–1918* (Boston MA: Little Brown) [**20**, *27*, *30*]

Light, P. & Simmons, B. 1983. The effect of a communication task upon the representation of depth relationships in children's drawings. *Journal of Experimental Child Psychology* **35**:81–92 [**14**, *10*]

Lindblom, C.E. 1990. *Inquiry and Change: The Troubled Attempt to Understand and Shape Society* (New Haven CT: Yale University Press) [**21**, *21*]

Lindgren, M. 1996 Några tankar kring Christopher Polhems teknikpedagogik (Some reflections on Christopher Polhem's approach to engineering education). In *Teknik i skolan (Technology in the School)*, edited by Lindgren, M. *et al.* (Lund: Studentlitteratur), [**17**, *9*]

Lloyd, E.A. 1992. Unit of selection. In *Keywords in Evolutionary Biology*, edited by Keller, E.F. & Lloyd, E.A. (Cambridge MA: Harvard University Press), pp.334–40 [**5**, *35*]

Long, P.O. 1985. The contribution of architectural writers to a 'scientific' outlook in the fifteenth and sixteenth centuries. *Journal of Medieval and Renaissance Studies* **15**:265–98 [**9**, *38*]

Longino, H. 1990. *Science as Social Knowledge: Values and Objectivity in Scientific Inquiry* (Princeton NJ: Princeton University Press) [**21**, *11*, *19*, *23*]

Lorenz, K. 1962. Kant's doctrine of the *a priori* in the light of contemporary biology, *General Systems* **7**:23–35 [**1**, *4*; **16**, *5*, *39*]

Lovelock, J. 1979. *Gaia: A New Look at Life on the Earth* (Oxford: Oxford University Press) [**4**, *14*]

Lovelock, J. 1991. *Gaia: The Practical Science of Planetary Medicine* (London: Gaia Books) [**21**, *30*]

Lumsden, C.J. & Wilson, E.O. 1981. *Genes, Mind, and Culture* (Cambridge MA: Harvard University Press) [**6**, *2*]

Macdonald, S., Lamberton, D.M. & Mandeville, T.D., eds. 1983. *The Trouble with Technology: Explorations in the Process of Technological Change* (London: Frances Pinter) [**18**, *28*]

Mackenzie, D. 1990. *Inventing Accuracy: A Historical Sociology of Nuclear Missile Guidance* (Cambridge MA: MIT Press0 [**16**, *3*, *21*]

Malerba, F. & Orsenigo, L. 1995. Schumpeterian patterns of innovation. *Cambridge Journal of Economics* **19**:47–65 [**19**, *3*, *6*]

Malone, T.W. 1997. Is empowerment just a fad? Control, decision making and IT. *Sloan Management Review* **38** (2):23–35 [**19**, *9*]

Margulis, L. & Sagan, D. 1987. *Microcosmos: Four Billion Years of Evolution from our Microbial Ancestors* (London: Allen & Unwin) [**9**, *8*]

Mark, R. 1982. *Experiments in Gothic Structure* (Cambridge MA: MIT Press) [**9**, *33*]

Mark, R. 1990. *Light Wind and Structure: The Mystery of the Master Builders* (Cambridge MA: MIT Press) [**9**, *37*; **16**, *31*]

Mark, R. & Clark, W.W. 1984. Gothic structural experimentation. *Scientific American* **25**:144–53 [**9**, *26*]

Marshall, A. (1890) 1930. *Principles of Economics* (London: Macmillan) [**5**, *1*]

Maynard Smith, J. 1972. *On Evolution* (Edinburgh: Edinburgh University Press) [**1**, *6*]

Maynard Smith, J. 1976. What determines the rate of evolution? *American Naturalist* **110**:331–8 [**5**, *15*]

Maynard Smith, J. 1986. *The Problems of Biology* (Oxford: Oxford University Press) [**2**, *2*; **3**, *5*]

Maynard Smith, J. & Szathmáry, E. 1995. *The Major Transitions in Evolution* (Oxford: Freeman) [**2**, *4*, *16*; **3**, *6*; **22**, *3*]

Mayr, E. 1963. *Population, Species and Evolution* (Cambridge MA: Harvard University Press) [**16**, *34*]

Mayr, E. 1982. *The Growth of Biological Thought* (Cambridge MA: Harvard University Press) [**2**, *3*; **3**, *4*]

McFarland, S.L. 1995. *America's Pursuit of Precision Bombing, 1910–1945* (Washington DC: Smithsonian Institution Press) [**20**, *18*]

McMahon, W.E. 1978. *Dreadnought Battleships and Battle Cruisers* (Washington DC: University Press of America) [**20**, *35*]

McNeill, W.H. 1982. *The Pursuit of Power* (Chicago IL: University of Chicago Press) [**16**, *25*; **20**, *4*, *9*, *11*, *14*, *22*, *24*, *28*; **21**, *36*]

Mead, M. & Métraux, R. 1957. The image of the scientist amongst high-school students. In *The Sociology of Science*, edited by Barber, B. & Hirsch, W. (New York NY: Free Press of Glencoe), pp.230–46 [**14**, *1*]

Meadows, S. 1996. A Study of Year 2 Pupils Learning Design Technology. Unpublished diploma, Educational Studies, Oxford University, Oxford [**14**, *3*, *12*]

Meyer, P., Linn, F., Heidmann, I., Meyer, H., Niedenhof, I. & Saedler, H. 1992. Endogenous and environmental factors influence 35S promoter methylation of a maize A1 gene construct in transgenic petunia and its colour phenotype. *Molecular and General Genetics* **231**:345–52 [**3**, *27*]

Michael, D.N. 1973. *On Learning to Plan – and Planning to Learn* (San Francisco CA: Jossey Bass) [**19**, *11*]

Midgley, M. 1996. *Utopias, Dolphins and Computers: Problems of Philosophical Plumbing* (London: Routledge) [**21**, *28*, *32*]

Mikula, B.C. 1995. Environmental programming of heritable epigenetic changes in paramutant r-gene expression using temperature and light at a specific stage of early development in maize seedlings. *Genetics* **140**:1379–87 [**3**, *29*]

Miller, G.F., Todd, P.M. & Hegde, S.U. 1989. Designing neural networks using genetic algorithms. In *Proceedings of the Third International Conference on Genetic Algorithms*, edited by Schaffer, J.D. (Morgan Kaufmann) pp.379–84 [**15**, *10*, *18*]

Miller, G.F. 1994. Exploiting mate choice in evolutionary computation: Sexual selection as a process of search, optimization, and diversification. In *Evolutionary Computing: Proceedings of the 1994 Artificial Intelligence and Simulation of Behavior (AISB) Society Workshop*, edited by Fogarty, T.F. (Berlin: Springer-Verlag), pp.65–7 [**15**, *15*, *24*]

Miller, G.F. & Todd, P.M. 1995. The role of mate choice in biocomputation: Sexual selection as a process of search, optimization, and diversification. In *Evolution and Biocomputation: Computational Models of Evolution. Lecture Notes in Computer Science*

899, edited by Banzaf, W. & Eeckman, F.H. (Berlin: Springer-Verlag), pp.169–204
[**15**, *24*]

Ministry of Education 1995. Technology in the New Zealand Curriculum. Wellington:
Learning Media [**21**, *41*]

Misa, T.J. 1985. Military needs, commercial realities, and the development of the
transistor, 1948–1958. In *Military Enterprise and Technological Change: Perspectives on
the American Experience*, edited by Smith, M.R. (Cambridge MA: MIT Press),
pp.253–88 [**20**, *20*]

Mitchell, M. 1996. *An Introduction to Genetic Algorithms* (Cambridge MA: MIT Press) [**15**, *8*]

Moehrle, A. & Paro, R. 1994. Spreading the silence: epigenetic transcriptional
regulation during Drosophila development. *Developmental Genetics* 15:478–84
[*3*, *25*]

Mokyr, J. 1990. *The Lever of Riches* (New York NY: Oxford University Press) [**1**, *3*, *5*; **7**, *3*, *4*,
5, *7*, *64*; **21**, *36*]

Mokyr, J. 1996. Evolution and technological change: A new metaphor for economic
history? In *Technological Change*, edited by Fox, R. (London: Harwood Publishers),
pp. [*5*, *6*, *19*]

Mokyr, J. 1997. Induced technical innovation and medical history: An evolutionary
approach. Paper read at Induced Innovation, at Laxenburg, Austria [*5*, *43*]

Monbiot, G. 1998. Science with scruples. In *The Values of Science: The Oxford Amnesty
Lectures 1997*, edited by Williams, W. (Oxford: Oxford University Press), pp.42–57
[*21*, *25*]

Morgall, J.M. 1993. *Technology Assessment: A Feminist Perspective* (Philadelphia PA: Temple
University Press) [**21**, *38*]

Morrison, J.S. & Coates, J.F. 1986. *The Athenian Trireme: The History and Reconstruction of
an Ancient Greek Warship* (Cambridge: Cambridge University Press) [**18**, *41*]

Morse, E.S. 1936. *Japan Day by Day. 1877, 1878–79, 1882–83.* 2 vols (Tokyo) [**7**, *26*, *27*, *29*, *30*,
31, *32*, *33*, *34*, *35*, *36*, *41*, *42*, *43*, *47*, *51*, *52*]

Mumford, L. 1947. *Technics and Civilization* [**7**, *6*, *9*]

Munz, P. 1993. *Philosophical Darwinism: On the Origin of Knowledge by Means of Natural
Selection* (London: Routledge) [**1**, *4*, *26*; **4**, *5*, *11*]

Nef, J.U. 1950. *War and Human Progress: An Essay on the Rise of Industrial Civilization,
Economic History Review* (Cambridge MA: Harvard University Press) [**20**, *3*]

Nelkin, D. 1995. *Selling Science: How the Press Covers Science and Technology* (New York NY:
Freeman) [**21**, *52*]

Nelson, R.R. 1962. The link between science and technology: The case of the
transistor. In *The Rate and Direction of Inventive Activity*, edited by Nelson, R.R.
(Princeton NJ: Princeton University Press) [*6*, *13*]

Nelson, R.R. 1995. Recent evolutionary theorizing about economic change. *Journal of
Economic Literature* (March):48–90 [**1**, *16*; **6**, *1*, *10*; **22**, *1*]

Nelson, R.R. & Winter, S. 1982. *An Evolutionary Theory of Economic Change* (Cambridge
MA: Belknap) [**1**, *20*; **5**, *8*, *32*; **6**, *9*; **12**, *8*; **18**, *2*, *6*, *26*, *27*, *43*; **19**, *1*]

Newell, A. & Simon, H. 1972. *Human Problem-solving* (Englewood Cliffs NJ: Prentice-
Hall)

Noble, D. 1984. *Forces of Production: A Social History of Industrial Automation* (New York NY: Alfred Knopf) [**21**, *2*]

O'Connell, C., F. 1985. The Corps of Engineers and the rise of modern management, 1827–1856. In *Military Enterprise and Technological Change: Perspectives on the American Experience*, edited by Smith, M.R. (Cambridge MA: MIT Press), pp.87–116 [**20**, *15*]

O'Connell, R.L. 1991. *Sacred Vessels: The Cult of the Battleship and the Rise of the U.S. Navy* (Boulder CO: Westview Press) [**20**, *34*]

Oeser, E. 1984. The evolution of scientific method. In Concepts and Approaches in Evolutionary Epistemology, edited by Wuketits, F. (Dordrecht: Reidel), pp.149–84 [**1**, *3*]

Ogburn, W.F. 1922. *Social Change* (New York NY: B.W.Huebsch) [**21**, *45*]

Ohlman, H. 1990. Information: Timekeeping, telecommunications and audiovisual technologies. In *An Encyclopedia of the History of Technology*, edited by McNeil, I. (London: Routledge), pp.686–758 [**5**, *22*]

Oliphant, L. 1859. *Narrative of The Earl of Elgin's Mission to China and Japan in the Years 1857, '58, '59* (London) [**7**, *24*, *57*]

O'Neill, B. 1991. Bridge design stretched to the limits. *New Scientist.* **132** (26 Oct.):28–35 [**9**, *40*]

Pacey, A. 1983. *The Culture of Technology* (Oxford: Blackwell) [**14**, *5*; **18**, *13*; **21**, *1*]

Patino, M.M., Liu, J.-J., Glover, J.R. & Lindquist, S. 1996. Support for the prion hypothesis for inheritance of a phenotypic trait in yeast. *Science* **273**:622–6 [**3**, *20*]

Pavitt, K. 1984. Sectoral patterns of technical change: Towards a taxonomy and a theory. *Research Policy* **13**(6):343–74 [**17**, *5*]

Penrose, E.T. 1952. Biological analogies in the theory of the firm. *American Economic Review* **42** (5):804–19 [**5**, *2*, *37*]

Penrose, L.S. 1959. Automatic mechanical self-reproduction. *New Biology* **92**:117 [**3**, *18*]

Perkins, D.N. 1981. *The Mind's Best Work* (Cambridge MA: Harvard University Press) [**12**, *9*, *14*]

Perkins, D.N. 1992. The topography of invention. In *Inventive Minds: Creativity in Technology*, edited by Weber, R.J. & Perkins, D.N. (New York NY: Oxford University Press), pp.238–50 [**12**, *12*, *16*]

Perkins, D.N. 1994. Creativity: Beyond the Darwinian paradigm. In *Dimensions of Creativity*, edited by Boden, M. (Cambridge MA: MIT Press), pp.119–42 [**12**, *12*, *16*]

Perkins, D.N. 1995. Insight in minds and genes. In *The Nature of Insight*, edited by Sternberg, R.J. & Davidson, J.E. (Cambridge MA: MIT Press), pp.495–533 [**12**, *12*, *16*]

Perkins, D.N. & Weber, R.J. 1992. Effable invention. In *Inventive Minds: Creativity in Technology*, edited by Weber, R.J. & Perkins, D.N. (New York NY: Oxford University Press), pp.317–36

Petroski, H. 1993. *The Evolution of Useful Things* (New York NJ: Alfred Knopf) [**5**, *7*; **17**, *3*]

Petroski, H. 1994. *Design Paradigms: Case Histories of Error and Judgment in Engineering* (Cambridge: Cambridge University Press) [**13**, *1*, *2*]

Petroski, H. 1995. *Engineers of Dreams: Great Bridge Builders and the Spanning of America* (New York NY: Alfred Knopf) [**13**, *1*, *3*, *4*]

Piaget, J. & Inhelder, B. 1958. *The Growth of Logical Thinking* (London: Routledge & Kegan Paul) [**14**, *11*]

Pickering, A. 1995. *The Mangle of Practice: Time, Agency and Science* (Chicago IL: University of Chicago Press) [**16**, *1*, *30*; **20**, *42*]

Pinker, S. 1994. *The Language Instinct* (London: Penguin) [**4**, *20*; **17**, *16*]

Plotkin, H.C. 1982. *Learning, Development and Culture: Essays in Evolutionary Epistemology* (Chichester: Wiley) [**6**, *5*]

Plotkin, H.C. 1994. *The Nature of Knowledge: Concerning Adaptations, Instinct and the Evolution of Intelligence* (Harmondsworth: Penguin) [**1**, *11*, *26*; **4**, *3*, *4*, *12*, *18*; **18**, *22*]

Poincaré, H. 1921. *'Mathematical Creation', the Foundations of Science* (New York NY: Science Press) pp.383–94 [**16**, *37*]

Polanyi, M. 1958. *Personal Knowledge* (London: Routledge & Kegan Paul) [**5**, *34*; **14**, *16*]

Polanyi, M. 1967. *The Tacit Dimension* (London: Routledge & Kegan Paul) [**18**, *12*]

Popper, K.R. 1968 [1963]. *Conjectures and Refutation: The Growth of Scientific Knowledge*. 3rd edn (New York NY: Harper Torchbooks) [**6**, *6*; **16**, *10*]

Popper, K.R. 1978. Natural selection and the emergence of mind. *Dialectica* **22**:339–55 [**16**, *10*]

Porter, M.E. & Miller, V.E. 1985. How information gives you competitive advantage. *Harvard Business Review* (July-August):149–60 [**18**, *14*]

Prahalad, C.K. & Hamel, G. 1990. The core competence of the corporation. *Harvard Business Review* (May–June):79–91 [**18**, *20*]

Prescott, G.B. 1884. *Bell's Electric Speaking Telephone: Its Invention, Construction, Application, Modification, and History*. reprint Arno 1972 ed (New York NY: D. Appleton) [**11**, *18*]

Purchas, S. 1938. *Purchas His Pilgrimes in Japan, Extracted from Hakluyts Posthumus …*, edited by Wild, C. (Kobe) [**7**, *14*, *37*]

Pursell, C. 1995. *The Machine in America: A Social History of Technology* (Baltimore MD: Johns Hopkins University Press) [**16**, *27*]

Quine, W.V.O. 1953. Two dogmas of empiricism. *Philosophical Review* **60**:20–43 [**16**, *11*]

Ramsden, P. 1992. *Learning to Teach in Higher Education* (London: Rouledge) [**21**, *47*]

Ramstad, Y. 1994. On the Nature of Economic Evolution. In *Evolutionary and Neo-Schumpeterian Approaches to Economics*, edited by Magnusson, L. (Boston MA Kluwer), pp. [**5**, *2*]

Rasmussen, S., Knudsen, C., Feldberg, R. & Hindsholm, M. 1990. The core-world: emergence and evolution of cooperative structures in a computational chemistry. *Physica D* **42**:111–34 [**4**, *25*]

Ravetz, J.R. 1971. *Scientific Knowledge and its Social Problems* (Oxford: Clarendon Press) [**18**, *44*]

Ray, T.S. 1992. An approach to the synthesis of life. In *Artificial Life II*, edited by Langton, C.G., Taylor, C., Farmer, J.D. & Rasmussen, S. (Reading MA: Addison-Wesley), pp.371–408 [**4**, *26*]

Rechenberg, I. 1973. *Evolutionsstrategie: Optimierung technische Systeme nach Prinzipien der biologischen Evolution* (Stuttgart: Frommann-Holzboog) [**15**, *4*]

Regamey, F. 1892. *Japan in Art and Industry, with a Glance at Japanese Manners and Customs* (London) [**7**, *48*]

Rennie, L.J. & McClafferty, T.P. 1996. Science centres and science learning. *Studies in Science Education* **26**:53–98 [**21**, *54*]

Ridley, M. 1985. *The Problem of Evolution* (Oxford: Oxford University Press) [**5**, *45*]

Ridley, M. 1996. *Evolution*. 2nd edn (Oxford: Blackwell Science) [**1**, *6*]

Riedl, R. 1984. *Biology of Knowledge: The Evolutionary Basis of Reason*, translated by Paul Foulkes. 3rd edn (Chichester: Wiley) [**1**, *4*]

Roemer, I., Reik, W., Dean, W. & Klose, J. 1997. Epigenetic inheritance in the mouse. *Current Biology* **7**:277–80 [**3**, *32*]

Rosenberg, N. 1982. Learning by using. In *Inside the Black Box: Technology and Economics*, edited by Rosenberg, N. (Cambridge: Cambridge University Press), pp.120–40 [**18**, *1*]

Rosenberg, N. & Vincenti, W.G. 1978. *The Britannia Bridge: The Generation and Diffusion of Technological Knowledge* (Cambridge MA: MIT Press) [**13**, *5*]

Rosenberg, R.A. *et al.* eds. 1991. *The Papers of Thomas A. Edison* (Baltimore MD: Johns Hopkins University Press) [**11**, *2*]

Rothschild, J., ed. 1982. *Women, Technology and Innovation* (Oxford: Pergamon Press) [**21**, *6*]

Rouse, J. 1987. *Knowledge and Power: Towards a Political Philosophy of Science* (Ithaca NY: Cornell University Press) [**9**, *13*]

Ruse, M. 1996. *Monad to Man: The Concept of Progress in Evolutionary Biology* (Cambridge MA: Harvard University Press) [**2**, *13*; **5**, *40*, *46*; **21**, *56*]

Sachs, T. 1988. Epigenetic selection: an alternative mechanism of pattern formation. *Journal of Theoretical Biology* **134**:547–59 [**3**, *33*, *37*]

Sapp, J. 1987. *Beyond the Gene* (New York NY: Oxford University Press) [**3**, *2*]

Sapp, J. 1994. *Evolution by Association: A History of Symbiosis* (Oxford: Oxford University Press) [**9**, *18*]

Saviotti, P.P. 1996. *Technological Evolution, Variety and The Economy* (Cheltenham: Edward Elgar) [**1**, *20*]

Schick, K.D. & Toth, N. 1993. *Making Silent Stones Speak: Human Evolution and the Dawn of Technology* (London: Weidenfeld & Nicolson) [**18**, *39*]

Schiebinger, L. 1989. *The Mind Has No Sex: Women in the Origins of Modern Science* (Cambridge MA: Harvard University Press) [**21**, *6*]

Schon, D. 1983. *The Reflective Practitioner* (London: Temple Smith) [**14**, *13*; **21**, *10*]

Schwefel, H.P. 1977. *Numerische Optimierung von Computermodellen mittels der Evolutionsstrategie* (Basel: Birkhauser) [**15**, *4*]

Searle, J.R. 1995. *The Construction of Social Reality* (London: Penguin) [**1**, *19*]

Sen, A. 1993. On the Darwinian view of progress. *Population and Development Review* **19** (1):123–37 [**5**, *42*]

Service, R.F. 1997. High-speed materials design. *Science* **277**:474–5 [**15**, *21*]

de Seversky, A. 1942. *Victory through Air Power* (New York NY: Simon & Schuster) [**20**, *38*]

Shapiro, J.A. 1995. Adaptive mutation: who's really in the garden. *Science* **268**: 373–4 [**16**, *33*]

Shelby, L.R. 1970. The education of medieval English master masons. *Medieval Studies* **32**:1–26 [**9**, *34*, *35*]

Shelby, L.R. 1971. Medieval masons' templets. *Journal of the Society of Architectural Historians* **30**:140–52

Shelby, L.R. 1976. The 'secret' of the medieval masons. In *On Pre-Modern Technology and Science*, edited by Hall, B. & West, D. (Malibu CA: Undena Publications), pp.201–22 [9, 32]

Shelby, L.R. 1981. The contractors of Chartres [review]. *GESTA* 20:173–8 [9, 36]

Simon, H. 1957. *Models of Man* (New York NY: Wiley) [10, 8]

Simon, H. 1969. *The Sciences of the Artificial* (Cambridge MA: MIT Press) [16, 29; 17, 4, 11]

Sims, K. 1991. Artificial evolution for computer graphics. *Computer Graphics* 25 (4):319–28 [15, 23]

Smith, C.S. 1981. *A Search for Structure* (Cambridge MA: MIT Press) [8, 1, 2]

Smith, M.R. 1985a. Army Ordnance and the 'American system' of manufacturing, 1815–1861. In *Military Enterprise and Technological Change: Perspectives on the American Experience*, edited by Smith, M.R. (Cambridge MA: MIT Press), pp.39–86 [20, 1]

Smith, M.R. 1985b. Introduction. In *Military Enterprise and Technological Change: Perspectives on the American Experience*, edited by Smith, M.R. (Cambridge MA: MIT Press) pp.1–37 [20, 12]

Sniegowski, P.D. & Lenski, R.E. 1995. Mutation and adaptation: the directed mutation controversy in evolutionary perspective. *Annual Review of Ecology and Systematics* 26:553–78 [3, 9]

Solomon, J. 1987. Science technology and society education: Tools for thinking about social issues. *International Journal of Science Education* 10 (4):379–87 [14, 4]

Solomon, J., Duveen, J. & Scott, L. 1996. Large-scale exploration of pupils' understanding of the nature of science. *Science Education* 80 (5):493–508 [14, 2]

Solomon, J. & Hall, S. 1997. An inquiry into progression in primary technology: A role for teaching. *International Journal for Design Education* [14, 10]

Sombart, W. 1913. *Studienzur Entwicklungsgeschichte des Modernen Kapitalismus, II. Krieg und Kapitalismus* (Munich) [20, 2]

Soros, G. & Giddens, A. 1997. Beyond chaos and dogma. *New Statesman* (31 October):24–7 [21, 37]

Spindler, K. 1994. *The Man in the Ice* (London: Weidenfeld and Nicolson) [18, 38]

Staudenmaier, J.M. 1995. Rationality versus contingency in the history of technology. In *Does Technology Drive History? The Dilemma of Technological Determinism*, edited by Smith, M.R. & Marx, L. (Cambridge MA: MIT Press) [9, 10]

Staudenmaier, S.J. 1985. *Technology's Storytellers: Reweaving the Human Fabric* (Cambridge MA: MIT Press) [14, 8; 21, 3]

Steadman, P. 1979. *The Evolution of Designs: Biological Analogy in Architecture and the Applied Arts* (Cambridge: Cambridge University Press) [9, 17, 21, 43, 44]

Stebbins, G.L. 1982. *Darwin to DNA, Molecules to Humanity* (San Francisco CA: W.H. Freeman) [5, 14, 18]

Sternberg, R. & Davidson, J.E., eds. 1995. *The Nature of Insight* (Cambridge MA: MIT Press) [12, 13]

Street, B. 1984. *Literacy in Theory and Practice* (Cambridge: Cambridge University Press) [21, 14]

Strong, G. 1996. Uncovered. *Sunday Age, Melbourne*:6 [9, 42]

Suchman, L. 1987. *Plans and Situated Actions: The Problem of Human-Machine Communication* (Cambridge: Cambridge University Press) [9, 20]

Sumida, J.T. 1993. *In Defense of Naval Supremacy: Financial Limitation, Technological Innovation and British Naval Policy, 1889–1914* (New York NY: Routledge) [**20**, *32*]

Suppes, P. & Atkinson, R.C. 1960. *Markovian Learning Theory* (Stanford CA: Stanford University Press) [**10**, *5*]

Swanson, E.B. 1988. *Information System Implementation: Bridging the Gap between Design and Utilization* (Homewood IL: R.D. Irwin) [**18**, *50*]

TAEM. 1985–. The speaking telephone interferences. United States Patent Office. Evidence on behalf of Thomas A. Edison. In *The Thomas A. Edison Papers: A Selective Microfilm Edition (Frederick, Md.: University Publications of America, 1985–)*, edited by TAEM (Frederick MD: University Publications of America) [**11**, *3, 17, 19*]

Thomas, B. 1991. Alfred Marshall on economic biology. *Review of Poliical Economy* **3** (1):1–14 [**5**, *1*]

Thompson, S.P. 1883. *Philipp Reis: Inventor of the Telephone* (London: E. & F.N. Spon) [**11**, *8, 9*]

Thunberg, C.P. 1796 [1793]. *Travels in Europe, Africa and Asia*. 3rd edn (London) [**7**, *18, 22, 23, 39, 44, 49, 56, 58, 59*]

Todd, S. & Latham, W. 1992. *Evolutionary Art and Computers* (San Diego CA: Academic Press) [**15**, *23*]

Toker, F. 1985. Gothic architecture by remote control: An illustrated building contract of 1340. *Art Bulletin* **67**:67–95 [**9**, *27*]

Toulmin, S. 1972. *Human Understanding*, Vol. 1 (Oxford: Oxford University Press) [**1**, *3*]

Travers, T.H.E. 1979. Technology, tactics, and morale: Jean de Bloch, the Boer War, and British military theory. *Journal of Military History* **51**:264–86 [**20**, *25*]

Turkle, S. & Papert, S. 1990. Epistemological pluralism: Styles and voices within the computer culture. *Signs: Journal of Women in Culture and Society* **16** (1):128–57 [**21**, *8*]

Turnbull, D. 1993a. The ad hoc collective work of building Gothic cathedrals with templates, string, and geometry. *Science, Technology & Human Values* **18**:315–40 [**9**, *14, 25*]

Turnbull, D. 1993b. Local knowledge and comparative scientific traditions. *Knowledge and Policy* **6** (3/4):29–54 [**9**, *15, 19*]

Turnbull, D. 1996a. Cartography and science in early modern Europe: Mapping the construction of knowledge spaces. *Imago Mundi* **48**:5–24 [**9**, *16, 19*]

Turnbull, D. 1996b. Constructing knowledge spaces and locating sites of resistance in the early modern cartographic transformation. In *Social Cartography: Mapping Ways of Seeing Social and Educational Change*, edited by Paulston, R. (New York NY: Garland Publishing Inc.), pp.53–79 [**9**, *16, 19*]

UNICTAD. 1978. *Manual for the Acquisition of Technology by Developing Countries* (New York NY: UNICTAD) [**18**, *36*]

Utterback, J.M. & Abernathy, W.J. 1975. A dynamic model of process and product innovation *OMEGA* **3**(6):639–56 [**18**, *37*]

Van Creveld, M. 1977. *Supplying War* (Cambridge: Cambridge University Press) [**20**, *28, 29*]

Vincenti, W.G. 1982. Control volume analysis: A difference in thinking between engineering and physics. *Technology and Culture* **23**: 145–74 [**16**, *9*]

Vincenti, W.G. 1990. *What Engineers Know and How They Know It: Analytical Studies from*

Aeronautical History (Baltimore MD: Johns Hopkin University Press) [**1**, *9*; **5**, *11*, *31*; **13**, *8*; **17**, *3*, *11*; **18**, *11*, *49*]

Vincenti, W.G. 1994. The retractable airplane landing gear and the Northrop 'anomaly': Variation-selection and the shaping of technology. *Technology and Culture* **35**:1–33 [**6**, *11*; **13**, *9*, *10*]

Vincenti, W.G. 1997. Engineering theory in the making: Aerodynamic calculation 'breaks the sound barrier'. *Technology and Culture* **38** (October):819–51 [**20**, *19*]

Voss, C.A. 1988. Implementation: A key issue in manufacturing technology: the need for a field study. *Research Policy* **17** (2):55–63 [**18**, *50*]

Vygotsky, L.S. 1978. *Mind in Society: The Development of Higher Psychological Processes* (Cambridge MA: Harvard University Press) [**14**, *7*; **21**, *47*]

Waddington, C.H. 1957. *The Strategy of the Genes* (London: Allen & Unwin) [**3**, *3*]

Waldrop, M.M. 1992. *Complexity: The Emerging Science at the Edge of Order and Chaos* (New York NY: Simon & Schuster) [**12**, *10*]

Watkins, S. 1990. Sky-high jigsaw. *Age, Melbourne*:7 [**9**, *28*]

Watson-Verran, H. & Turnbull, D. 1995. Science and other indigenous knowledge systems. In *Handbook of Science and Technology Studies*, edited by Jasanoff, S., Markle, G., Pinch, T. & Petersen, J. (Thousand Oaks CA: Sage Publications), pp.115–39 [**9**, *15*]

Weber, R. 1992. *Forks, Phonographs and Hot-Air Balloons: A Fieldguide to Inventive Thinking* (New York NY: Oxford University Press) [**12**, *25*]

Weber, R.J. & Dixon, S. 1989. Invention and gain analysis. *Cognitive Psychology* **21**:283–302 [**12**, *25*]

Weber, R.J. & Perkins, D.N. 1989. How to invent artifacts and ideas. *New Ideas in Psychology* **7**:49–72 [**11**, *13*]

Weber, R.J. & Perkins, D.N., eds. 1992. *Inventive Minds: Creativity in Technology* (New York NY: Oxford University Press) [**12**, *27*, *28*]

Wesson, R. 1991. *Beyond Natural Selection* (Cambridge MA: MIT Press) [**12**, *23*, *24*]

White, L. 1978. *Medieval Religion and Technology* (Berkeley CA: University of California Press) [**5**, *19*]

White, L.J. 1962. *Medieval Technology and Social Change* (New York NY: Oxford University Press) [**20**, *10*]

Wilson, D.S. & Sober, E. 1994. Reintroducing group selection to human behavioral sciences. *Behavioral and Brain Science* **17**:585–654 [**2**, *6*]

Wilson, S.W. 1986. Knowledge growth in an artificial animal. In *Adaptive and Learning Systems*, edited by Narendra, K.S. (New York NY: Plenum) pp.255–66 [**4**, *27*]

Wolffe, A.P. 1994. Inheritance of chromatin states. *Developmental Genetics* **15**:463–70 [**3**, *26*]

Womack, J.P., Jones, D.T. & Roos, D. 1990. *The Machine that Changed the World* (New York NY: Rawson Associates) [**18**, *34*]

Wood, D. & Dempster, D. 1969 (1961). *The Narrow Margin: The Battle of Britain and the Rise of Air Power, 1930–1940* (New York NY: Paperback Library) [**20**, *37*]

Woodward, J. 1980. *Industrial Organisation: Theory and Practice*. 2nd edn (Oxford: Oxford University Press) [**18**, *29*]

Wright, S. 1932. The roles of mutation, inbreeding, crossbreeding and selection in evolution. Paper read at Sixth International Congress of Genetics, 1986 [**12**, *7*]

Wuketits, F. 1984. Evolutionary epistemology: A challenge to science and philosophy. In *Concepts and Approaches in Evolutionary Epistemology*, edited by Wuketits, F. (Dordrecht: Reidel), pp.1–33 [**1**, *3*]

Wynne, B. 1991. Knowledges in context. *Science, Technology and Human Values* **16** (1):111–21 [**21**, *27*, *51*]

Ziman, J.M. 1978. *Reliable knowledge* (Cambridge: Cambridge University Press) [**16**, *14*, *26*]

Ziman, J.M. 1984. *An Introduction to Science Studies* (Cambridge: Cambridge University Press) [**16**, *38*]

Ziman, J.M. 1995. *Of One Mind: The Collectivization of Science* (Woodbury NY: AIP Press) [**1**, *8*]

Ziman, J.M. 1996a. Darwin and/or Lamarck: Selection and/or design: Technological innovation as an evolutionary process. *Times Higher Education Supplement* [**12**, *1*, *20*; **18**, *2*, *3*, *5*, *33*]

Ziman, J.M. 1996b. The marriage of design and selection in the evolution of cultural artefacts. In *The Certainty of Doubt: Tributes to Peter Mun*, edited by Fairbairn, M. & Oliver, W.H. (Wellington NZ: Victoria University Press), pp.269–303 [**12**, *1*, *20*; **18**, *2*, *3*, *5*, *33*]

Ziman, J.M. 2000. *Real Science: What it is and what it means* (Cambridge: Cambridge University Press) [**16**, *17*, *23*]

Index

Page numbers in italics refer to an illustration.

Names of authors cited in the text are to be found in the bibliography. Names of people discussed in the text are to be found in the index.